THE SENSES IN LATE MEDIEVAL ENGLAND

THE SENSES IN LATE MEDIEVAL ENGLAND

C.M. WOOLGAR

YALE UNIVERSITY PRESS
NEW HAVEN AND LONDON

For information about this and other Yale University Press publications, please contact:

U.S. Office:	sales.press@yale.edu	yalebooks.com
Europe Office:	sales@yaleup.co.uk	www.yaleup.co.uk

Set in Minion by J&L Composition, Filey, North Yorkshire

Library of Congress Cataloging-in-Publication Data

Woolgar, C. M.
  The senses in late medieval England/C.M. Woolgar.
    p. cm.
  Includes bibliographical references and index.
  ISBN: 978--300-20605-0
  1. Senses and sensation—England—History—To 1500.   I.   Title.
  BF233.W75 2006
  152.10942'0902—dc22

2006017766

10 9 8 7 6 5 4 3 2 1

CONTENTS

ILLUSTRATIONS

FOREWORD

I have been interested in the history of the day-to-day for many years, inspired initially by Pierre Goubert's work on Beauvais, which I found by chance in a second-hand book shop in Guildford in 1974. I then pursued the subject through medieval history and archaeology at university. The sources for medieval England offer different possibilities to those for seventeenth-century France and, in working on the records of the great household, I came to believe that there was much more to be learned about the senses and that the subject might be considered more widely across medieval society. The premiss of this book is that daily life in the past cannot be understood without first comprehending how perception functioned. Evidence for this study has not been lacking: in fact the problem has been the opposite. As almost anything could touch on the senses, the question was to identify those records that would consistently produce dividends. I was encouraged in this project by Dr Robert Baldock of Yale University Press, who has allowed me to present my thoughts in this form. My colleagues at Southampton have helped in many ways, not least in responding generously to papers I presented, as did audiences at Leeds and Oxford. I am also grateful to Yale's two readers for their comments and suggestions.

In bringing this work to fruition I have received considerable assistance from the University of Southampton. In particular, I have had two periods of leave from my normal duties in the University Library: neither would have been possible without the support of Dr Mark Brown, the University Librarian, and my colleagues in the Special Collections Division, who have readily confirmed my belief that it is better for large institutions to be more robust and systematic in their arrangements than a reliance on individuals could ever allow. I am also grateful to the Inter-Library Loan section, which has always responded admirably to my requests. Together, they have made Southampton an exceptional base from which to work. Coming at the end of a period of major building work that has transformed the University Library and our working lives, this project has provided me with a welcome perspective on the purposes and qualities of libraries.

A grant from the Arts and Humanities Research Council through its research leave scheme allowed me to bring this book to a conclusion: I am grateful to the AHRC and to those who supported my applications for funding. Prolonging my research leave in this way created an opportunity for me to concentrate on this task for a longer period of time than I have ever before committed to writing.

As that writing was largely carried out at home, acknowledging the conveniences of my own day-to-day domesticity is not mere form. The author's family – and mine is no exception – forswears a great deal. Sue, Tom, Matt and Dan have lived with much more than discussion of whether the dog means what it says when it barks and other recherché aspects of my conversation. That this work has been completed at all owes much to the distractions, relief and humour they have provided. Although they may wonder at what it is that I have provided in return, I proffer them this book with gratitude.

Acknowledgements

In using the libraries, archives and collections of a number of institutions, I have incurred debts to many archivists, curators and librarians, and their reprographic services. If I do not acknowledge them individually, it is because they are very many and it would be invidious to single them out. The following have kindly given permission for me to use quotations and illustrations: the Bridgeman Art Library; the British Library; the Trustees of the British Museum; Cadw Photographic Library; the Conway Library, Courtauld Institute of Art, University of London; English Heritage Photographic Library; Glasgow City Council (Museums: the Burrell Collection); A.F. Kersting; the Museum of London; Norfolk Museums and Archaeology Service; the Dean and Chapter of Salisbury Cathedral; the Master and Fellows of Trinity College, Cambridge; V&A Images/Victoria and Albert Museum; the National Library of Wales; and the Dean and Chapter of Westminster Abbey.

A note on the text

For ease of reading, in all quotations capitalisation and punctuation have been normalised to modern forms. I have also standardised i/j and u/v to reflect modern usage. All abbreviations and contractions have been expanded silently. All biblical references are taken from the Vulgate: *Biblia sacra secundum vulgatam versionem* ed. R. Weber *et al.* (4th edition, Stuttgart, 1994). Books of the Latin Bible are cited according to the conventions given at the end of the list of abbreviations. The references to the Cantilupe canonisation proceedings have been taken from the Vatican manuscript in preference to the version in the *Acta Sanctorum*.

Christopher M. Woolgar
University of Southampton
26 February 2006

MAP

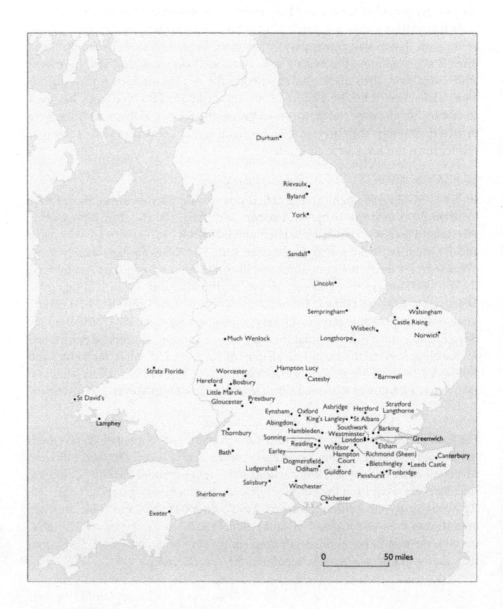

The principal places referred to in the text

Chapter 1

INTRODUCTION

Medieval sources are replete with evidence of a passion for sensory experience: the smell and taste of spices and herbs, the pungency of burning aromatics, the bright and gaudy colours. But how were these sensations interpreted? If our beliefs about perception are rooted in science and philosophy, from the Enlightenment to neuro-science and cognitive psychology, the starting point for medieval man was different. Twenty-first-century medicine or physiology, for example, does not necessarily help us appreciate the consequences of Galenic medicine and Aristotelian philosophy – or local superstition – as a cultural basis for perception. Science now works with an expanded range of senses and, at the same time, a conviction that we are born with innate sensory predispositions, such as a sweet tooth and a desire for salt. Equally it is recognised that there is much that is acquired empirically, particularly during the first year of life.

A considerable element in perception is cultural: it comes to us from those around us and carries many meanings that may be obscure but are part of attitudes that now form custom and practice. Many sensory references, medieval or older in origin, are no longer clear in meaning, for example, the 'odour of sanctity',[1] 'stinking sin',[2] or the colloquial 'nose for the nasty'. In the late medieval period the lustre of a colour might be more important than its hue, cacophony might be the work of the Devil, while ecclesiastical music might resemble that of the angels. These are reflections of a culture that thought differently about the senses. Here appearances mattered, appearances that might reflect the inner soul. Moral and spiritual quali-ties were apparent through external signs – a pleasant smell might indicate sanctity for some, or a good or noble person more generally; to dress in white might be not only a claim to purity, but establish that state; and, conversely, a bad smell or black could indicate or create evil. It is perhaps impossible to overstate the significance of sensory perception in understanding how medieval people behaved in their daily lives. We do not have access to their physical senses in a way that would allow a cognitive scientist or physiologist to address this question – this is little to be regretted as, *prima facie*, there was probably almost complete symmetry between their physical sensations and ours – but we do have a substantial body of informa-tion about how sensation affected and dictated their behaviour, and that material forms the core of this book. It is not a study of biological universals, but of the cultural attitudes that constitute and accompany perception.[3]

The senses are to us much more circumscribed than would have been understood in the Middle Ages or in Antiquity. Since the Enlightenment, their operation has commonly been seen in the high culture of the West as a narrow and analytical process of physiology, biology and electrochemistry, with an ever increasing preci-sion of definition of what we might reasonably expect a sense to do. The senses are receptors. Our assessment of the information that they bring produces 'output' in terms of behaviour. In medieval Europe, sensory perception was a much more open process, not just a form of transmission of information about objects, but one which enabled tangible qualities and, indeed, spiritual or intangible qualities, to be passed from one party or object to another. It was also a two-way process: the senses gave out information or affected others directly, as well as receiving information or serving as a conduit that might change an individual. In this way much that we might now seek to explain as miraculous or magical took place as part of the natural order. When, in St Luke's Gospel, a sick woman touched the border of Christ's garment and was cured, Christ asked straight away who had touched him, 'for I perceive that virtue has gone out of me':[4] his remark was understood in the medieval period as part of the ordinary process of perception. Sound – whether the song of the sirens, bad language, that is, the work of the Devil, or the music of the angels – transmitted much more than the literal message of the text. From Antiquity, the lethal powers of the basilisk were conveyed by sight, although they were later believed to pass by its breath, touch and hiss as well.[5] The impact of perception, however, might vary from individual to individual. To the encyclo-paedist, Bartholomew the Englishman, writing in the first part of the thirteenth century, the effects of smells were based on their composition, in terms of the four elements, and their similarity to the complexion of the perceiver.[6] These beliefs indicate that we should expect from the senses a range of behavioural responses and uses significantly different to our own.

The Middle Ages inherited from Classical Antiquity the notion of five senses and we may be conditioned, culturally, to accept the existence of the same five – sight, hearing, smell, taste and touch. Modern science has confirmed the foundation they constitute for sensory perception, but has added others to the sensorium, such as proprioception and the workings of the vestibular system, which allow us to sense where the parts of the body are and to ensure balance.[7] A historian, however, needs to consider a different range of possibilities. Comparative anthropology indicates that other cultures had and have different numbers of senses, ordered and order them in different hierarchies and had and have a variety of models for their opera-tion. Sensory experience, equally, can be a matter of training: chefs or musicians, for

1 Man's body, like all things, was composed of earth, water, fire and air: the four elements here personified from left to right. This copy of Bartholomew the Englishman's encyclopaedia, *De proprietatibus rerum*, in its French translation by Corbichon, was made in Bruges in 1482, probably for Edward IV.

example, might have heightened perceptions of taste or sound. Our view of the senses as mainly passive receptors is not necessarily shared with the past, or with contemporary cultures. The boundaries of life – both the span of our existence as sentient beings and the range of things that might be considered animate – may need to be drawn in different places.[8] With a view to broadening the framework for analysis, a whole sub-discipline of archaeology – cognitive archaeology – has been devoted to the interpretative matrix that comes between physical evidence and human thought, a process that is further complicated in the case of our hominid precursors, whose senses may have functioned biologically in a different way.[9]

Historical studies of the senses are not lacking for Antiquity and seminal works, such as Alain Corbin's *The foul and the fragrant*, have made major contributions to research from the early modern period onwards.[10] These works document changes in the relative importance of some senses in Western environments, for example the devaluation of the sense of smell.[11] A second category of research has looked at sensory environments – soundscapes, or smells and sights – some, for the medieval period.[12] In many ways, neither analysis penetrates far enough, rarely looking beyond the classical five senses. While there are studies of the theory of the senses in terms of philosophy, theology, medicine, literary and artistic works, of hierarchies of sense, the primacy of sight, or the importance of olfaction and touch,[13] an assessment of the practical operation of the senses and their cultural impact, as opposed to a theoretical or purely literary one, is lacking. In addition, derived from sources that are overwhelmingly clerical and Latinate in origin, they propagate a comparatively uniform notion of sensory perception in the medieval culture of Western

Europe, whereas, on the analogy of anthropology, one might expect a much less even weave.

By making explicit what, in late medieval England, was implicit in the messages conveyed by the senses, this study addresses these deficiencies. It starts by examining beliefs about perception and looks at the range of senses commonly recognised. There was a very strong interplay of theoretical analyses of perception, the product of a learned culture and derived from theology, natural philosophy and medicine, on the one hand, with popular beliefs and practices on the other. Subsequent chapters scrutinise individual senses, considering them from both a popular standpoint and a learned perspective. If the first part of the book focuses on ideas and beliefs, the second part puts some of these into a social context, to see the effects on the realities of daily life. With the aim of revealing different textures, three chapters concentrate on sensory environments that were closely related and for which we have a good deal of information about day-to-day living and material culture. All centre on great households in the late medieval period: those of bishops, the Queens of England and the aristocracy at the end of the Middle Ages. These environments, ostensibly similar in many ways, can be shown to have differed significantly: manners of perception varied from individual to individual, and there was also change over time. In conclusion, these sensory nuances, identifiable on a micro-level, changes in daily reality, have to be set alongside others that have been signalled as major changes in perception, such as the shift from an early and high medieval way of seeing everything as allegory, through to a perception of landscape and surroundings that was more 'natural';[14] or the reappraisal of man and his sensory abilities that is apparent in theology, medicine, philosophy and many practical aspects of everyday life from the twelfth and thirteenth centuries.[15]

This book is based on a wide range of sources that illuminate beliefs in the senses, largely incidentally, from narrative accounts, the lives of saints, miracle and medical collections, sermons and devotional works, coroners' records and heresy proceedings, to domestic financial records, material culture and buildings. Every endeavour has been made to see beyond the Latin, clerical culture that was responsible for creating many of these records to the realities of everyday life, but at times there is much that is opaque. The period covered by the study is broadly from the mid-twelfth to the mid-sixteenth century. Before this period, evidence for practical responses to the senses, as opposed to philosophical and theological positions relating to them, is very scanty. At its conclusion, the Renaissance, Reformation and Enlightenment were to change fundamentally the way people thought about the senses and perception. The focus of the study is England, but with a recognition that many of the ideas were more broadly based, from the universities of northern Europe to the school of Salerno in southern Italy and Arabic learning. Equally there were within England different groups, receptive in distinctive ways to continental or local influence.

Chapter 2

IDEAS ABOUT THE SENSES

Throughout the Middle Ages, intellectual and popular ideas about the senses were closely intertwined and were a source of mutual inspiration. The range of senses and the manner of their operation indicated by demotic accounts all have similarities with higher, literate views of the subject. Yet there remained important distinctions, some of which can be seen from the vocabulary used to describe perception. The higher culture of cognition, laid out in encyclopaedias or works of theology and natural philosophy, or in learned disquisitions on physiology and medicine, would have been comprehensible to few, although these ideas underpinned notions that were crucial to the day-to-day operation of the Church. These concepts had to be translated into forms that were readily intelligible, in sermons or at confession, in ways that mirrored common beliefs or assimilated those practices to Christianity. Beliefs about the senses were diverse, even if the picture we have of many of them from clerical sources tends to homogeneity. This chapter examines some of the underlying ideas to show both their variety and the relationship between popular understanding and practice on the one hand, and the theoretical on the other.

We are dependent for much of our knowledge of perception in late medieval England on a small group of words that describes the senses and their operation in each of the three principal languages then in use – Latin, Anglo-Norman French and Middle English.[1] The Latin words for the senses – *visus* (sight), *auditus* (hearing), *olfactus* (smell), *gustus* (taste), *tactus* (touch) and *affatus* (speech) – had held their meanings since late Antiquity and their interpretation in medieval sources usually presents little ambiguity. They were well understood by the literate, that is, the clerical community. There was relatively little modification to the range of their meanings and that of related words. For example, *gustare* (to taste) had the sense in classical Latin of 'to taste a little', and it also came to mean 'to listen to' or 'overhear'; in medieval England it also had the connotation of 'to experience'.

The vocabularies of Anglo-Norman French, the vernacular used by the upper classes until the early fifteenth century, and of Middle English, used by all others, were much more fluid as far as sense terms were concerned. In the thirteenth-century *La lumere as lais*, the five senses were listed as those by which the wicked would suffer unremitting torment in Hell:

> En veue [sight], en oie [hearing], e en guster [taste],
> En fleurur [smell], en sentir par taster [sensing by touch] . . .²

The last two verbs in this group, *sentir* and *taster*, held a range of meanings. *Sentir* – here, 'to sense' – could also be employed for 'to think', 'to taste' or 'to smell' (*scentir* was a variant form). Both *taster* and its analogue in Middle English, *tasten*, could refer to the sense of touch, as intended here, or to the sense of taste. In the contemporary *Mirour de Seinte Eglyse* the five senses were described as looking with the eye (*garder de oyl*), hearing with the ear (*oyr de oreilie*), tasting with the tongue (*guster de lange*), smelling with the nose (*sentir de nes*), and feeling with the sense of touch (*tucher de tast*).³ Further, the *Mirour* defined the four levels of existence with reference to sensation: being (*estre*), like stones; being and living (*vivre*), like trees and plants; being, living and having sensation (*sentir*), like animals; and being, living, having sensation and reasoning (*resuner*), like man and the angels.⁴

Veer, 'to see', held wider meanings, commonly associated today with 'seeing', of observation, realisation, recognition, study and consideration; whereas *oir*, 'to hear', 'understand' or 'comprehend', *guster*, 'to taste', and *fleurir*, 'to smell', were less widely elaborated. *Parler*, 'to speak', was sometimes enumerated among the senses in Anglo-Norman sources, a point to which we shall return. *Tucher*, 'to touch', also had the sense, as it does in modern English, of 'speaking about something'. In addition, *odurer* was used for 'to smell', and *puir* was used for bad smells, with the adjectives *puant*, for 'stinking', and *pulent*, for 'fetid' or 'nasty'. *Escut, escout* or *escult* (cf. modern *écouter*), 'listening', was usually combined with a verb: 'to receive a hearing' (*aver escut*), 'to give ear to' (*doner escut*), 'to pay attention' (*faire escut*) or 'to be all ears' (*estre tut en escut*).

The use of sense words in Middle English is marked by some important differences from the modern language. If *heren*, 'to hear', *listen*, 'to listen', *seien*, 'to say', *sen*, 'to see', *smellen*, 'to smell', and *speken*, 'to speak', all had meanings similar to the present, other words exhibit significant variations and/or wider meanings. *Sense* was not generally employed before the sixteenth century to mean 'a faculty of perception': texts speak of the 'five wits'. *Sensibilite*, however, did have this meaning earlier, of sense as a faculty, as did *sensible*, 'perceivable'. Besides the dual use of *taste*, for both touch and taste, *savour* was used for both taste and smell, as were *smak* and *smacchen*. The last two also had the meanings both of smelling or inhaling of smoke, and of giving off an odour. *Flavour* referred to odour or fragrance, or again to the smell of fire, but not to taste. *Odour* could refer to the sense of smell; the smell of an object; specifically to a sweet or fragrant smell, or to a bad one, a stench; or to a mystic, spiritual fragrance. *Scent* was employed in the late medieval period for the sense of smell or an odour, but not before the eighteenth century to mean cosmetic perfumery. *Touch*, besides denoting the faculty of touch, might also mean 'to partake of food' or 'taste'. *Think* is an amalgamation of two similar, but distinct, words in Old English, *þincan*, 'to seem', and *þencan*, 'to think' or 'to reason', perhaps even 'to seem to oneself', the forms of which merged linguistically in Middle English *thinken*, which conveyed the meanings of both earlier words.

These words show areas where linguistic discrimination was limited in the vernaculars, or was so unless texts chose explicitly to make distinctions. Close associations between words for taste and smell have often been the subject of comment, and it has been argued that there is a general law that 'taste' words often move into the 'smell' arena, but not vice versa, as well as a tendency for words related to touch to transfer to taste.[5] The crucial point, however, is not one of general semantic change, but that words used in common for more than one sense imply no primary division of the ideas relating to these senses in general consciousness. Perception may not generally have been as precisely allocated to a sense as the listings of the sensory faculties imply. There is much of the formulaic about the appearance of five senses in a list, albeit not always the same five. The use of some words to refer to more than one sense, and of *sentir* to refer to almost all the senses, suggests that this distinction may not have been as important in practice as it was in literary convention. Where the horizons for discrimination were limited – or expanded, as with the addition of speech to the sensorium – one has to be careful both to distinguish the context and to ask whether the writer intended a distinction to be made between sensory faculties. Ambiguity in the text, or the use of a word that may represent more than one sense, may represent ambiguity in terms of understanding perception. Although the first part of this book is organised by sensory faculty, many in medieval England may not have thought exactly in this way.

What were the characteristics of the senses commonly recognised by medieval society? Cases of sensory deprivation are helpful in delineating some of these. There is much evidence for these in miracle collections, where deficiences were made good. Alditha, who was cured at the tomb of St William of Norwich in the 1150s, had been going deaf for a long time: she could only make out the words of those who put their mouth very close to her ear. She was effectively confined to her house, fearing the derision and disgrace that her deafness would bring if it were detected by others.[6] The friends of Simon de Cantelu, concerned at his growing deafness, gradually had less conversation with him lest they brought shame on one they loved. He prayed to Thomas Becket.[7] Their cure relieved both Alditha and Simon of a moral affliction as much as a physical one.

Punishment by mutilation of the sense organs inflicted moral opprobrium as well as physical pain. In the marketplace at Shefford, on 22 February 1275, Thomas Kek of Oxford was caught immediately he cut the purse of Walter Sparuwe. He was put in the pillory. The town's catchpole later took Thomas out and cut off his ear, before leading him to the boundary of the town to send him on his way: two days later he was found dead in suspicious circumstances on the road to Shillington, three miles away.[8]

Loss of senses was often reported as a consequence of serious illness or madness. Petronilla, a nun of Polesworth, suffered epilepsy and loss of all 'human senses', until cured by the good offices of Thomas Becket.[9] The daughter of John the Fisherman, who lived near Seaford, was partly disabled after sleeping in the open one summer, and became mad, but was restored to her full senses after visiting Reading Abbey and the hand of St James there, around 1173.[10] Rose Tholus, cured of a cancer in her eye with the spiritual assistance of Simon de Montfort, had been deprived of her senses by the suffering in her head.[11] In the late 1490s, William Wotton, of Lingfield in Surrey, had his leg kicked badly by a horse and the injury subsequently led him to lose his senses to madness, from which he recovered miraculously.[12]

Other miracles referred to partial impairment of the senses. A series of witnesses, including Master Thomas of Chobham, the subdean of Salisbury and author of a noted guide for confessors, attested to the cure of one Simon, through the intervention of St Osmund. The event must have occurred late in the reign of King John, while England was under interdict. Simon was paralysed from the waist down and had spent more than a year in this condition at the house of Sampson the Skinner, an elderly and poor man, but it had been too much for Sampson and his wife to sustain him and Simon then moved to the gate of the castle at Old Sarum where the sick used to congregate. While he was living with Sampson, however, Sampson's wife, perhaps tried by their resident, had tested his infirmity by sticking a knife into his shins – Simon had not felt this, nor had he shed blood, nor could he feel the heat of a fire.[13] In a similar case, a woman whose legs had been crippled, her feet drawn up to her waist, for a period of three years, could feel nothing in the lower part of her body; but, cured while keeping vigil at the tomb of St Hugh of Lincoln, felt the extension of sensation and the cracking of her bones.[14]

Complete loss of sensation was one of the signs of death. In July 1306 Gilbert, the infant son of a London goldsmith, was found in a container of water that had been left outside the door to the hall of the house where he lived. Beatrix le Mareschal, Gilbert's nurse, reported that it was believed that Gilbert was dead: his body was completely cold and blue, his mouth was closed and he was not breathing, his eyes were fixed like those of the dead and he appeared to have no sense or movement.[15] A fifteenth-century translation of Guy de Chauliac's treatise on surgery advised what to do in the event of swooning, known widely as 'the little death' and a serious matter, as it was regarded as leading to death itself. Treatment involved stimulating the senses, pulling hair, the ears, the nose, rubbing the extremities, striking the patient and calling him by name, and dousing him in rose-water, or, failing that, cold water.[16] Margery Kempe, approaching Jerusalem, was so overcome with the

2 Old Sarum. An enclosure, to keep away dogs and pigs, was constructed at the gate of the castle by a canon of the cathedral for one Simon, prior to his cure by St Osmund. The cathedral was laid out to the north-east of the castle.

sweetness and grace of God, that she was on the point of falling off her ass, but was helped by two German pilgrims. One, a priest, believing she was sick, put spices in her mouth.[17] Strong stimulus to the senses assisted revival.

Loss of senses also occurred in sleep and at times it was hard to distinguish this from death. The vision of Thurkill, an Essex peasant, took place in October 1206. He remained for two days and nights insensible and immobile, as if in a great sleep or stupor. His wife could not raise him, even on Sunday morning before church, or when she returned, despite shaking him and calling his name. He was woken on the Sunday evening, when his mouth was forced open and a priest was on the point of pouring in holy water.[18] A late fourteenth-century sermon noted that men had five senses and that stopping these wits brought sleep. At this point, man was half dead, unable to work or defend himself against enemies, particularly those that attacked the soul.[19]

These examples point to the strong moral connotation that attached to the senses; and this was true at both a popular level and among theologians and philosophers. The wits were the gift of God and they might be taken away by the Devil. The fifteenth-century sermon cycle, *Jacob's Well*, regarded the tavern as the well of gluttony, the Devil's schoolhouse and chapel. Here the Devil did his miracles: 'He takyth awey mannys feet, þat he may noȝt go, & his tunge, þat he may noȝt speke, alle his wyttes & his bodyly strengthe.'[20] Those without particular senses, either temporarily or more permanently, were often seen as evil, evidence of the Devil at work – and they had powers that were evil. There was something shameful in disability, the loss of limbs or other faculties, as well as the literal loss of the senses in madness. The moral counterpoint to these deficiencies was their miraculous cure, although Sampson's wife showed curiosity, perhaps a desire to expose a charlatan as well as to test the operation of the senses. Practical tests or attempts at revival were of particular urgency, not only from a human point of view, but because of this moral environment.

Vision, hearing, smell, taste and touch are to be understood on most occasions 'the five senses' are referred to, especially by clerical authors. A Becket miracle, recorded by William of Canterbury, described a victim deprived of the five senses in an accident in which the tongue of a bell fractured his skull.[21] Walter Daniel's account of the death and burial of Aelred of Rievaulx (d. 1167), reported that he retained the integrity of his five senses without diminution until his death, although his speech was very short and disjointed.[22] William and Walter probably had little doubt in their minds about which senses were involved: in clerical literature the canonical five were very rarely supplemented, or reduced, and usually only to suit a specific exegetical purpose.[23]

Beyond the five senses of Antiquity, there are two prime candidates for inclusion in the late medieval sensorium: speech, as noted above in the context of Anglo-Norman; and the operation of virtues and qualities of holiness in a manner closely akin to touch, which are considered together in the next chapter. In the case of speech, there was inherent confusion between the senses and the organs of sense; but we should not expect a clear distinction that matches our model of the senses. To our notions, speech is immediately different, giving out information, rather than receiving it, but we must set aside that preconception: in medieval terms, too, other senses could act in this way.[24]

References to speech in some miracles exclude it from the senses, but this evidence is not as straightforward as it may at first seem. The report of the cure of Robert, son of Liviva of Rochester, a boy who fell into the Medway while playing, stated that when found he had neither voice nor senses;[25] a mad youth who was brought to Worcester for the intercession of St Wulfstan was similarly deprived of his voice and senses;[26] and a little child – later resuscitated by the intervention of St Osmund – who fell into the well of William West at Salisbury, was also without voice or senses.[27] These miracles either occurred or were reported between *c.* 1180 and 1231, the last two examples in documents probably compiled with the saint's canonisation in mind. At this period the process of canonisation became more rigorous and, as far as England was concerned, it was the Papacy that was the arbiter of sanctity. Accordingly, the documentation was carefully prepared.[28] It might be considered that these reports of miracles presented sensory perception in ways that their authors believed would find favour with the Papacy. In this case, however, the common phraseology – *non erat vox neque sensus* – was more than accidental, repeating the words of the Vulgate preceding Elisha's revival of the Shunammite's son.[29] In reading these miracles, therefore, the allusion was not necessarily to a framework of sensory perception, but to one that compared the merits of a prospective saint with an established prophet.

Other texts are suggestive, but equally not definitive. Around 1499, a farmer from Malpas lost the ability to speak, as if his tongue had been clipped or pinned by a brooch. Prior to the restoration of his speech, by the miraculous intervention of Henry VI, it was noted that 'he who had been created a man was not like a man', that is, speech was one of man's defining characteristics.[30]

There is stronger evidence connected to the practice of confession. The Third and Fourth Lateran Councils of 1179 and 1215 introduced a programme that was to reform radically the practical operation of Western Christianity. The legislation of 1215 instituted confession, for all, at least annually, to the priest of the parish. The priest, as part of the process of confession, was to examine the penitent on the principal elements of the faith and on his sins. This change spawned an instructional literature advising priests how to take confession and, in parallel, a no less vigorous production for penitents, on how to live their lives. These are of interest as one of the most common ways of examining the penitent was to look at his actions through the medium of the senses, to establish how he might have committed sins: the senses were the gateways between the external body and the internal soul.[31] Treatises with schemas mapping the sins to the senses became common from the start of the thirteenth century.[32] Although questions about the senses were absent from penitential interrogatories before the start of the thirteenth century, confession, based on the senses, had a much longer pedigree in both public rites for penance and in the literature covering the temptations of the senses, dating back to the early Middle Ages.[33]

Robert Grosseteste, Archdeacon of Leicester before teaching in Oxford and becoming Bishop of Lincoln, addressed to parish priests, to monks, and possibly to his own household, several works on penance and confession which are typical of the areas covered, listing the five senses of the body – vision (*visus*), hearing (*auditus*), taste (*gustus*), smell (*odoratus*) and touch (*tactus*) – and the sins that might be associated with them.[34] These confessional formulas had enduring influence throughout the later Middle Ages. Without doubt all would have encountered the five senses in sermons,[35] or at confession, as outlined for example in John Mirk's *Instructions for parish priests*, aiming to elicit a wide range of venial sins.[36]

From time to time, texts in this genre encompass speech, either explicitly or implicitly, as a sense: sometimes as a sixth sense, sometimes combined with taste as 'the mouth'. It has been argued that this conflation of the senses with the activities of the sense organs originated in confusion derived from the attention given to the sins of the tongue, frequently enumerated at great length in penitential treatises. The phenomenon does not start at this point and it is driven in particular by texts in the vernacular, which may owe their model of the senses more to popular tradition than to the classical and clerical inheritance.[37] The fourteenth-century *Book of vices and virtues* (derived from Lorens d'Orléans' *Somme le Roi* of the thirteenth century) emphasised the need to govern the five wits of the body by reason, so that each served its office without sin and mistaking, the eyes to see, the ears to hear, the nostrils to smell, the mouth to taste and to speak, and all the body to touch.[38] Henry of Lancaster's devotional treatise *Le Livre de seyntz medicines*, completed in 1354, opined that all the senses that God had given him had been diverted from their true purpose and towards sin – senses of hearing, vision, smell and speaking, and other senses which he did not have time to mention. Henceforward his five senses (which he does not itemise) would be like fox-hounds, working for good, rooting out and destroying sin.[39] In the morality play the *Castle of Perseverance*, Mankind is shriven of the sins of the five wits, of whatever has been done, seen with the eyes, heard with the ears, smelled with the nose, spoken with the mouth and all the body's bad working.[40] John Burell, examined for heresy in April 1429 in relation to the nature of the sacrament, noted that his brother and two others had taught him that no priest had the power of making the body of Christ on the altar, that God had created all priests. He went on to list the sensory characteristics of a human body, saying that each priest had a head and eyes to see, ears to hear, a mouth to speak and all the limbs of a man; and to note that the sacrament, the true body of Christ, did not have eyes, ears, mouth, hands to feel and legs to walk, but was bread made of wheat flour.[41] This popular enumeration of the senses appears, at about the same time, in the sermons of *Jacob's Well*. Describing virginal and clerical continence, it was noted that a man should be dead to the world, as a man who had lost his five (in fact, six) wits, 'as syȝte, heryng, swelwyng, smellyng, felyng & speche', and alive to God. In its discussion of chastity, it set out the common designation of the senses as the gates of the soul:

> Kepe þi syȝte fro leccherous syȝtes & þin erys from leccherous heryng and þi mowth fro leccherous speche and þi nose fro dyshonest smellyng and þi lymes handys & mowth & þin oþere membrys fro leccherous towchyng. Þese ar þi v [five] wyttes & ȝatys [gates] through whiche þe feendys entryth into þe herte.[42]

These examples also show that perception was widely considered to be a two-way process. *Jacob's Well* noted that men were often overcome by the fiend because they did not keep well the sensory gates to the body. Samson, Solomon and David were overcome by women's wiles – that is, the Devil, partly through the power of women's speech, through its bad qualities, overcame good.[43] What was transmitted through the gates of the body was not just the sense of the words, but the moral and spiritual qualities that were associated with the speaker. An early fifteenth-century popular devotional text, *The doctrine of the hert*, copied its thirteenth-century continental exemplar (attributed variously to Hugh of St Cher or Gerard of Liège) in listing the five sensory gates of the body as 'tastyng, touchyng, seying, hiryng, and smellyng', noting that by these gates the soul went out to outward things and outward things also came into the soul. The senses both operated as receptors of information and gave out tangible and intangible information about the individual, which was capable of changing the character of the perceiver and the perceived.[44]

These popular and pragmatic views of the senses need to be set alongside ideas about their operation, as understood by higher, literate culture. An overview of these can most usefully be drawn from encyclopaedias. The philosophical understanding of the senses and cognition was advanced by debate among scholastics, conducted in the universities. It was disseminated by the movement of both teachers and students across Europe, and frequently into ecclesiastical posts; but the finer points of the study of cognition would not have troubled many outside these groups. Aside from the works of spiritual writers and medical works of a practical nature, medieval ideas about the world were an amalgam of classical philosophy (largely Aristotle, Plato and the Stoics) with elements of Christian and pagan tradition. Knowledge grew by systematisation and accretion, with new material and thought built into or set alongside the ideas of earlier authorities, and with finely wrought explanations of its inherent contradictions.[45] In the medieval encyclopaedia the reader had before him a range of possibilities and an indication of the complexities of debate, much as similar works today make accessible the cutting edge of science, otherwise intelligible to a restricted range of individuals alone.

Before the twelfth century, there were very few works that were encyclopaedic in nature. The most important of these, the *Etymologies* of Isidore of Seville (d. 636), was interested not in the practicalities of life, but in explaining what words meant based on their etymology, by which he largely meant a system of homophones. Thus sight – *visus* – had that appellation because it was livelier (*vivacior*) than the other senses; taste – *gustus* – because it was associated with what came from the throat (*a gutture*).[46] The work of Rabanus Maurus, *De universo, c.* 840, repeated much of Isidore but gave it an allegorical meaning, an explanation in terms of Christianity and its morality.[47]

These early encyclopaedias were based on three essential principles, ultimately derived from St Augustine: that they might demonstrate the association or presence of morality in all things; that their purpose was to help elucidate Scripture; and that the study of things that had been created helped one to understand the Creator.[48] An educated reader would have looked to these texts to understand how the world worked, not primarily for a scientific explanation as we might conceive it, based on experience, but for an explanation drawn from metaphysics and theology, based on a systematic arrangement rather than an alphabetical one. Bartholomew the Englishman's encyclopaedia, *De proprietatibus rerum*, for example, started with God and the angels, before going on to describe all aspects of Creation, from humans (and all parts mentioned in Scripture, including the senses), the earth, heavens and all elements, lands, jewels, metals, plants and animals, concluding with a book on 'accidents' – colours, tastes, liquors and smells; that is, an attribute or quality of an object, but not part of its essence.[49]

From the numbers of manuscripts that survive – there are about 900 copies of Isidore's *Etymologies*, and about 100 of Bartholomew's work, besides early printed editions and translations into English and French – it is apparent that these texts circulated widely, were used extensively in teaching in the schools of Western Europe and were consequently very influential in the diffusion of ideas.[50] Given their importance, one strand of examples in the present book is drawn from the works of the encyclopaedists.

Bartholomew's early thirteenth-century work was translated into English by John Trevisa between 1394 and 1399, although this translation was little used in the later Middle Ages: most relied on the Latin original or the French translation of Corbichon.[51] Bartholomew worked through the senses from those located highest in the body, proceeding downwards: sight, hearing, smell, taste and touch. The messages from the physical senses were received and interpreted in the brain, in the front ventricle, by what was known as 'the common sense'.[52] In Book 5, Bartholomew considered in more detail each of the features of the body. In his examination of the nose, his first source was Isidore. Isidore indicated that the nose – *nares* – was the instrument of smelling. It had this name because by it we perceive things that smell and spiritual things, and judge between sweet and stinking smells. Wrathful and rowdy men, rash and ignorant, were said to be, 'without a nose'.[53] Isidore used the similar-sounding word *gnarus* (knowing), and its opposite, *ignarus* (ignorant), to indicate that the action of smelling was literally 'knowing with your nose'.[54] Elsewhere Isidore explained that *spiritus* – with which the spiritual things were related – was the soul after death, what would be called *anima* in a living being. Here, then, was a physical sense which could assess what we might call an intangible quality.[55]

Bartholomew proceeded to explore the physiology of smell, for which he turned to the work of Constantine the African (d. before 1099), a monk of Monte Cassino known for his translations of Arabic medical texts. Constantine held that air was drawn into the brain through the nose, one nostril for incoming air for the brain, the other for ejecting superfluities. Bartholomew indicated that this, in his view, was not quite right; that the instruments of smell were two teats hanging inside the nostrils; they drew in the air and a sinew came from the brain, giving them the spirit of feeling (sometimes referred to as the 'animal spirit', that is, coming from the *anima* or soul). Returning again to Constantine, he noted that a 'fumosity' or dry smoke (as opposed to a vapour, a wet one) from the air that had been drawn into the nose was joined to the spirit of feeling, which took a likeness of the smoke and presented it in the brain for the judgement of the soul.[56] From physiology, the discussion has moved to the transmission of odour. Bartholomew focused on disputed territory between Platonic and Aristotelian natural philosophy as developed in the Middle Ages. Aristotle and his followers held that odour was an immaterial quality (known as a *species*) that belonged to an object and radiated from it, which was then captured by the senses and imprinted on the common sense in the brain. Bartholomew and Constantine saw smell as a smoke or vapour, as Plato had done, having a physical quality.[57]

If the transmission of odour was in dispute, so was the physiology of olfactory perception. Galenic medicine had a different view of the operation of this sense. The nose was a passage to carry smell to the brain; the bone separating the nose from the brain was porous and the air passed through for smell to be sensed by two projections from the front ventricles of the brain: in the case of smell, the brain itself was a sense organ.[58] What Galen's followers did have in common with Bartholomew was the view of the operation of the senses as a form of touch, an area that will be explored in more detail later.

Smell was mapped on to another framework of classical and medieval analysis. To medieval theorists, all matter was composed from one or more of four elements, each with its own characteristics: fire (hot and dry), water (cold and wet), earth (cold and dry) and air (hot and wet). Four related humours – choler or yellow bile, phlegm, black bile and blood – were crucial to the existence of the human body. External influences, known together as the six non-naturals – air; food and drink; sleep and waking; movement and rest; consumption and excretion; and the emotions – might in turn affect the humours.[59] To Bartholomew, those smells that were similar in complexion to one's own make-up in terms of elements and humours were those that were likely to be found pleasing; those that were dissimilar were likely to be displeasing or even noxious.[60] At various times the mapping of smell against the elements was different, as the sense did not easily fit against air, earth, fire or water. Plato held that none of the four elements had smell: it arose between them as vapour, or mist, from material in a state of change. Galen, on the other hand, pointed to different characteristics in odours: the odour of roses, for example, he understood as cooling; odours produced by burning, for example asphalt, cassia, myrrh, frankincense, he saw as dry and hot.[61]

In this way, Bartholomew's analysis of smell – and the other senses – encompassed a moral or spiritual standpoint, systems of natural philosophy and of physiology. We now need to turn to these areas of analysis to look in more detail at how each appraised the senses.

In mapping human existence, the Church turned to the six days of Creation to provide a framework for its theologians, in discussions that would encompass both man's soul and his body.[62] Between 1228 and 1235, drawing on the work of Bede and other commentators, Robert Grosseteste used the days of Creation as a parallel to the life of man. The light of the world illuminated the outward senses of the babe (*infans*) as it came into the world at birth. The five senses were fixed in the body in infancy and childhood, equivalent to the first two days of God's work; and the development of the body progressed in a way that mirrored (and derived its knowledge from) the progress of Creation overall, gaining strength and memory in childhood (*puericia*: the second day); in adolescence (*adolescentia*: the third day), completing its physical growth; in manhood (*iuventus*), attaining wisdom and understanding; in its fifties and sixties (*etas senioris*), facing bodily deterioration of the outward senses, but growth in internal powers; and finally, coming to old age (*senectus*), with the further weakening of the body and death, but the strengthening of the mind, which grew more beautiful and gained control over the harmful, physical senses and desires of the flesh. To Grosseteste, this was the epitome of Creation, advanced age bringing forth the living soul and completing the process of making man in the image and likeness of God.[63]

While Grosseteste here placed the acquisition of senses at birth, there was debate among philosophers over the point at which the foetus gained both a soul and sensation, a progression from vegetative to animal, and then to human, intellectual soul. The writer of a late fourteenth-century sermon on the conception of John the Baptist noted that the Baptist danced in the womb of his mother, Elizabeth, during the visit of her cousin Mary – and therefore in the presence of Christ. To the writer's mind, this indicated that one should not believe that it was six months into a pregnancy before the soul was coupled with the body, having before that the 'soul' of a plant and then of a beast, and that the Baptist was alive before Christ was conceived.[64]

Once the senses were acquired, theology offered man a choice between the temptations of the flesh through the senses, leading to perdition, and their proper use in devotion to God, or at least in a morally acceptable way, leading to salvation. Discussion of the ethics of sensory perception had a long pedigree from patristic texts, such as Augustine's *Confessions*, onwards.[65] The senses might be the gateway between the inner and the outer man, but, by the end of the first millennium, they were regarded with rank suspicion by theologians. St Anselm noted that 'delight coming from the senses is rarely good: more often it is truly bad.'[66] At the start of the thirteenth century, Adam of Eynsham, the biographer of St Hugh of Lincoln, recalled the aptness of the hymn of St Ambrose, *Deus creator omnium*, which, as a

Carthusian monk, Hugh had sung often: 'Once divested of the hazards of the senses, the innermost parts of the heart dream of thee.'[67] Recounting Hugh's diet and his behaviour as Bishop of Lincoln, when entertaining in his household or as a guest elsewhere, Adam recalled his great detachment from acting, music or the delights of the table, 'as if only his exterior senses were charmed by these sweet delights'.[68] The snares of the external senses – and attempts to spurn or deny them, or to limit their influence – formed a prominent strand in religious thought and daily life through the later Middle Ages. The Council of London of 1268 aimed to control closely the movements of nuns in order to preserve their innocence. After specifying those places within the monastery to which the nuns might go, it was enjoined that at other times they should remain in the cloister, 'which might enclose them and the senses of her body . . . Thinking about and looking at God alone with the eyes of the mind, they might, in contemplating its sweetness, have a foretaste of eternal life.'[69] *The cloud of unknowing* noted that even the extreme position of destroying the sensory organs would fail to save the individual from sin.[70] The focus of confession on the five senses largely maintained this negative way of viewing sensation – and depriving a man of his senses could be a punishment from God.[71]

Although it was common among theologians and moralists, not all treatises on confession took this deleterious standpoint. The attitude of Grosseteste to the senses was positive, recognising that these were the channels, given by God, by which one could discover good and conduct oneself well.[72] The senses might be put to good use. It was probably a representation of the good use of the senses that Langland intended by the five sons of Inwit by his first wife, Sir See-well, Sir Say-well, Sir Hear-well the courteous, Sir Work-well-with-thine-hand, and Sir Godefray Go-well. Their task was to save the Lady Anima (the Soul), until Kynde (the Creator) came for her.[73] A sermon, in a collection composed in the late 1380s or 1390s in the Wycliffite tradition, provided a long discussion on the senses and how they might be used for good: sight, to see God's law, that is Christ; hearing, to hear God's word; smell, which indicates to God whether a man is good or evil; taste, for by man's speech we may understand who tastes of God's sweetness, tasting it on the tongue; and touch, to feel Christ's yoke, which is sweet and soft and which draws men upwards and not down to Hell.[74]

There was a long tradition of responding through the senses to the physical presence of the sacred. The early pilgrims to the Holy Places recounted how, in Judaea, they had physically seen the divine, touched it and tasted it. St Jerome made the transcendental connection to the holy, linking the events of a biblical past with his present experience, by the 'eye of faith'. Fourth-century theologians accepted that the Incarnation had legitimised sensory experience as a way of knowing God.[75] At about the same time, Augustine argued that only senses that transcended the corporeal could know the divine – making use of a set of internal senses that reflected the external ones.[76] These ideas found their correlate in the theological and devotional practices of the later Middle Ages, especially in mystical works. The Cistercian, Baldwin of Ford, successively Bishop of Worcester and Archbishop of Canterbury in the 1180s, in his treatise *De duplici resurrectione* (on Revelation 20–1), matched the progress of the ascetic to the progress of the soul, indicating the mystical signifi-cance of the five spiritual senses.[77] *Jacob's Well* put names to the virtues that the spir-itual senses described, the five watergates of the soul: understanding (the sight of the eye of the soul), desire (the ear and hearing of the soul), delight (the mouth and swallowing of the soul), the mind (the nose and olfactory abilities of the soul), and consent (the soul's ability to feel).[78]

In mystic experience, the sensations of the spiritual senses became physically manifest to the believer. Those who viewed sensory devotion as their object linked their spirituality to the sensual language of passages in the Old Testament, particu-larly the Song of Songs. But saints are not easy companions and, as one tract alleged, those that tried this spirituality 'travayle þeire ymaginacion so undiscreetly, þat at þe laste þei turne here [their] brayne in here [their] hedes'. The Devil then had the power to 'feyne sum fals liȝt or sounes, swete smelles in þeire noses, wonderful taastes in þeire moþes, & many queynte hetes & brennynges in þeire bodily brestes or in þeire bowelles, in þeire backes & in þeire reynes [kidneys], & in þeire pryve membres'.[79]

In terms of natural philosophy and the study of cognition, from the thirteenth century the emphasis of debates about the nature of the world changed to scientific explanation. This was a consequence of the availability of the full range of Aristotle's works, from *c.* 1260–80, present in authoritative Latin translations from the original Greek, as opposed to the limited group of works transmitted to Latin scholars through the medium of Arabic. This transition was of major importance, from an educated point of view, in the understanding of cognition and the operation of the senses.[80] As far as the senses were concerned, there was a good deal to draw on in Aristotle's work, in the *Libri naturales*, and particularly in *De anima* and *De sensu et sensato*.[81] It was immediately productive of increasing and lively debate in the schools and universities of north-western Europe, particularly in the period to *c.* 1350, resulting from the contradictions between Aristotle on the one hand, medical and theological viewpoints on the other, and the development of rival theories of cognition. These debates centred on how the sense organs worked, the functioning of the common sense – that is, the internal sense that received the infor-mation from the external senses – and how information was transmitted to them, as well as how imagination and memory created or recalled sensations. As indica-tive of the general debate that would have been captured by the encyclopaedists and those trained in the schools, these arguments were important, particularly for the development of late medieval philosophy, but it is not necessary here to do more than sample a range of viewpoints, especially those that find resonances in the workings of perception as it was more generally understood.

Two major areas of debate were the functioning of what were known as the internal senses, and the means by which information was transmitted to the senses, both external and internal. The internal senses, or 'inward wits', were the link between man and the external world on the one hand, and the intellect and spiritual truth on the other. Ideas about this came from both philosophy and medicine, but philosophy accepted the evidence of the latter and constructed its theories around medicine.[82] Medicine derived its ideas from fragments of Greek medical works, especially those of Galen and his followers, from Greek philosophy and from the works of Arabic writers, producing a scheme of cognition based on three key powers – imagination, thought and memory – located in the three main ventricles of the brain, although with some elaboration and variation.[83] Early medieval philosophy had supplied a single common sense to receive the information from the external senses, but Arabic philosophers, particularly Avicenna and Averroës, had elaborated the number and function of the faculties that constituted the internal senses. By the thirteenth century, the internal senses were commonly described as containing, in the first ventricle of the brain, the common sense, the function of which was to receive sense impressions directly from the bodily senses. The establishment of this location for the common sense had itself been a matter of debate. Robert Kilwardby, Archbishop of Canterbury from 1272, writing probably in the 1250s, had attempted to reconcile the view of Aristotle that the heart was the organ of common sense with that of medicine and St Augustine, which argued for the brain.[84] The common sense did not retain sense impressions; instead it passed them through – the internal senses were often portrayed in diagrams in a linear sequence – to two senses in the middle ventricle, commonly indentified as two forms of 'imagination', cognitive and estimative. The former we might now recognise as imagination, a sense of fantasy based on experience and reason; the latter operated more in the realm of instinct, allowing one to make judgements beyond the information contained in the sense impression, for example what might be dangerous or otherwise. The fourth area, located in the rear ventricle, was memory. The number of internal senses that might be identified and their functions were disputed. Roger Bacon and Thomas Aquinas both enumerated five internal senses, but with differing capacities.[85]

Among the earliest Englishmen to engage in detail with this new learning was John Blund (c. 1175–1248). He studied probably first at Paris, then taught arts at Oxford until the interdict of 1209–14, when he may have returned to the Continent. Writing close to 1200, his own treatise *De anima* examined the five senses and the common sense, working from Aristotle's *De anima* and the commentary upon it by Avicenna. In his discussion of taste Blund noted that, according to Avicenna, it was a force or virtue arranged in a broad nerve to receive flavours coming from the outside. To this Blund countered that the nerve was a bodily thing and bodily things cannot perceive; only the soul can. Likewise it was said that taste was a kind

3 The internal and external senses, from a thirteenth-century English manuscript of St Augustine's *De spiritu et anima*. The five external senses are each marked by the name of the sense; the internal senses are listed across the brain, starting with the common sense at the front.

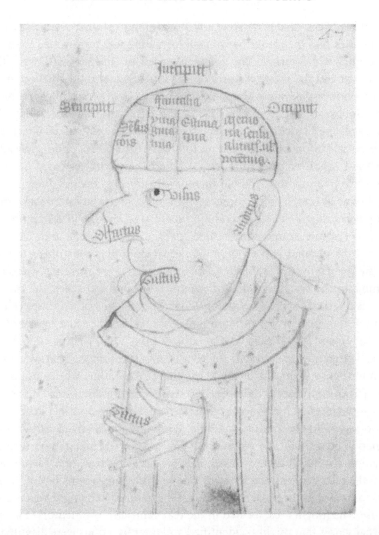

of touch, as it functioned only when it touched something with flavour; but if this were the case, Blund argued, then it would be a form of touch, that is, not a sense by itself. Some said that taste was distinguished from touch in that taste detected flavours, whereas touch distinguished hot, cold, wetness, dryness, weight or lightness. Blund then analysed the action of taste by analogy with hearing and smell: air was the medium that brought both sound and odour; and according to Aristotle, it was saliva that acted as the medium bringing taste through its humidity.[86]

Theories of how information was transmitted to the sense organs were of considerable importance and were reflected in popular beliefs about the operation of the senses. The work of a later writer, Walter Burley (1275–post 1344), an important commentator on the works of Aristotle, was typical of the continuing interest in questions of cognition and sensory perception. In his *De sensibilibus* (On perceivable things) of *c.* 1337, he set out the view that there were five internal senses corresponding to the five external ones. The common sense distinguished the information they supplied; cognition distinguished the accidents – colour, number and shape. Burley then discussed the imagination and memory, their location in the brain, and a series of views of the operation of individual senses. He described six different opinions about the way sight functioned. The first argued that the presence of an object alone was necessary for it to be seen. To this he countered that if one looked at the sun, then closed one's eyes or turned away from it to some dark place, there would still be an image of the sun, albeit less bright, then subsequently purple and fading, an after-image. A second view was that vision was the result of the action of the object, being suffered or impressed on sight. A third opinion was that vision resulted from rays going out from the eye to the object; a fourth, that sight resulted both from the presence of the object and rays from the eye mixing together; fifthly that a repeated image (*species*) coming from the object created a visible image (*species*) or act of seeing in the eye; and sixthly, similarly, except that the visible *species* was translated into a spiritual (or intelligible) *species* for the purposes of cognition.

Burley went on to discuss how the senses worked. Sight was comprehended not by the eye, but within the front part of the brain, which was linked to the eyes by the two optic nerves. The perception of sound was not located in the ear, but within an air pocket in a membrane in the brain; smell was also located in the front part of the brain, in a web of nerves. If the operation of these three senses was located in the brain, for the other two it was not. The sense of taste was not in the flesh of the tongue, but in a fine and subtle net – that is, of nerves – within the tongue. The sense of touch was equally not in the flesh, but in a similiar net stretched over the whole body.[87]

Burley's discussion of sight touched on the two principal theories of its operation – extramission and intromission – which underlay more widely recognised notions of perception. Extramission was more commonly understood in the earlier Middle Ages as explaining the action of sight. The eye sent out rays of its own, gathering strength from ambient light, which were then returned back to the eye from the objects they touched with an impression of that object. The theory had its origins in Neoplatonism, reworked by St Augustine, who believed the ray was constituted from a kind of fire that gathered behind the pupil and brought back to the eye light or fire from the object. Intromission worked in a way that was in part similar. No ray came from the eye, however. Nothing was seen except as a result of light arising from the object in view, transmitted to the eye through the air by a process known as the multiplication of *species*.[88] In theory, information was transmitted from an object to the body and into the mind by *species*, a likeness, image or representation of the original. There were three kinds of *species*: *species in medio*, which multiplied all the way from the object to the sense organs of the body; *species* in the sense organs (sensible *species*), and *species* in the common sense (intelligible *species*). The *species* took with them the accidents of the original, for example its colour. This theory of cognition was grounded in Aristotle and was developed by the 'perspectivists' (as they were known: this theory was especially appropriate to sight), notably Roger Bacon (*c.* 1220–92) but developing the ideas of Grosseteste, Alexander of Hales and Robert Kilwardby, and is also especially associated with Thomas Aquinas (1225–74), although it was generally accepted. Counter-theories, of Peter Olivi (1247/8–98) and William of Ockham (*c.* 1285–1347), held that *species* were not necessary, that cognition was intuitive, the result of an impressed quality on the senses, that it was not dependent on *species* and their transmission.[89] Some held that the visible object was able, in this way, to assimilate the eye and soul to its nature and essence – a tangible contact and change, and one that was commonly understood at a popular level.[90]

A further, important dimension in the understanding of the senses came through medicine and physiology, particularly from Arabic medical works. These became more widely known through the translations of Constantine the African and the medical school at Salerno, among the first to make use of the new translations of Aristotle's *Libri naturales*. The school at Salerno developed the means of exploiting medicine based on the principles of natural philosophy, through teaching and investigation. It employed a series of questions – *questiones phisicales* – as a means of teaching. They had a significant influence in England, for example on Alexander Neckham, and on the scientific works of the school based in Hereford; and in the mid-thirteenth century, at Paris, on Englishmen such as Roger Bacon and Robert Kilwardby. Beyond this, they had an enduring presence through the Middle Ages and into the early modern period.[91]

Many of the questions were couched as problems, rather than describing, as the philosophical texts did, how the senses worked; and the responses, where we have them, are informative. One of the prose Salernitan questions asked why the man who used cumin – which is both hot and humid by nature – remained pale. The spirits ought to have been roused by the heat and the subject ought therefore to have had a good colour. In this case the cumin was probably taken orally, rather than used purely as an aromatic, and the range of answers is an interesting illustration of how this problem might have no definitive resolution in the scheme of thought. One answer suggested that cumin made the person pale from its substance and not from its qualities. Cumin was pale in colour and, on entering the stomach and liver, it made humours of a similar essence, which nourished the limbs and made them pale following its nature. A second solution argued that cumin was naturally a subtle or fine substance and reduced the spirit – and as it was the spirit that coloured the body, the skin was discoloured. A third argued that cumin was above all wind-inducing, and as the excess of wind is expelled, so the vital spirit is diminished with it, with paleness and discoloured skin following as a consequence.[92] Some questions related more directly to sensation: why, for example, did the brain not have feeling, when all the senses came from it? The soul, having its seat in the brain, put the virtue of sensing into the limbs through the nerves that came from the brain; the nerves operated as if they were a part of the brain. How were the nerves capable of feeling, if the brain was not? The brain did not receive movement because of its softness; but the nerves which proceeded from the brain and which were not separated from it were of a mean between hardness and softness, whence they could receive change.[93] Collections of these problems, originally destined for learned readers, circulated more widely, leading to the development of more popular interests in science and medicine.[94]

As the senses were central to both cognition and higher truths, it was of some moment to place them in rank. This was a task for philosophers and students of physiology, as much as those of theology. Discussions about the place of vision in Western culture and the dominance of this sense over others have their roots in these exchanges; but one needs to recognise that these debates have their own temporal qualities and context and are conducted within a restricted framework, the high culture of cognition, which would have been understood by few. The debates were also a natural consequence of the great demand within medieval thought for systematisation and enumeration.

A very common order of rank placed vision at the head of the list, followed by hearing, smell, taste and touch. This mirrored the order of the sensory organs, describing them from the top of one's head downwards. It had as its authority Aristotle, in *De anima* and *De sensu et sensato*, the first Latin translations of which were available from the second half of the twelfth century. Although he had not argued explicitly for this ranking, for example recognising the importance of hearing for conveying information and understanding, and touch, Aristotle's treatment of them implied an order.[95] It was also the order employed by Cicero in his *De natura deorum* and some early Christian authors.

Another order was also current, placing taste before smell. This appeared in the works of Gregory the Great, Rabanus Maurus and, in the twelfth century, in the *Cosmographia* of Bernard Silvestris. To these medieval authors, the significance of the senses – all things had a teleological significance, in this case not very different from those assigned by Aristotle and Plato – was that they helped the inner body to move towards spiritual truth.[96]

Although this order of rank was very common, it did vary if it suited the purpose of the discussion. An early thirteenth-century treatise on taste argued for the primacy of that sense, an argument that would have found favour with many physicians.[97] Taste might also, but perhaps unusually, have an especial significance in relation to divine revelation, through the physical consumption of the Eucharist.[98] An author might vary the order, again depending on purpose. Aquinas, for example, excluded taste and smell from the senses that might appreciate things on aesthetic grounds; here he placed hearing above touch, but reversed the order when elsewhere discussing the volume of information a sense might convey.[99]

The representation of the senses in images could convey an order, but this was unusual. Given the wide range of references to the senses in medieval writing and their significance in the progression of the soul to its salvation, it is both striking and surprising that the senses feature so little among iconographic evidence. The religious cosmology of the Middle Ages was filled with allegorical relationships, such as the Constellations or the Ages of Man. Although the senses had an important place in this view of the universe, they were not usually indicated by anything other than words within a diagram, even if other elements were drawn out.[100] Across Europe, the senses appeared as the subject of works of art very rarely, and even more rarely in England. That this is not just an accident of preservation can be seen from the analysis of the decorative cycles of wall paintings for the palaces of Henry III, based on the extensive evidence of instructions given in administrative documents. This shows other conventional summaries of knowledge, such as *mappe mundi*, for example in the Painted Chamber at Westminster and in the King's Hall at Winchester; but depictions of the senses were absent.[101]

In so far as we can tell, the senses were not a major feature of decorative art in Antiquity. This meant that few models for depiction passed through to the Middle Ages. Representation therefore took a wider variety of form, although there were some themes that recurred.[102] The earliest occurrences of the motif are all secular, for example on the ninth-century Fuller Brooch, where each of the senses is depicted as a man in the central roundel.[103] It is also possible that the figure on the Alfred Jewel may represent sight.[104] Literary works – particularly secular works and those in the vernacular – had a longer tradition of linking the senses with a sensuality traditionally associated with women and especially with love poetry. The early fourteenth-century poem 'Annot and John' has five verses working through the senses, comparing the beloved to paragons of sensory perception.[105] In the play *Wisdom*, probably written *c.* 1465–70, the five internal wits are played by five prudent virgins dressed in white kirtles and mantles, chaplets and kerchiefs. They enter, however, singing a sensuous passage from the Song of Songs.[106] But in iconographical terms, the senses did not make a signficant contribution to secular works until the end of the Middle Ages.

Although they were always exceptional, from the thirteenth century, when depictions of the senses occur more often, there were three principal genres of representation: depictions of human figures performing actions associated with a sense; illustrations of the sense organs, sometimes on the body, or labelled diagrammatically; or the use of animals to represent senses that might be particularly associated with them. The recovery and dissemination of Aristotle's works, particularly the *Parva naturalia*, offered new possibilities for illuminators. Texts were sometimes

4 The Fuller Brooch: a ninth-century disc-brooch, subsequently remodelled as a pendant. Made of silver and niello, it is perhaps associated with the court of Wessex. The central figure represents Sight with, top left inner circle, a figure representing Taste (with his hand in his mouth); top right, a figure with his hands behind his back, Smell, sniffing a leaf; bottom right, Touch, a figure rubbing his hands together; and bottom left, Hearing, a figure with his hand cupped to his ear.

accompanied by a series of human figures, each performing a function clearly associated with a sense, such as playing a musical instrument or smelling flowers; but it was only in a minority of books that the five senses were depicted.

Scholastic works constituted a second category. Some of these, such as the *Anticlaudianus* of Alan of Lille (d. 1203) – a poem in which the five senses were horses drawing the chariot of Prudentia towards Heaven – were illustrated by a drawing of the sense organ; others positioned both the internal and external senses on the body.[107] These typically labelled an illustration of the body with the names of the senses.

The third group, which employed animals to represent the senses, had different roots, associated with moralising literature. Encyclopaedias referred to those animals whose sensory features were regarded as better developed than those of man. Thomas de Cantimpré (d. 1270 or 1272) identified the boar for its sense of hearing, the lynx for sight, the ape or monkey for taste, the vulture for smell, and the spider for touch.[108] The earliest illustrations based on these associations occur about 1240. The most famous English example, the wall paintings in the Longthorpe Tower near Peterborough, probably executed between 1320 and 1340 but possibly as late as 1350, contains on the east wall a similar image with the beast standing for each sense located around a wheel. Here a boar represents hearing; a

5 In this fourteenth-century manuscript of the translation of Aristotle's works the figure blowing a trumpet represents sound and the sense of hearing.

cock, sight; a monkey or ape, taste; a vulture, smell; and a spider in its web, touch. Behind the wheel stands a king, with his hand resting on a spoke. He is probably to be interpreted as the common sense, or the higher reason of the soul, judging the perceptions that come from the corporeal senses.[109] The poor state of the bottom right-hand corner of the Longthorpe painting has led to some discussion about the identity of the cock and whether the creature should be interpreted as a basilisk or a cockatrice, but there has been no doubt that it represents the sense of sight. The positioning of the senses on the wheel – and the place of touch as the highest – may reflect a relative ranking of the senses. Touch was held by some to be more perfect in man than in other animals – in contrast to the views of Thomas de Cantimpré and others. Touch was also a model for the way in which sensory perception worked in man, with the internal senses touching the external ones.[110] There is little dispute, however, that the picture stands in general as a model for the interpretation of man's sensory perception, with the external senses governed by the kingly figure.

6 Longthorpe Tower, near Peterborough: the wheel of the senses.

Until the later fifteenth century the senses in artistic works were personified as men, although it is just possible that the women featured in a series of tapestries owned by Henry V – 'D'un dame qi harpe ung note' – may have a sensory context.[111] Thereafter, the senses were more frequently depicted as women.[112] In high art, the first definite female association appears with the famous sequence of six tapestries known as the Lady and the Unicorn, made for the Le Viste family of Lyons,[113] and the model then spread quite rapidly. Henry VIII possessed a painting of 'a woman playinge upon a lute with a booke before her and a litle potte with lillies springinge oute therof', which may have been intended to represent sound, sight (reading the book) and smell.[114]

These discussions and illustrations of the senses reflect only a part of the range of perception, the five senses drawn from Antiquity. The use of different vocabularies by the three languages of late medieval England – Latin, Anglo-Norman and Middle English – shows how paradigms for perception might be shaped diversely. Speech was popularly recognised as part of the sensorium, even if it was excluded in some ecclesiastical contexts. The human body and its senses were not regimented, enclosed and exclusive in their perceptions. Many things might have effects on the body and the body might equally affect them. These effects were often explained in the medieval period as 'virtues' or 'powers' of substances and they were as much a part of the process of sensation as vision or hearing, operating in a similar twofold way. They were part of a model of perception very different from our own, what we might label 'extra-sensory perception'; and it is this that we must next discuss, along with a detailed examination of touch and of cosmic forces that might influence the body and its senses.

Chapter 3

TOUCH, VIRTUES AND HOLINESS

Touch had a pivotal role in many analyses of the senses. A common metaphor centred on the spider. In the middle of its web, feeling all movement both within and without the web, it was likened to the soul – in the middle of the heart, feeling all, without spreading itself, but giving life to the whole body, controlling the movement of all the limbs.[1] If not touch itself, it was something closely akin to it that imprinted the messages of the external senses on the internal ones and within the brain. Virtues and holiness might also make a considerable impact through touch or general proximity: this was one of the most common ways by which moral or intangible qualities might pass between beings or to and from objects.

Bartholomew summarised thinking about this sense. Regarded as the least subtle of the wits – by which he meant it was the most earthy and beast-like sense – it was touch that let the soul know hot and cold, dry and wet, soft and hard, and rough and smooth. Unlike the other senses, it was present in all the physical parts of the body, although it was most commonly associated with the palms of the hands and the soles of the feet. Sensation could be lost through illness or physical disability, 'in lymes þat ben croked and contract', or temporarily, through sleeping in an awkward position or through cold, and bodily pain might severely affect it. If the sense were completely destroyed, so was the individual. Bartholomew also recognised that this sense underpinned all the other senses: 'Þe wit of groping inprentiþ his felinge in alle þe lymes of oþir wittis and so doþ none of þe oþir wittis.'[2] It was commonly noted that touch was both an active sense – in the case of 'touching' – and that it was also present in a passive mode, in being touched.

Different properties attached to the two hands or, more generally, to things close to or touching the two sides of the body. Bartholomew recounted Isidore's belief that the right hand was to be used for the giving of peace, faith, trust and salvation, deriving the Latin for right, *dextera*, from *dare*, 'to give'. He held that the left hand, *sinistra*, came from *sinere*, 'to suffer' or 'to depend on someone or something', that is, the right hand.[3] Bartholomew also recounted Aristotle's argument that the blood was hotter on the right side than the left and that therefore the right hand was more

7 Longthorpe Tower:
the spider and its web,
a metaphor both for
the sense of touch and
for the soul, in the
centre of the body,
feeling all movement.

ready and able to work than the left; and from Isidore, that the right side was more agile, while the left side – *laeva* – was better able to lift (*levare*) loads. It was for this reason that the shield, buckler, sword or knife was carried on this side. The head of the heart or 'sharp end' was set on the left side of the body – therefore there was more spirit of life on that side and it was the stronger side.[4]

There were also general moral or spiritual divisions between the two sides, associating the right with pity or mercy, honour and virtue; and the left with justice, infamy and vice. This division had a foundation in Greek philosophy, in the works of Pythagoras, and biblical authority: at the Last Judgment, the good, for their works of mercy, were to be placed on the right; the damned, on the left. From these general divisions, much followed. In religious art, for example, the crucified Christ was almost always represented with his right foot above the left, with the wound from Longinus' spear on the right, Christ's wounds symbolising pity.[5]

In terms of popular belief much significance lay in whether the right or left hand, or the right or left side of the body, was used. When the daughter of the smith of Postwick came to the tomb of St William of Norwich to pray for the cure of her illness, she held her candle in her right hand.[6] Seeking mercy, in 1430, Baldwin Cooper of Beccles read his abjuration for heresy while touching a book of gospels with his right hand.[7] The left side was associated literally with the 'sinister', a meaning that the word had borne since Antiquity. The blessing of Titivillus in the play *Mankind*, probably written between 1465 and 1470, was given with his left hand, sending his three companions to the Devil.[8] Part of the penance of seven Kent

8 The crucifixion, from the Gorleston Psalter. Christ's right foot is above his left and the wound from Longinus' spear is on the right-hand side. A later addition to the Psalter, probably painted *c.* 1320–30, possibly after it had passed into the possession of Norwich Cathedral Priory, the image has a strong Italian influence and is based on a Siennese model.

heretics sentenced by Archbishop Warham in May and June 1511 was for each of them to wear an image of a faggot surrounded by red flame on the left upper arm of their outer garment, openly and without concealment.[9]

The effectiveness of the virtues of some stones was dependent on the side of the body on which they were carried. Aspites (perhaps the Arabian stone *aspilates*, recorded by Pliny), a red and shining stone, was considered particularly powerful if carried on the right-hand side;[10] whereas diamond and the two kinds of stone known as chelidonius, both allegedly removed from the gut of a swallow in August, were carried on the left.[11] An Anglo-Norman poem on the art of love, dating from the second half of the thirteenth century, centred on a cherub or cupid, part of whose task was to cure the lovesick with treacle (theriac, a special ointment that was a cure-all against the strongest poisons), held in a special box hung from his left-hand side.[12] In chiromancy, however, the practitioner was required to consider both hands, but it was believed that the principal indications for a man would come from the right hand, for a woman, from the left.[13] This division, based on gender, was one of many that attributed sinister things to women, and the consequences of physical contact with the sinister were high among the reasons for not touching females.

The fingers each had names, functions and common associations that might give a particular emphasis to touch. The *Etymologies*, rehearsed by Bartholomew, illustrate the long-term currency of these ideas. The thumb – *pollex* – had the most virtue and strength: the origin of the name was ascribed by Isidore to *pollere*, 'to be strong or powerful'. The forefinger, known to Isidore as the index (or pointing) finger and the *salutaris*, or 'greeting' finger (its meaning in Antiquity was 'beneficial' or 'useful'), was referred to additionally in Trevisa's translation of Bartholomew as 'the lick-pot' and 'the teacher', from its demonstrative functions. The middle finger was known as *impudicus*, literally 'unchaste' or 'shameless', according to Isidore because of its association with insulting gesture. Next came the ring finger, *anularis*, also known as 'the leech', 'leechman' or doctor, *medicus*, as it was this finger that was used to administer a salve around the eye. The little finger was known as the *auricularis*, or 'ear-finger', from its use for cleaning the ears.[14]

These names for the fingers appear regularly in medieval records and must have been widely understood. In the late twelfth century, Emma of Halberton, who had the lack of respect to work on Whitsunday, had an accident sewing, the thread passing through her middle finger – which, William, a monk of Canterbury recounted, was deservedly known as the *impudicus*, on account of this scandalous activity – and on into her ring or 'medical' finger.[15] In late December 1265, Richard of Eltisley, of Eaton Socon, was knocked to the ground with a willow staff by William Moring of Staploe, who leapt on him and bit the index finger of his right hand, maiming him. In November 1271, Emma, the wife of John of Bretville of Great Barford went to the Bedfordshire county court to appeal against Simon, the son of Roger of Cainhoe. Simon had struck John's left hand with a sword, cutting the sinews of the auricular finger and breaking the bones of the *medicus* next to it.[16] An early fifteenth-century translation of John Arderne's treatise on the fistula referred to the index finger as the 'schewyng fynger'.[17] If the case of Emma of Halberton indicates how moral qualities might attach to the fingers for the worse, then for a bridegroom to place a ring on his spouse's ring finger was especially beneficial: it was said that there was a vein coming from the woman's heart to that finger 'and therfore the ringe is putt on the same finger, that sche shulde kepe unite and luff with hym, and he with hyr'.[18]

The association of names with the fingers formalised some of the ideas that were commonly associated with gesture. The notion that external actions revealed the inner man went back at least to Cicero on the one hand and to Augustine on the other – and the possibilities were particularly important in relation to touch. Gesture, especially with the hands, sometimes combined with touch directly, sometimes more obliquely, was used in specialised ways to convey information, qualities or a moral, spiritual or even legal force. Signs and touch brought much more than bare signification.[19] It was held in the tenth and eleventh centuries that one needed to make the sign of the cross when yawning, to prevent the Devil entering the body (perhaps the origin of the present-day gesture of covering the mouth when yawning).[20] In ceremonies of fealty, the supplicant or vassal, with hands held in the modern position of prayer, palms together, placed his hands in between those of his lord, swearing to become the lord's man; a similar process took place at manumission, which relaxed the bonds of servitude or serfdom; and handfasting – clasping hands – formally contracted marriage.[21] In 1381, prominent among the demands of the peasants led by Wat Tyler were those that no one should be a serf or owe homage or service to any man, but pay rent for their land. In counterpoise to the power conveyed by these accepted gestures of touch, Tyler approached his King, dagger in hand, and half kneeling, took the King's hand to lecture him forcibly in this cause. The *Anonimalle Chronicle* reinforced this scandalous misfeasance with its description of the Kentish rebel's gestures: given water and ale, he quaffed it in ugly and villainous fashion, also mounting his horse, even though he was in the presence of the King. This was behaviour well outside the accepted order. Yet the significance of gesture could vary markedly between countries. Clasping hands does not seem to have been common in medieval England as a form of greeting – Tyler's gesture was not made with this intent, but was a deliberate flouting of the conventions of touch associated with lordship. Those literary texts that mention hand-clasping as a greeting are, in the main, translations and replicate the customs of other countries and times.[22]

9 Handfasting as part of a clandestine marriage ceremony, conducted by a friar rather than a priest. From the *Omne bonum*, the encyclopaedic compilation of James le Palmer (d. 1375), possibly written in the late 1330s and 1340s, and later owned by Thomas Arundel, Archbishop of Canterbury.

Touch was required to make an arrest. In 1324, the bailiff of Arnald le Pouwer, the steward of the liberty of Kilkenny, attempted to arrest the Bishop of Ossory to take him to prison, but was apprehensive of manhandling him – touching was a form of violence or assault. The Bishop, however, was quite clear: more was needed. He would not consider himself under arrest unless the bailiff had touched him, or the reins of his horse, with his hand, or at least with his wand of office; he knew of no other method of capturing or attaching someone in the realms of the King of England. This delicacy the bailiff realised he could avoid, pointing out that he might raise the hue and cry: he had armed assistants ready with their horns to summon the people against the Bishop as they would against a felon.[23]

Even if touching was not pursued in the courts, it might appear in confession, where all assaults and threats against the clergy were to be divulged. Touching women was of great concern to moralists.[24] There were two strands to this belief – first, that it might lead to sexual incontinence; and secondly, that by touching, the very nature of a woman would pollute a man, especially a cleric or one of a holy life. Peter Idley, in his *Instructions to his son* of the mid-fifteenth century, reported of the first:

10 A beadle making an arrest, touching the man with his wand of office. From the *Omne bonum*, possibly late 1330s and 1340s.

Also in Englond ys now a commun custome,
Kyssyng off woman and þem handle and touch;
And off yt myche synn doth arysse and come . . .

He gave as a model the example of St Jerome, who attempted to exclude all women, even his own sister, not only from his presence but also from his thoughts.[25] The fifteenth-century English version of the *Alphabet of tales* recounted how a clergyman, on a journey with his mother, had to carry her over some water. Before he did so, he wrapped his hands in his clothing so that he might not touch her directly, because of the peril that might come from touching any woman.[26] Those in a privileged position, who had to have contact with women, were counselled to behave circumspectly. Surgeons were not to speak in a foolish way to women in a sick man's house, not to ogle the women in a great household and certainly not to kiss them or touch their hands, their breasts or their genitals.[27] The greatest reservation about touching women came from the association with menstruation and an abhorrence of touching menstrual blood which, it was believed, had extraordinary powers, causing trees to become barren and herbs to wither.[28]

Touch, proximity and association had significant consequences. Those that kept the company of fools might be adjudged fools; those that wished to be considered saintly should associate with the holy.[29] Families, lovers, friends and companions were in an environment where physical touch was the expectation and unavoidable. Sharing beds was commonplace. In a didactic work of 1396, aiming to expand the Anglo-Norman vocabulary of the student, the concern of one of two companions visiting an inn was the cold feet of the other, with whom he was about to share a bed; and, once they were in bed, one complained that the other should not touch him – they had sufficient hazard in the form of the insect life.[30] Unlike Benedictine monks, who were to sleep separately, although in common dormitories, the scholars of Winchester College, in its first statutes, were permitted to share a bed until the age of fourteen, but with no more than two per bed.[31] In peasant households, spouses might share their bed with an infant child, while most of the other children slept together in a second bed.[32] It was not only lovers who might sleep naked:[33] in 1356, following Edward III's campaign in Scotland, William Douglas surprised Sir Robert Herle, who had returned to his manor with about twenty of his colleagues in order to relax better than they might with the army. The English were caught naked in bed, Sir Robert and his chamberlain scarcely escaping.[34] Elsewhere in society, others, such as the monks of Eynsham Abbey, were to remain clothed and with the bedcovers drawn up to the elbows.[35]

The press of crowds made touch unavoidable and it may have been this that made the combined moral force of people assembled together much stronger than that of an individual. Numbered here are the crowds that gathered at shrines, for example, clustered at the oratory for St Erkenwald on his feast day, but frustrated at its closure for painting – and at the painter, Teodwin, whose insistence on painting on this day led him to be struck down;[36] or the unusually large gathering thronging the streets of Norwich on the way to the cathedral on Maundy Thursday 1144, frustrating the Jews who wished that day to dispose of the body of St William.[37] But the press of the late medieval town might simply overwhelm individuals, as was the case in 1322 at the distribution of alms for the soul of a London fishmonger at Blackfriars, when fifty-five people died in the crush.[38]

Conversely distance – not being touched or being out of proximity – created space beyond which moral infection as much as physical disease might not travel. Lepers, who had both a physical and a moral affliction, Jews, heretics and excommunicants, all of whom were a source of moral contagion, were kept separate from the ordinary community. Legislation in the thirteenth century enforced a division between Christians and these parties. Taking up the theme of the Fourth Lateran Council (1215) and following royal writs of 1218, the Council of Oxford of 1222 ordered the Jews to wear badges on their upper garment in a colour that contrasted with it, so that Christians could identify them. Jews were not to have Christian servants. Royal writs of the same year and of 1235 forbade the sale of foodstuffs and other necessities to them, except through royal officials. Statutes for the diocese of Chichester, dating from between 1245 and 1252, forbade Jews and Christians to live together; and those of 1257 for Salisbury again forbade the Jews to have Christian servants, nurses or midwives. A royal statute of 1253 forbade the Jews to use Christian nurses, and an ordinance of 1271 prohibited Christians from acting as nurses, bakers, brewers or cooks for the Jews. Jews were also imprisoned separately.[39] The abjuration of John Eldon of Beccles in 1430 indicated that he had been familiar with heretics and that, from that time henceforward, he would not receive them nor 'wittyngly Y shal felaship with thaym, ne be hoomly with thaym, ne gyve thaym consel, yeftes, sokour, favour ne confort'.[40]

Setting an individual apart for the purposes of punishment created a moral distance. The Rule of St Benedict envisaged the punishment of defaulters at meals by seating them aside from the common table: the spirit of the injunction was followed in English Benedictine houses, with local variations. More serious offences were treated by further isolating the miscreant.[41] At Strubby, Lincolnshire, in 1342, Gilbert, a clerk, who was the brother of the vicar, was alleged to have committed adultery with two women. He was required to stand at the font in his surplice, reciting the psalter.[42] In 1511, two Kent heretics were punished by separation from their communities. Joan Lowes of Cranbrook was to reside in the nunnery at Sheppey and was not to go more than a mile from it; Julian Hills of Tenterden was sent to the nunnery of the Holy Sepulchre immediately outside Canterbury and was not to go beyond the suburbs of the city for the rest of her life without archiepiscopal dispensation.[43]

Punishment, however, might take a physical form. While it might be the regular fate of children, in and out of school, for adults it had a different character. The pain of punishment and mortification were integral parts of the sense of touch, carrying both physical and moral implications. The scourging of Christ formed an exemplar for the administration of discipline for the refractory as much as a model for imitation by the penitent. Punishment took place in public, an important component in its moral force, as was the humility with which it was received. Corporal punishment was part of monastic discipline in chapter: the individual would be beaten with a single staff while prostrate but clothed, or finer rods would be used on the bare back. In early Cistercian practice, no one might beat a superior; punishment was to be inflicted by equals or by those of higher rank.[44]

Beatings might be inflicted by the clergy on the laity as penance, which was then delivered in a highly ritualised fashion.[45] Typical was the sentence which Richard Allot merited in 1336 for fornication: to be beaten three times around the parish church of Friesthrope.[46] The notoriety of a crime might require public humiliation in addition to beating: in 1338, Richard de Croxby, sentenced for his fornication with Margaret Joye, was to go on three successive Sundays round his parish church clad only in his breeches and similarly round the market at Rasen, and Margaret was subsequently sentenced to be beaten around the church and marketplace there;[47] in 1428 Thomas Wade was sentenced for his heresy to be beaten three times around the cloister and market at Norwich, as well as six times around his parish church.[48] Less usual was the insistence that rank was no exception. Roger de Clifford, a baron, on account of the extortion and conflict he had engendered with the Church, was made to do public penance by Thomas Cantilupe, Bishop of Hereford, who spurned his offer of £100 in lieu. Roger went barefoot in a tunic without a hood in procession through Hereford Cathedral and its cloister, followed by Cantilupe, who beat him with a rod to the high altar.[49]

The moral component in beating made it especially suitable for driving out demons. Two individuals suffering from madness were brought to Worcester for St Wulfstan to cure. Beating did not rid the first, a woman, of her demon, although she was subsequently relieved. The second, a youth, had a vision of Wulfstan who struck him with his pastoral staff on one arm and then the other, releasing him from the ill effects.[50] In a similar way, others possessed – such as heretics – might be beaten out of town, the fate of the group of Cathars, largely from Flanders or Germany, driven out of England in 1166.[51]

Capital punishment, in the hands of justice administered by the state, or as a result of ecclesiastical sanction, was similarly intended to carry with it a high moral charge. Stories of martyrdom might inspire those dying for their faith to disdain the moral force of their oppressors. The late twelfth-century Anglo-Norman life of St Lawrence suggests that on the grill he felt no heat, just coldness and repose, and the comfort of God.[52] Failed hangings might be attributed to divine intervention and the common belief that if one had once been hanged, but had revived or escaped through accident, one should not again be subject to the penalty, was closely linked to a perception of moral virtue in the incident.[53] But it was not always seen like this. Cristina Cray, imprisoned and hanged at Hereford for stealing a pig, was cut down at vespers, as was the custom there, and revived. Apparently she had been measured to Thomas Cantilupe on entering prison – a practice that involved preparing a candle to the length of her body and offering it to the saint. She remained in ecclesiastical sanctuary for three weeks lest she was taken again, before abjuring the realm: despite the miracle, local feeling against her ran high.[54]

Mortification of the flesh – wearing hair shirts, going barefoot, minimal washing, extreme dietary abstinence – might be self-inflicted or might be imposed as penance.[55] A man wearing iron bands or fetters on his arms, as a penance, came asking for alms at the feast held on the day of the weaning of St William of Norwich. The bands shattered as the baby toyed with them.[56] Practices such as creeping to the cross on Good Friday would normally have been carried out barefoot, although monastic texts make clear that it was not to be carried to extremes. While the Benedictine monks to whom Lanfranc's constitutions were addressed might be expected to remain barefoot for that day's service, the Abbot had discretion to allow them to put on their shoes in case of extreme cold; and at the end of the day the monks had their feet washed in warm water and were allowed to put on their day shoes, to make up for the chill.[57]

It was mitigation like this that distinguished the normal run of monastic practice from extreme asceticism. Those following this course of their own volition, however discreetly, might find their practices identified as markers of sancity. Margery Kempe made sure that the hair undergarment or lining to her kirtle – probably made from matting placed under malt when it was dried in the kiln – was concealed from everyone, including her husband, although she slept with him every night and bore him children at this time.[58] The moral component of touch extended in this way to clothing. It was originally the practice of the Benedictines not to wear under-garments to separate the coarser, woollen cloth of their outer clothing from their body, but this was subsequently mitigated by practices that allowed woollen under-garments. Linen, however, was generally forbidden to ecclesiastics: it was considered soft and effeminate, although it was worn by the sick, the Premonstratensians and the Augustinian canons. Its use became a common plaint in the war against sensualism.[59] A Cistercian monk at Rievaulx went to the abbot, Aelred, seeking to leave the monastery and complaining of the coarse clothing and the irritation it caused his flesh – to which Aelred was sympathetic, preferring to offer the monk the continued benefit of the claustral life.[60] The coarseness of the clothing was one of the austerities cited by an elderly monk of Grand Chartreuse when attempting to persuade the future St Hugh of Lincoln – whose sensitivity he suspected – from joining the order.[61]

Throughout the later Middle Ages the link between touch and sensuality extended to 'softness' generally, to delicate clothes, soft bedding, whether straw or something more luxurious, soft furnishings, and behaviour typifying idleness – light labour, ill-discipline, lying in bed, or lounging around and bathing.[62] The canons of Barnwell were not allowed in choir to rest on their elbows, cross their legs, stretch their legs out or sit with their legs wide apart.[63] Saddles used by monks and nuns were to be appropriate or 'regular', along with reins and under-saddles: some texts suggest that it was ornament that was to be excluded, but others indicate that it was a matter of deportment.[64] In the litany of targets that might include pride and covetousness, *Jacob's Well* attacked under sloth the tenderness of the flesh that led a man to avoid penance and strictures of the body – the Devil rests in this man as in his soft featherbed.[65] Featherbeds, too, were among the many temptations that the Cistercian monk travelling outside his monastery had to avoid.[66] In confirmation of this association with the households of the upper classes and the iniquities that materialism might bring, St Alban, cast into prison, in Matthew Paris' metrical life of the saint, probably of 1230–40, felt the moral benefits of austerity and aspersion, his bed a dark grey rock, as hard as steel.[67]

This moral inflection needs to be borne in mind throughout the discussion of touch, with respect to listings of everyday furnishings as much as actions. Care was taken to explain why invalids had soft beds or why measures were employed to mitigate pain: they were doubtless necessary, to our minds, from a humane point of view, but this was something exceptional and had to be justified. Walter Daniel, writing of the arthritis that afflicted Aelred of Rievaulx, described how he was carried around in a linen sheet, both to move him from bed to bed and to relieve him from pain.[68] After a vision of Thomas Cantilupe, Margery de Hommer, who was so severely disabled she could not bear to touch anything or be touched unless there were pillows, cushions or linen sheets interposed, had to wait two months before her children were sufficiently convinced that it was wise to take her to Hereford to Cantilupe's tomb.[69] And it is just possible that we are intended to understand some virtue in the case of another Cantilupe miracle, the resuscitation of the infant Gilbert Russel, who was run over by a wagon while sleeping on hard, flat and solid ground.[70]

These views were the inverse of more general social habits, for example the desirabilty of creating a comfortable and warm environment, and an appreciation of bodily comforts sensible through touch. This can be seen from inventories for royal households, such as that of Henry V, which listed no fewer than forty-five featherbeds,[71] or that for Henry VIII, with its listings of featherbeds and beds of down.[72] Equally, in peasant households, straw pallets might be made as comfortable as possible with blankets and where the family had more wealth, beds and bedding might feature among the few possessions of these households listed in wills and inventories.[73]

Kissing constituted a special form of touch: as well as physical contact, the gesture might convey much more. The custom was common in England, in a variety of circumstances.[74] Aelred of Rievaulx looked at the qualities of the kiss, from the carnal to the spiritual, from the human to the divine. When two people kissed, their 'spirits', that is their breath, mixed and were joined, from which a sweetness of the mind was born: it bound together the minds of those kissing, so it was a threefold kiss – physical, spiritual and intellectual. Given these connections, Aelred went on to discuss the circumstances in which kisses might be given. A physical kiss could only be offered or received on defined, honest occasions, for example as a sign of reconciliation between friends; as a mark of inner peace, by those who were to take communion in church; or as a mark of delight, between spouses, between friends after a long absence; or as a sign of catholic unity, as when receiving a guest.[75] The practice of kissing can be followed in other early Cistercian examples. It was used both liturgically and ceremonially, for example in the service of Maundy, kissing the feet of the poor, kissing the cross during its adoration on Good Friday, kissing the altar at various points in services,[76] and, in meditation, the novice in choir contemplating and metaphorically kissing the wounds of Christ on the cross.[77]

All these had their parallels outside the cloister. In the first of the turbulent meetings between Henry II and Thomas Becket at Northampton in October 1164, the King deliberately declined to allow the Archbishop to kiss after mass – that is, it was the common expectation in England that he should have done so.[78] After Arnald le Pouwer was forced to seek pardon from the Bishop of Ossory, all those present, both prelates and those of the King's council, kissed each other to make peace between themselves.[79]

The act of kissing might bring further spiritual benefits. Hugh of Lincoln made a practice of entering leper houses as he passed. Having separated the men from the women, he used to kiss all the men, whatever their deformities. His chancellor likened him to St Martin, who had cured a leper by kissing him – to which Hugh immediately replied that Martin's kiss cured the leper, and the leper's kiss had indeed healed Hugh's soul.[80]

Other kisses might recognise forgiveness or convey blessing. In the play of Joseph's doubt, in the N-town cycle, after Joseph had been persuaded by God and an angel that he was in the wrong for his anger at Mary's pregnancy, he attempted to kiss her feet. Mary declined:

> Nay, lett be my fete, not þo ȝe take;
> My mowthe ȝe may kys, iwys,
> And welcom onto me.[81]

In the Towneley cycle, following the story of Genesis 27, the elderly and near-blind Isaac asked his son Esau to kiss him, without realising that it was his second son, Jacob, who, with his mother, deceived him in order to receive his blessing. The deception was exposed and Jacob fled, but not before kissing both his father and mother. On his return he met his brother Esau, as the stage direction indicates, kissing and embracing him.[82] Henry of Lancaster noted three types of kiss. The first was that wound of the mouth, the lecherous kiss, that had frequently poisoned him and in which the Devil was strongly implicated. He had always more willingly kissed an ugly, poor girl, lewd in her habits, than a good woman of high rank, however beautiful she was. The second kiss was that offered in friendship, but insincerely. The third was the kiss he gave to the Lord every day at mass, which symbolised the peace that followed the conflict that had arisen between Henry and Christ right from the age that he could sin.[83]

Kissing objects conveyed an intention of reverence or respect and beyond this at least a desire for the transfer of further powers. There was here a close connection between the spirit of the person kissing and the virtue or power of the objects kissed. In July 1274, the seven-year-old Prince Henry was taken by his grandmother, Eleanor of Provence, to Canterbury to meet his parents on their first return to England after Edward I's accession to the throne. While he was there he was taken to kiss the relics of St Augustine.[84] Friar Adam lay paralysed at Guildford in the new friary Eleanor of Provence had sponsored there until, with great ceremony, she had the arm of St Richard brought. The relic was carried to Adam's chamber, where he was cured after the bare bone touched his mouth: he was presumably too severely paralysed to kiss it himself.[85] It was the practice at Walsingham for the pilgrims to kiss the feet of the image of the Virgin there – whose beneficence might then transfer to the pilgrim. A local Lollard, William Colyn of South Creak, examined for his heresy in 1429, said that he would prefer to kiss the feet of Lady Lestrange;[86] and others of Lollard persuasion doubted the benefits of kissing stones and images.[87]

There was a strong tradition in Christianity – the Lollards excepted – from the miracles of Christ onwards that close proximity to the holy, touching it or kissing it, transferred the benefits or virtues of sanctity.[88] The object itself might have acquired virtue directly as the remains of the person of the saint, or by its proximity to the saint or his relics. This expectation was widespread in late medieval England. About 1155–6, Mauger Malcuvenant, who was probably the sheriff of Sussex, became very ill and wanted to take the habit of a monk before he died. The monks of Reading Abbey were reluctant to receive him, in case they also incurred his financial obligations; nonetheless, as he reached the conclusion of his life, they did promise to accept him. They came to give him the habit as he was laid out on the floor in anticipation of death, bringing with them an ampulla containing water in which their greatest relic, the hand of St James, had been washed. Against their expectations, Malcuvenant swallowed three drops from the ampulla and revived.[89] Sarah de Wileby (probably from Willoughby, in Warwickshire), who had suffered from leprosy for fifteen years and whose body was dreadfully afflicted with the disease, went to nearby Catesby Priory, where the pallium of Archbishop Edmund Rich had been sent after his death. Because of her condition, she was received secretly at the nunnery. Taking off her own squalid clothes, she was dressed in the Archbishop's pallium – and her own flesh was miraculously restored so that it was like that of a child. The sanctity of the Archbishop had transferred its virtue to his pallium and was in turn transferred to Sarah.[90]

Although treatises on the Ten Commandments give extensive coverage to similar practices that the Church considered magical and idolatrous, noting that many of these beliefs were a legacy of pagan customs – and the differences were not always clear[91] – the idea of virtues was inseparable from beliefs about holiness, sanctity, the miraculous and evil. A wide range of miraculous cures was ascribed to the virtue of touch itself. The sermon for the feast of St Thomas the Apostle, in Mirk's *Festial*, recounted how Thomas had found truth by touch, putting his hand in Christ's wounds. Accordingly the hand that had touched Christ – which was not buried with Thomas in his tomb – was responsible for many miracles; it was used to bless the sacraments and closed up when those that had not been shriven of their sins approached it to take the host. When the hand was set between parties in dispute, it would turn to indicate who was in the right.[92] *Jacob's Well* recounted the story of a gardener who, instead of giving away his surplus, hoarded it, became lame and was to have his leg amputated. An angel appeared to him on the morning that this was to happen and the gardener, confessing his sin and asking for mercy, was cured by the touch of the angel.[93]

The touch of a holy man or saint – or to touch one – was widely recognised as conveying virtue. It was this expectation that led men and women at funerals to try to touch the body or covering vestments: this was the occasion of miraculous cures when the body of St Erkenwald was brought to his grave.[94] While the body of Hugh of Avalon was in Lincoln Cathedral, awaiting burial, the face was uncovered, as was customary. A knight, whose arm was affected by a cancer, rested it on Hugh's face and was immediately cured.[95] It was to this end that crowds attempted to touch the cloths that covered the body of St Richard as it was carried back to Chichester in 1253.[96]

The same powers might also be transferred by touching the tomb of a holy man. Tombs of saints were constructed with openings (*foramina*) to allow pilgrims to get even closer to the relics. At Salisbury, Gervase Brode, a friar, recalled before the 1424 commission into the sanctity of St Osmund how, some thirty years earlier, a madman named Thomas had been brought, manacled, by his friends to the saint's tomb. Thomas had placed his hands in the apertures of the tomb, to withdraw them miraculously unbound and with his madness cured.[97]

Beyond relics, virtue might reside in other holy objects, particularly the sacraments. The sacraments conferred benefits directly upon the recipient, through touch and proximity, as much as through ingestion and anointing. Great care was taken with consecrated elements, especially the bread and wine of the Eucharist, as the sacred qualities could be contaminated by touch (it was for this reason that

11 The early thirteenth-century tomb on the south side of the Trinity Chapel (Lady Chapel) of Salisbury Cathedral where the remains of St Osmund are believed to have been deposited on their translation to the new cathedral. The *foramina* or openings are similar to those of the shrines of other saints.

vestry equipment contained fans for driving away flies)[98] and they could transfer to things they touched. If this happened, whatever the sacrament had fallen on had to be taken up and disposed of carefully: wood, carpet, scrapings from the earth were to be disposed of in the piscina, as was the water from the washing of the chalices and corporals (the cloths on which the consecrated elements rested).[99] John Mirk's *Instructions for parish priests* advised that if consecrated elements touched things that might not be burned, such as stones, those items should be placed with the relics. Flies or spiders that fell into the chalice were either to be drunk up by the priest or were to be burned with the residue of the wine.[100] Synodal statutes of the thirteenth century required the consecrated host to be placed in a pyx – sometimes described as one of unusual beauty – with a clean, linen cloth; consecrated hosts were not to be kept for more than seven days, lest they became damp and mouldy, and look and taste unpleasant; care was to be taken not to confuse consecrated hosts with unconsecrated ones; and the host was to be conveyed to the sick in a silver or pewter container.[101] In a fifteenth-century sermon, the communicant was advised not to chew the host, but to swallow it cleanly so that none remained in the teeth or other part of the mouth – and the water or wine that was given with it was intended to make it easier to swallow.[102]

The quasi-magical powers of the host made it a target for misuse. One tale from a thirteenth-century preaching collection recounted how a man from near New Ross in County Wexford, who had been on a pilgrimage, bought a cask of wine for resale in his tavern. It seemed to his wife that he had purchased it expensively, for there was no dearth of wine. She was afraid that they would lose badly by the transaction. When speaking with other women, one suggested that when she took communion at Christmas, she should keep part of the host in her mouth; and when she came home, she should place that part of the host in the wine. She would then be able to sell the wine as she wished. The plan was executed, but when the woman came to tap the wine cask, it was empty. The host was retrieved with the aid of the staff her husband had used as a pilgrim; and the cask was only replenished after the woman had fully confessed her sin.[103] Other tales recounted how unworthy celebrants might be magically deprived of the host, how their state of sin would not affect the power of the host and how the host had to be made of the purest flour, as well as the consequences of contamination.[104]

Holy water – water that had been blessed or had some association with sanctity, for example water used for washing relics – was equally potent. The washing away of sin was a well-established and familiar biblical trope onwards from Elisha's instructions to Naaman, the leper, captain of the Syrian army, who was sent to wash himself seven times in the River Jordan and who was cured of the disease.[105] In this way, one of the duties of the infirmarer, set out in Lanfranc's monastic constitutions, was to sprinkle holy water each day after compline on the beds of all the sick.[106] Between 1158 and 1165, following an outbreak of plague that had affected both humans and cattle around Bucklebury, Roger, Abbot of Reading, blessed the area from a high point and used water that he had blessed, and in which the reliquary containing the hand of St James had been dipped, immediately relieving the neighbourhood of the infection.[107] It may have been especially effective against demons in possession and ghosts.[108] The holy water or oil contained in lead ampullae, small flasks that were a popular acquisition at the shrines of saints, was well known for its healing properties. An ampulla discovered in Yorkshire in 1965 still contained a liquid with aromatics that doubtless would have conveyed the odour of sanctity as well as the prophylactic benefits of its sacralised contents.[109]

The powers of sanctity, virtues of relics and sacraments, also had an inverse effect, if one came into contact with them in disbelief or if attempts were made to use them maliciously, or by those in a state of sin. At Winchcombe Abbey, a monk whose fornication on the eve of the feast of St Kenelm had gone unconfessed, found his hands stuck to the psalter of Cwoenthryth, the saint's sister, which he had to carry in procession the following day. He was released on admitting his guilt.[110] A popular story, derived from the Gospel of Pseudo-Matthew and probably originating in the first quarter of the seventh century, concerned two midwives who had come to attend the Virgin Mary at the birth of Christ. Not only were they completely super-

12 An ampulla that would have contained Canterbury water, from the shrine of Thomas Becket and possibly dating from the thirteenth century.

fluous, but one of them, Salome, had the audacity to doubt the virgin birth and to touch Mary to investigate her claim. Her hand withered; and it was restored only after she had been allowed to touch the edge of the garment of the infant Jesus.[111] A second tale associated another midwife with the birth of Christ, equally miraculously but more positively. Anastasia, whose hands had been cut off by her father, had them restored on touching the infant.[112]

Popular tales recounted the consequences of touching the sacraments in a state of sin or of abusing the vessels in which they were contained. The pregnancy of a nun was discovered when she touched a corporal cloth that had been washed, revealing a spot of blood. The steward of Julian the Apostate, ordered by his emperor to urinate in holy vessels the Emperor had gathered, had – doubtless to the delight of the medieval audience – all the waste products of his body pass out through his mouth for the duration of his life.[113] The Croxton *Play of the sacrament*, probably composed in the Thetford area after 1461, reworked a theme more common on the Continent, of an attempt by Jews to capture and defile a consecrated wafer, paralleling the events leading to the passion of Christ. The worst offender was badly mutilated in the process: his arm remained attached to the wafer when it was thrown into an oven but was restored to him when he begged forgiveness from Jesus.[114] The power of a virtuous stone could be affected by the sins of those who carried it, requiring 'reconsecration' of its powers.[115]

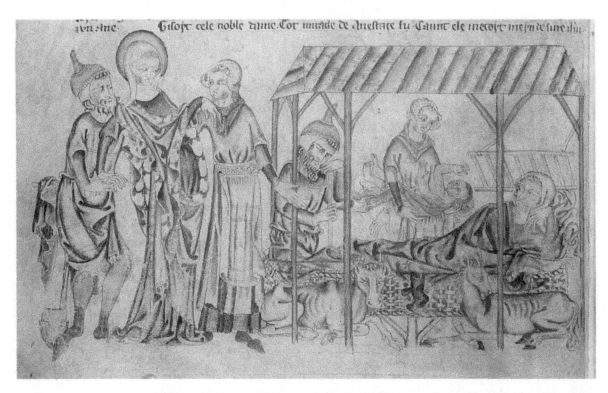

ivn Ane Gisoiχ cele noble dinie.Cor innade dr dineftaχe fu Cannt cle ineχoiχ mefn'iχ fure ihi

13 Anastasia leads Mary and Joseph to the stable; in the second scene, her hands are restored and her prosthetic hand can be seen hanging down beside her as she holds the infant Jesus. From the Holkham Bible Picture Book, *c.* 1320–30.

Closely related were the practices of swearing while touching relics or holy books, and trial by ordeal.[116] The virtue of the object that was touched revealed the truth or falsity of the parties involved. Touch was a guarantee of the verdict – literally a *veredictum* or 'true saying' – reached by a jury. Harold's doom was settled at Bayeux, well before the battle of Hastings: it was here that he swore fealty to Duke William on two reliquaries, and he is portrayed in the Bayeux Tapestry with a hand on each of them. The word *sacramentum* – an oath – had a plurality of meaning. In Antiquity it was the oath of allegiance, in origin a military oath; but by the fourth century it had already acquired the meaning of 'sacrament', that is, signs of the divine.[117] Taking oaths on the Gospels was a standard expectation in judicial process.[118]

Trial by ordeal represented a variant in the process of establishing truth by exposure to virtue. In this case, the trial required that either the iron which was to be heated, or the water in which the accused was to be immersed, first had all demonic force exorcised from it and, secondly, was blessed, so that the instrument of judgment had the virtues of holiness. The same applied to trial by battle, where the outcome was deemed to rely on the will of God. The ordeal by iron required the accused to hold the iron for a set number of paces. The hand was subsequently bound, and judgment, after three days, depended on whether the wound was

UBI HAROLD:SACRAMENTVM:FECIT:⁖ HIC HAROLD:DVX:⁖
VVILLELMO DVCI:⁖

healing cleanly. But other virtues might interplay with the iron. The servant of
Flothold, who was subject to ordeal by iron at Winchester, had the assistance of
St Swithun. When his hand was unbound after three days it appeared to his enemies
unscathed, but to his friends it appeared to be dreadfully burnt.[119]

The other sacraments might convey similar powers. Unction – anointing with
sacred oil – was prominent among these, nowhere more than at the coronations of
kings of England and France. Anointing showed the divine nature of the kings and
changed them, investing their kingship with a sacral power. While the French kings
had the oil that was brought from Heaven for the baptism of Clovis, the situation in
England remained more ambiguous until the late fourteenth century, at which
point it was established that the oil to be used at the coronation was that given by
the Virgin to Thomas Becket, while in exile, contained in a phial in an eagle of
gold.[120] This sacral power manifested itself in many ways that allowed the divine
virtue so acquired to be transferred. In England, from the reign of Edward I, there
is good evidence for the practice of touching for the king's evil (scrofula): it may
have started under Henry III, probably in imitation of Louis IX of France.[121] The
making of cramp rings – rings that held a medicinal benefit – from the king's
offering on Good Friday was well established by the reign of Edward III. In 1353,
6s. 8d. was offered and then 5s. worth of the offering was redeemed so that the rings

14 Harold, with his
hands on two
reliquaries, swears
fealty to William at
Bayeux. From the
Bayeux Tapestry.

15 The anointing of David, from the Rutland Psalter, *c.* 1260.

could be made. The sum became regularised as three gold nobles and five silver shillings (25s. in all) by the 1370s and continued as that amount on a customary basis throughout the fifteenth century.[122]

The consequences of unction can be seen further in practices in the royal household that gave special regard to the King and his possessions. By the reign of Henry VII, only the esquires of the body might touch the King in person. On Twelfth Night, when the King formally wore his crown and his most 'royal' – that is, prestigious – clothes, these had to be blessed and no secular person was to touch them. They were brought to him wrapped in a kerchief for him to dress himself. At the making of the King's bed, an esquire of the body was to sprinkle the bed with holy water.[123]

Anointing played a central role in other sacraments, especially baptism, unction of the sick and of the dying. These sacraments were conveyed by touch, either by the priest or, in the case of baptism, by the laity, who might perform the ceremony out of urgency and where the actions of ordinary individuals created something that was no less holy. Early thirteenth-century synodal statutes specify that if water is used in the process and this happens away from the church, the surplus is to be tipped into a fire, or brought back to the font or baptistry in the church. Any container used to carry the water is also to be burned or to be handed over to the church for sacred use.[124]

Jacob's Well explained the practice to a rural congregation in southern England in the fifteenth century. Six gifts were imparted to the child by the Holy Spirit. Placing salt in the child's mouth transferred the gift of the wisdom 'for to have savour afterward in God whan he comyth to dyscrecyoun'. In anointing the ears and nose with spittle the child was given the counsel in order to have the circumspection and the vigilance (*warhed*) to govern his five wits. When the sign of the cross was made on the child's forehead and chest, the child received understanding, allowing him to perceive virtues. Through the anointing with oil on his chest and back the child received pity; in the dowsing with holy water from the font he received strength against the Devil and sin; and in anointing the nape of the neck, he received the gift of knowledge in order to tell good and evil apart. A further element was symbolised through touch: at the christening the child was to be wrapped in a white cloth, signifying the cleanness of body and soul that was required for salvation. A seventh gift of the Spirit was transferred by touch at confirmation, when the bishop anointed the individual's forehead, conveying the gift of dread of the Holy Ghost.[125]

Anointing the sick was an important element in miracles. Alice de Lonesdale, a young girl from the diocese of York, who had injured her foot in a fall on the road at Stamford and whose infirmities had spread while she was sleeping rough in Southwark, was taken around Whitsun 1303 to Hereford Cathedral where she stayed three days at the tomb of Thomas Cantilupe. Here the saint appeared to her in a dream and anointed her chest, arms and legs with a milky ointment, curing her of her disability.[126]

Twelfth-century Cistercian practice shows the care taken in the disposal of items that had touched holy materials involved in unction of the sick. After wiping away the oil, the cloths employed for this purpose were to be burned in the piscina allotted for the task.[127] Although it was unusual, in special cases unction might take place after death. In January 1167, after the body of Aelred of Rievaulx had been washed in the customary manner, a small phial of balsam that he had used as medicine was brought. His friend and biographer, Walter Daniel, anointed the two fingers and thumb of his right hand, those he had used for writing, for his spiritual works. Others argued for anointing his tongue and his face – and miraculously the ointment went much further than the monks could possibly have imagined.[128]

Anointing the dying – extreme unction – was one of the practices disdained by the Lollards. Hawise Moon of Loddon, Norfolk, confessed in her abjuration of heretical practices in 1430 that she had believed it was but a trifle to anoint a sick man with consecrated oil, for it would suffice for every man at his end only to think of God.[129]

16 Rievaulx: the chapter house, where Aelred was buried, seen from the cloister. The shrine to his predecessor as abbot, William, was located immediately to the left of the door.

The framework of virtue accorded by sanctity and touch might extend to places. The sanctity of a place might be assured by its formal consecration and by the presence there of other sacred elements: cemeteries, for example, on account of the 'many bodies of saints and the saved resting there'.[130] Consecrated ground would protect the dead Christian from the fiend, keeping him safe until the resurrection, unless the Christian had died in a state of sin – for it was commonly believed that the Devil might disturb a body in this condition. Two further precautions were taken to identify the bodies of the virtuous: wrapping them in a white sheet to indicate that the body was shriven, and placing a wax cross on the body to symbolise that the person died showing charity to God and man. Without confession and absolution, the dead man was not entitled to the Church's prayers, and he was to be buried with a broken stick, to symbolise that he had a long journey to make; but the priest would nonetheless sprinkle the grave with holy water to keep the fiend at bay. The distinction was sharply maintained: those not in communion with the Church or who died in mortal sin, for example by suicide, were not to be buried in consecrated ground. There were further gradations. Mothers who died in childbirth – that is, in a bodily state of corruption, not having been churched, but still in communion with the Church – might be buried in the churchyard, but only after the dead child – which had never received the sacrament – had been taken out of the body to be buried outside the churchyard.[131] Similarly, those that had been excommunicated were not to be buried in consecrated ground; cemeteries were to be enclosed to protect them from animals that might cause pollution; and a long list of disreputable activities that might let in the fiend – ranging from holding law courts or playing games, especially those that might involve bloodshed, to dancing and 'dishonest singing' – were not to take place there.[132]

Other places attracted devotions, although they were not always sanctioned by the Church. In July 1240, Bishop Walter Cantilupe of Worcester condemned the practices at a holy well in North Cerney and another near Gloucester.[133] In contrast, the well at the battlefield at Evesham, the powers of which became established in the aftermath of the events of 1265, seems to have attracted formal support quite rapidly in its translation into a brief and minor cult for Simon de Montfort.[134]

The virtue of holy ground might require respect in other ways. Both the N-town play of Moses and the Towneley play of Pharaoh reiterated God's injunction to Moses to remove his shoes as he approached the burning bush, as the ground was holy.[135] The ground itself might acquire the association of virtue through the merits of an individual of sanctity. Some of the ground where Simon de Montfort had fallen at Evesham was dug up and taken away by William, the rector of Werrington. Mixed with water, it was used to revive a man who was on the point of death.[136]

Beyond the framework of Christianity, there were beliefs in the virtues of many other things that conveyed benefits by touch. Their influence might extend well beyond our usual frame of time and space. A vast repertoire of herbs, roots and stones carried about the person was traditionally regarded as efficacious against specific ailments. Hung around the neck, peony root was a protection against the falling sickness in children, dock root against boils and swellings, and the root of a brassica, provided it had not touched the ground again once it had been pulled up, against a sore throat.[137] Macer's *Herbal*, which provides these examples, was written in France in the second half of the eleventh century and was frequently copied alongside Marbode of Rennes' treatise on stones, written about the same time. Both were clearly aimed at medicinal practice.[138] Lapidaries – treatises on stones and gems – were especially popular in England. Marbode's account of 'adamas' – which might refer to diamond, quartz or corundum and other stones – noted its especial virtue for magic, making the bearer unbeatable, combating spirits and curing insanity.[139] Some of these notions can be traced back to Greek medicine, but the practices remained current in the Middle Ages.

There was no clear division between categories of object. Some were very closely tied to religion, through association with individuals, or inscriptions, or relics. The cup bequeathed by St Edmund of Abingdon to his chancellor, Richard, later Bishop of Chichester – and a saint – must have conveyed with it strong personal associations, but we know nothing more of it.[140] On the other hand, in the later Middle Ages a group of materials was associated with St Dunstan. In 1388 Westminster Abbey had the liturgical vestments that belonged to him, as well as two censers and two bells; Glastonbury had other vestments and liturgical apparatus, along with a ring said to have been taken from Dunstan's finger on his translation from

Canterbury; and Abingdon had bells. A further ring, of gold set with a sapphire, allegedly made by Dunstan, was to be found among the rings of abbots and bishops held by the royal wardrobe in 1300, and, subsequently, in the possession of Piers Gaveston. It passed to Edward II's queen, Isabella, and was with the jewels in the secret coffer in her chamber at her death, returning at that point to the wardrobe of Edward III.[141]

Rings containing stones that had a particular virtue were prized highly.[142] Sapphire was held to be especially worthy for kings to wear on their fingers. It was held to set aside envy, to bring to accord men in dissent, and to protect against witchcraft.[143] Because of their virtues, stones and jewels were held to be valuable enhancements to shrines, reliquaries and monastic or royal treasuries.[144] Henry II, hearing of the travails in labour of Aquilina, the wife of Gilbert Bassett, sent her all the gems and stones he had that were believed to assist with birth.[145] Many stones like this were exceptional items and their pedigrees could be traced. Ralph fitz Bernard, an English noble, gave a topaz for the shrine of Thomas Becket at Canterbury, a stone which had originally come from a Count of Flanders, Robert II of Jerusalem (d. 1111);[146] and the jewels of St Albans and their origins were catalogued by Matthew Paris in 1257. In 1307, the shrine of Thomas Cantilupe at Hereford had innumerable belts and jewels worn by women, with no fewer than 450 gold rings, seventy silver rings, fifty-five gold necklaces (*monilia*) and thirty-one of silver, as well as assorted precious stones, all brought by the faithful.[147] As well as their virtues, these ecclesiastical collections of stones recalled the description of the heavenly Jerusalem, with its walls constructed of jewels, of sapphires and emeralds.[148]

Engraved stones, particularly those that were antique cameos or intaglios, had an especial significance. Anglo-Norman lapidaries of the early thirteenth century described some of the effects these jewels might have. A cameo with Mars seated on an eagle, with a rod in his hand, if set in a ring of copper mixed with brass, would allow the wearer to overcome his enemies and all men would obey him, provided the cameo was worn on a Sunday before sunrise. The wearer should also wear white linen clothing and abstain from eating doves. Another gem, a crystal engraved with a naked woman with long hair, with a man, if correctly mounted in a ring twelve times the weight of the stone, with amber, aloes and pennyroyal set below the stone, would make the wearer attractive to all – and if he were to touch a woman with the ring, without doubt she would do his bidding; if it were placed under her head when she was sleeping, she would dream whatever the man wished.[149]

For their virtues, engraved gems were frequently used in signet rings, for example that of Richard I or Isabel, Countess of Gloucester (d. 1217).[150] An early Christian council had banned their use in episcopal rings, but they were sought after and their adverse effects might be mitigated by proper dedication. Archbishop Hubert Walter

17 The jewels of St Albans, described by Matthew Paris.

(d. 1205) was buried with, among other things, four antique, engraved gems, one set in a gold ring and three in his crozier.[151]

Three further examples illustrate the range of virtues that might be attributed to stones, jewels and related substances. The symbolic extinguishing of all lights in the church and throughout the monastery on Maundy Thursday was matched by the kindling of new fire on Easter Saturday (although it sometimes happened on the two preceding days as well). To create the fire, it was the practice to use a shaped beryl or other crystal to focus the rays of the sun, rather than striking a flint: the fire was generated by Heaven and not by earth. Both Westminster Abbey and York Minster had beryls for this purpose.[152]

18 A signet ring used as a counter-seal of Countess Isabel of Gloucester, 1216 or 1217. The ring contained an engraved gem, a helmeted bust of Nike, with an eagle, below, between two standards. The legend reads EGO SV' AQILA: CVSTOS DNE MEE ('I am the eagle, the guardian of my lady').

The philosopher's stone – which gave the ability to read men's characters in their faces – was a popular feature of alchemical texts. It was this, according to the *Secreta secretorum*, that gave Alexander knowledge in order to conquer the world.[153] Others might seek to create substances with similar effects. The production by alchemy of the quintessence, created by establishing harmony between the four elements, brought an elixir that would both prolong life to its natural duration and ensure soundness, or incorruptibility, of body and mind. It was sought by royal courts at various points in the late Middle Ages, for example in 1455–7, as a cure for Henry VI's insanity.[154] It was this that was shown to Henry VII in January 1499 by a 'stranger from Perpignan', for which demonstration he was given a reward of 40s.[155]

The practice of taking the assay, testing food and drink for poison, commonly made use of a unicorn's horn – that is, the horn of the narwhal. Fragments of the horn were carried by servants of the great household, to touch the lord's food and drink, defeating any poison. They were sometimes set on the inside of drinking vessels to the same end. Poison was also detectable by the fossils known as serpents' tongues.[156]

Amulets worked in a similar fashion, by physical contact and proximity. They might be directly associated with Christian religion, perhaps as reliquaries, such as the Middleham Jewel; by texts engraved on them, such as the Coventry Ring;[157] or directly with certain saints, such as the badges and emblems of pilgrims.[158] Others, such as handsels, the New Year's gifts that were deemed to have an auspicious effect through the coming twelve months, might be condemned from the pulpit.[159] Religious texts might add efficacy. Rolls based on the measurement of Christ and his wounds, whose texts guaranteed protection against a panoply of disasters – the avoidance of sudden death, especially from a weapon, protection from poison or false witness – were also valued by women for their beneficence in childbirth.[160] A

19 The Coventry Ring, late fifteenth century, has an amuletic text paying homage to the five wounds of Christ, with the names of the Three Kings and the words 'ananyzapta' and 'tetragrammaton', which had a magical force.

further variation in the use of text might include its consumption, cut up into small pieces in a drink – literally eating the words of the Magnificat, although the treatise on childbirth that advocated this, admittedly for aborting a stillbirth, also recommended the presence of a midwife.[161]

The virtues of amulets and other objects might be enhanced in various ways, for example by the employment of gesture, particularly the sign of the cross, which had a power all of its own. A charm brought back by a knight in the suite of Lionel of Clarence, at the time of his marriage in 1368 to the daughter of the Signore of Milan, cured spasm and cramp. Its text was punctuated by the sign of the cross, made as a gesture at the same time:

In nomine patris + et filii + et spiritus sancti + Amen.
+ Thebal + Enthe + Enthanay + In nomine Patris + et Filii + et Spiritus sancti + Amen. + Ihesu Nazarenus + Maria + Iohannes + Michael + Gabriel + Raphael + Verbum caro factum est +.

20 A fifteenth-century roll, with amuletic properties, offering protection in childbirth and against death in a range of circumstances, included the measurements of the cross and the wound in Christ's right-hand side.

Importantly, however, the text was to be kept secret and the charm made secretly, so that it should not be known to all 'lest perchance it should lose the virtues given by God'.[162]

The words of charms were only a part of the range of virtues that might be conveyed by, or associated with, text written on an object.[163] Inscriptions could generate a power all of their own. On the negative side, branding heretics on the forehead or the cheek with a letter H indicated clearly that these individuals possessed an evil virtue; and the letter F – for *fauxine* (falsity) – was similarly used under the statute of 1361 for labourers who broke their terms of employment.[164] Names were imbued with inherent virtues. Techniques for predicting the future used calculations derived from the letters of a person's name and curses might be prepared using the name of the victim.[165] To the ordinary person, as much as the student of exegesis, all names had meaning and their interpretation could not be left to chance. Lists of *nomina sacra* explained the full meaning of the name, beyond its philological interpretation: in St Jerome's catalogue of Hebrew names from the Old Testament, arranged book by book, in 4 Kings, Massefath meant 'reflection' or 'contemplation', Marodach, 'causing bitterness'.[166] Naming a child after a saint not only invoked the protection of that saint, but also imbued the child with the characteristics associated with the choice.[167]

Names were also given to objects in much the same way, to ensure that they conveyed the virtues, or values, inherent in the name. Among the goods of the privy wardrobe in the custody of John de Flete was a cup, probably made out of a coconut

21 Floor tiles in the nave at Rievaulx. Although they may not now be in their original location, there were extensive campaigns of tiling in the abbey. The words AVE MARIA created a prayer that was continuously present, an important component in the sensory environment of the building.

or similar nutshell, called 'Craddock'. This word may be derived from the Welsh *cariadog* or *caradoc*, 'dear one', 'lover', or 'friend', and the item may once have been owned by Llywelyn ap Gruffudd, other parts of whose property were to be found in the wardrobe and household of the English Crown; or it may be a direct reference to Craddock, one of King Arthur's knights, who was the only one to prove that his wife was faithful. The implication may therefore be that it was a form of loving cup. The same inventory also had a great mazer with a silver foot, called 'Edward', which would have brought with it the beneficence of the King.[168] Little store seems to have been set on an object as a result of antiquity alone, unless it had a special power.[169] Besides a name, an object might have an inscription – much as bells proclaimed their virtue in short inscriptions, such as *Vox domini* ('the voice of the Lord') or *Pestem fugo* ('I drive out the plague'), protecting the surrounding area with their sound.[170] Late fourteenth-century jugs sometimes exhorted drinkers to kindly or decent behaviour. The mottoes and letters emblazoned on many late medieval objects – from tapestries and clothing to rings and other jewellery – would have conveyed something more than the mere words of the text.

22 A late fourteenth-century jug, with the royal arms of England. The legend urges the drinker to stand away from the fire in order to let new arrivals near to it.

If these were the ways that touch and proximity played a part in conveying a wide range of virtues and powers, their role in medical practice was in some ways similar, but also at the same time one that we might recognise. Medical preparations, including poultices and 'plasters' (a solid or semi-solid mixture, typically bound on with fabric), were commonly applied directly to the affected areas, although they might also treat internal disorders. Plasters for headaches frequently included mixtures of aromatic herbs and plants, such as camomile, pennyroyal, rue and laurel, cooked in water or with wine, mixed with salt and honey, or thyme, ground with vinegar and oil of roses, all applied directly to the head generally, or to the forehead.[171] The more sophisticated medical recipes based their reasoning on the theories of Greek medicine, explaining how the plants employed would balance the humoral deficiencies that were the cause of the ailment. Garlic, which was hot and dry, was useful against an array of bites and stings, ranging from those of dogs and adders to venomous beasts generally.[172]

There was little explanation of how properties transferred from the curative, for example through absorption into the skin, although directions sometimes required ointments to be rubbed on affected areas. The presence of the healing or mediating substance was sufficient. There was nothing unusual, therefore, in the treatment that Henry of Lancaster described for fevers or frenzy, the result of general ill health or a wound. A red cockerel was to be killed, gutted and the carcass spread out: hot and bloody, still attired in its feathers, it was to be placed on the head of the frenetic. It comforted the spirits, allowed the senses to recover, cast out all delusions and, by its virtue, allowed the man to recover.[173] The direct application of other substances had gynaecological consequences. Macer's *Herbal*, allegedly following Pliny, advocated placing self-heal along with beef on the genital area of a woman to ensure that a pregnancy produced a male child; a fumigation of nettles and myrrh would stop menstruation; and nettle leaves and juice might be applied directly to a prolapsed uterus.[174]

Henry of Lancaster recounted how it was necessary to protect wounds to prevent them deteriorating. They should be bound and dressed with clean cloth, to keep out the air, dust and dirt, and other things, such as flies, that by their touch might hinder the healing process or cause disease. These were items that were traditionally regarded as evil and avoiding the transfer of their properties by touch was as important as ensuring that the virtues of a medicine were kept in contact with the affected area. The wound should only be touched by the correct medicine and the bandage kept the plaster or ointment in place.[175]

Disease might spread by touch, directly, or by bad air, especially that exhaled by the sick. Those that touched lepers were believed to attract the infection,[176] and those with itching and scabious infections – the result of an excess of salty humour, acquired by eating too much salted or piquant food (*scharpe metis*) or strong wine, or from manual work, lack of bathing or not wearing linen clothes – might spread contagion by touch.[177]

Touch and virtues might also operate at a 'supernatural' level, well beyond the bounds of life as we might define it. The earth was populated with fiends as well as angels, apparitions and some, at least, of the recently dead: all might have powers that could influence the living, particularly through touch or close proximity. To meet them was a common expectation. In 1196, after disturbances in Buckinghamshire by an extremely troublesome spirit, Bishop Hugh of Lincoln was consulted. Disturbances of this kind were a common occurrence in England and the solution was usually to dig up the body and burn it, but Hugh preferred a more dignified mode of proceeding, and had a charter of absolution affixed to the body.[178] The activities of another spirit, probably in northern England, an exile from the diocese of York, polluted the whole area through the bad air it engendered.[179] Some practices that may now defeat explanation were probably connected with the powers of spirits. According to the synodal statutes of Bishop Richard Poore for the

23 Scenes from the life of St Guthlac, mid-thirteenth century, west front, Crowland Abbey. In the right-hand lobe of the quatrefoil, Guthlac cures the sickness of Egga, a Mercian nobleman – and the cause of the illness, an evil spirit, can be seen fleeing towards the top right of the image.

diocese of Salisbury, between 1219 and 1228, and subsequently for the diocese of Durham after his translation there, a mother who died in childbirth could be cut open if it was believed the infant might live, provided that the mother's mouth were left open, possibly to let her spirit escape.[180]

Other, cosmic forces might affect the body. The influences of heavenly bodies might be read in astrology: they were reckoned to be strong at the moment of an individual's birth and might have an effect that lasted a lifetime.[181] The use of astrology to predict the future was more controversial. The first horoscopes for political purposes to survive from medieval England seem to have been cast in the 1150s and 1160s, to have required immense learning for their calculation, and to have been associated with Adelard of Bath.[182] The knowledge required to make planetary calculations became more widely disseminated in the thirteenth century

24 A horoscope probably cast by Adelard of Bath, 1151.

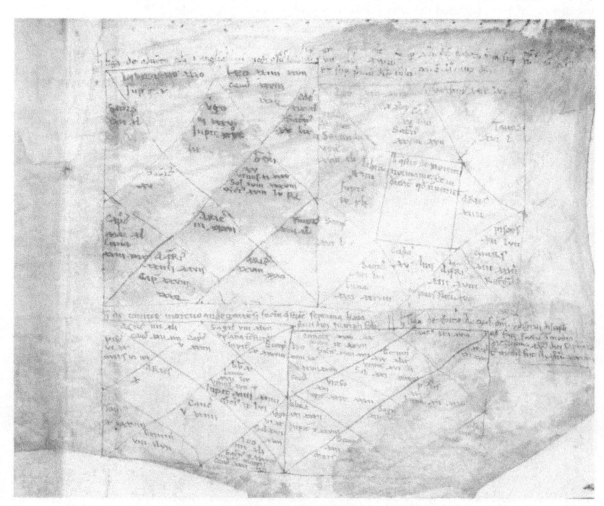

and a close link developed between astrology and medicine, in establishing propi-
tious moments for medical procedures.[183] Astrology had an influence at all levels of
society. In 1229, Bishop William of Blois instructed priests in the diocese of
Worcester to admonish their parishioners against practices linked to times of the
year, such as declining to contract a marriage or move into a new house unless there
was a crescent moon.[184] A primer for the Anglo-Norman language, written in 1396,
introducing the novice to words for the parts of the body, noted that the body was
divided into twelve parts matching the twelve signs of the zodiac. As the moon
moved through the signs, they each ruled the related part of the body.[185]

Homiletic literature proscribed a great many practices, taking as its cue the First
Commandment. The lists of predictive and magical practices are often repetitive
and it is plausible that many existed only in clerical imagination. If the practice of
scapulamancy operated in west Wales – and Gerald of Wales assures us that infi-
delity was exposed through consulting the shoulderblades of rams, an assertion
reiterated by Ranulph Higden without critique – the practices that came before English
ecclesiastical courts in the fourteenth and fifteenth centuries were more limited in
their description, often described solely as sorcery (*sortilegium*).[186] Nonetheless,
these commonly seem to have included the use of charms, or prayers asking for evil
to befall others. It was alleged before a court in June 1462 that John Smyth and his
wife had defamed Margaret Walbott of Wisbech in this way. The following year,
another Wisbech inhabitant was accused of enchanting fishing nets.[187] At the upper
end of the scale, an astronomer – that is, an expert in the theory of the knowledge
of heavenly bodies, as opposed to the practitioner, the astrologer – was paid 20s. by
King Henry VII in the week before Easter 1499.[188] Common to all these examples
was the belief that external influences and powers could affect the human body and
the course of events, and that these might be revealed either through the study of
those things that caused the influence, or in such activities as chiromancy or the
study of physiognomy, in studying the body itself. To these activities were attached
explanations in more or less learned guise, depending on the circumstances.[189]

If touch is a sense that is familiar to us, it was understood by medieval people in
different ways. Much might pass by way of this sense, not only physical information,
but also moral and spiritual qualities. The distinctions and consequences of gender
were prominent: to touch a woman, or to be touched by one, might bring severe
repercussions. The ways in which individuals shaped their own immediate environments, their choice of clothing or of soft furnishings, might indicate moral qualities, also made patent to the outside world by the things they chose to touch or to
let touch them. Beyond this, other forms of sensory perception operated in related
ways. The sanctity of individuals and objects, or indeed their power for evil, might
operate through direct touch or by more general proximity, transcending any way
in which we might understand the operation of the sense. Charms, magic, astrology
and the whole range of medieval science, from alchemy through to medicine and

pharmacology, all made an impact on perception. This evidence – and, to us, more conventional descriptions of the body and its operation – demonstrates that in looking at perception we need to consider a wide range of activities. Morality, spirituality, ideas of order, manners, courtesy, taboos and religion, all had a close link to beliefs about the way the senses functioned in daily life. Nowhere was that link stronger than in the case of sound and the two senses the Middle Ages connected with it, hearing and speech.

Chapter 4

SOUND AND HEARING

In exploring medieval notions of sound, hearing and speech, one might look to philosophical texts for their relationship to signification, cognition and grammar, for example, in the works of Bacon.[1] Similarly there were in the works of Aristotle and Boethius ideas about the creation, transmission and perception of sound that had very great influence during the later Middle Ages.[2] From a theological point of view, from Augustine onwards, there was interest in the word as 'the messenger of reason': it was a concept notably resonant in the pastoral literature of the twelfth and thirteenth centuries. Speech was a special form of sound, an ethical act; the speaker was an instrument of the Holy Spirit.[3] The detail of these theories features in this chapter in so far as it is essential to understanding the operation of the senses. The discussion has been divided between hearing and sounds on the one hand, with speech and language reserved for the following chapter.

Hearing is as subject to cultural influence as any other sense. It is a commonplace that we cannot fully understand the music of another period without appreciating how its composer expected it to be played, but this is only one aspect of the matter. Recreating authenticity in performance does not address the separate question of how sound was perceived – and we should not assume that music was heard in the manner it now is.[4] The same considerations apply to language and all other sounds. Contemporary Western scholarship has focused on the use of language in signification; and the strength of the link between the word as signifier and the object it signifies has had a central role in linguistic theory.[5] In the later Middle Ages while sound had this function, it operated in other ways as well. It might be an immaterial substance in itself: in some circumstances what mattered was not that a sound or words represented anything or conveyed information about it, but that it was that thing. Sound might physically embody a force or virtue that might be good, evil, healthy or diseased.[6]

By the thirteenth century, there was a measure of consensus on the physiology of hearing. Sound was air that had been struck: its qualities and strengths might be affected by atmospheric conditions. The 'animal spirit' within the ears received an impression of the sound, conveyed through the air, by way of the ear-drum, which was linked to the brain by two sinews and which conveyed the likeness to the soul (*anima*) for interpretation. Bartholomew the Englishman noted how loud noise could impede or destroy the sense; and that the ears were susceptible to various ailments that could also hinder or prevent hearing, or induce a ringing sensation in the ear. He described how hardness of hearing often afflicted old men: age shrank the sinews. Aristotle had held that the sense was affected by too much love-making, which disturbed the spirits that made hearing perfect. He was also Bartholomew's source for remarks on the size of the ears and their consequences for character.[7]

Medieval theory had two dominant explanations for the transmission of sounds through the air. One view was that sound, like sight, was conveyed through *species*, the one touching the next and causing it to vibrate. Sound had no material substance itself, but was a perceivable quality, caused by the operation of a physical 'suffering' at one point, that is, the striking or shaping of the air; the sound was then transmitted through the air. This reasoning derived from Aristotle and Avicenna, and was present in the *De anima* of John Blund. The alternative school of thought held that sound was a material substance, coming from breath or *spiritus*, and as a material substance – a subtle form of air – it might be perceived as a form of touch. This was a fitting explanation for those, such as John of Salisbury, who sought to explain the psychological effects of music: as a spiritual substance, or as the vehicle for one, it might have a strong impact on the soul.[8]

Although humans might not be able to hear it, sound was omnipresent: the universe was not silent. The music of the spheres – the sound of the planets moving the air – was one of the three types of music described by Boethius, along with human music (which bound body and soul) and instrumental music. The last was the only one man could perceive.[9] Music, binding body and soul, provided the potential for creating harmony within ourselves, a point made substantively by the use of music for healing. Eadmer's biography of St Dunstan reveals how his skill as a musician brought his friends the medicine of celestial harmony. The infirmarer of St Augustine's Abbey was allowed, according to its customary, to bring a sick monk into a chapel where he was to use a psaltery to arouse the man's spirit by its sound and harmony, although he was to do this with the door shut lest the sound penetrate further and disturb the infirmary hall or the monks' chambers.[10]

The sense of hearing was given further significance within the framework of Christianity. At a time when almost all reading was conducted aloud and most people would only ever have heard the word of God in sermons, it was through this sense above all that knowledge of God was conveyed. Biblical injunctions to listen to the word of God were reiterated from the pulpit. A Lollard sermon turned to St Luke: 'He þat haþ eris of herynge, here he !'[11] Others recalled Psalm 77: 1 (Vulgate), 'My people, direct your attention to my law; turn your ears to the words of my mouth'; Revelation 2: 7, 'He who has ears to hear, let him hear what the Spirit says to the churches', and other texts. A late fourteenth-century sermon writer of a Wycliffite persuasion cited the words of St Paul: that hearing was given to man in order to know his belief, and that therefore belief was hearing and hearing was Christ's word.[12]

If this gave hearing primacy among the senses, man's sense of hearing did not have pre-eminence when set against the natural world. Faculties of certain animals were regarded as more effective. The boar was sometimes used to represent the sense on account of its reputed acuity of hearing, but texts also referred to the abilities of the mole (following Pliny), the bear (especially in Germany and Scandinavia) and deer. Beasts with large ears were accorded priority.[13]

To understand fully the place of hearing, sound has to be set in its natural environment. The soundscapes of the medieval world were very different from those of today. Henry of Lancaster was ready to hear the word of God, although his ears had been closed to all good and only too prepared to listen to the delicious songs of a man or woman, the nightingale or other bird, or to musical instruments or the cry of dogs. Not that he regarded it as a sin to listen to songs or any other thing that God had made for man's benefit: the natural world was there for him to hear, provided it did not provoke inappropriate sensuality.[14]

25 Longthorpe Tower: the boar.

In terms of absolute levels of sound, both town and countryside were very much quieter than today: much would have been audible that is now obscured by background noise. In early modern England – and the same must have been true for medieval England – sounds above 60 decibels were rare: a handful of natural sounds, such as storms and thunder; the cries of some animals, for example barking dogs close at hand; and a few man-made sounds – shouting, bells, music in a confined environment or the sound of some wind and percussion instruments and, exceptionally, explosions caused by gunpowder.[15] Such sounds as there were, many of which today might be considered among the quieter sounds, would have been heard with greater intensity. The sounds of the natural world, of bird-song, for example, would have been prominent. When the outlaw's *Song of trailbaston* was composed, in the 1310s or 1320s, its final stanza reported – much as a charter would have recorded – that it was written in a wood, under a bay tree, to the sound of blackbird and nightingale and the cry of the sparrowhawk, before the script was published as a parchment left on the highway.[16]

Not only were the sounds more intense, but they would have been more readily identifiable and familiar. In a typical village, the bark of a dog or the sound of cattle would have been known not only to the owners, but to all. By their very recognisability, many sounds would have received more attention than we might give them.

The spaces of medieval life were acoustically very different from ours. The walls of houses, constructed of wattle, daub and timber, with their window and door spaces, would have offered little acoustic barrier to conversation. The few stone buildings would have provided a markedly different environment, with resonances that were exploited or diminished according to circumstances. The cavernous, echoing pattern of the greatest churches created an auditory experience without parallel. Not only might they be specially designed, from the point of view of matching celestial harmonies and proportions, but further attention was given to sonorousness and amplification. The new Gothic style of the thirteenth century changed the proportions of churches, so that the largest buildings had naves much higher than they were wide, increasing the amount of reverberation within the building. It is exactly at this moment that one finds the development of polyphony, executed by small groups of singers (rather than large choirs), who might combine skilled voice production with a knowledge of the acoustics of the church. Reverberation meant the prolongation of musical notes, leading to overlapping tones and a continuous background of sound. The amount of sound reflection is greater in the middle and lower frequencies; higher pitches can be articulated more clearly and, for example, the use of descants might be especially effective.[17]

Other features that might enhance sound ranged from acoustic pots introduced into choir stalls at Fountains Abbey; resonance chambers, at Coventry Whitefriars or St Gregory's Priory, Canterbury; or 'whispering galleries', as at St Peter's Abbey,

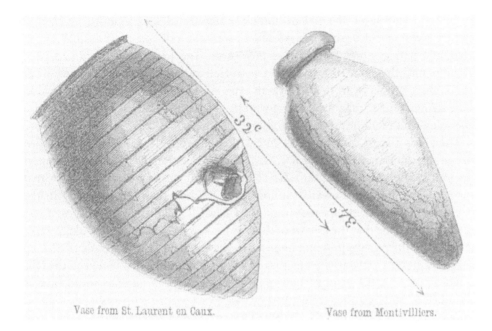

26 A thirteenth-century sounding pot from St Laurent en Caux, France, with a further French example of the seventeenth century, from Montivilliers.

Vase from St. Laurent en Caux.   Vase from Montivilliers.

Gloucester (now Gloucester Cathedral).[18] These features witness a keen interest in the acoustics of buildings and their character had very considerable implications for the structure of the liturgy and the development and effects of ecclesiastical music.

In a secular context, the greatest buildings, constructed of stone or brick, with tiled floors and sometimes wood-panelled walls, would also have had a marked, in-built resonance.[19] One needs to take account of the different purposes of rooms and their furnishings to see how the acoustic properties of the space were moulded. Although, as we shall see, the aim of many medieval soft furnishings was to display the splendour and wealth of the owner, wall hangings, floor coverings and carpets for tables would all have diminished the reflectivity of the hard surfaces. Canopies and carpets, which were marks of status, would have deadened sound in their immediate vicinity, perhaps helping to keep the conversation of the greatest from those around. Medieval beds, often the only or principal furniture in an aristocratic chamber and used throughout the day, were fitted with testers (hangings at the head of the bed), celures (a textile canopy), curtains, and furnished with cushions and pillows as well as mattresses and coverings, all of which would have absorbed sound. There seems to have been less use of curtains for windows, to judge by the few itemised among Henry VIII's possessions in 1547, but these, typically of sarcenet or satin, lined with buckram, would have further diminished sound.[20]

The great hall was one of the most resonant spaces within the household, contrasting noticeably with smaller chambers. One of the reasons for enjoining silence or permitting nothing more than moderate conversation at meals may have been the overall level of ambient noise: the dignity of silence added much to a formal occasion.[21] The substantial kitchens were typically portrayed as hellish, on account of their noise as much as their heat. The banging of pots and pans accompanied the entry of Adam and Eve to Hell in the mid-twelfth-century work *Jeu d'Adam*, probably of Anglo-Norman origin.[22] The moral associations of sound were strong. The word 'noise', replete with pejorative associations, may itself derive etymologically from *noxia*, 'harmful things'.[23]

There was a great contrast between enclosed and open spaces. Medieval garden design, the construction of arbours, the use of hedges and alleys, all had an impact on sound; the streets of the medieval town might be productive of noise that travelled into buildings, overwhelming many of the sounds coming out. In the countryside, open fields, meadows and woodland all had different acoustics. It was in the countryside that the impact of the seasons and the time of day would be most noted, with the relative presence or absence of moisture in the atmosphere making a significant difference to the qualities of sound. Crops also had an effect: the tall-standing corn of the Middle Ages must have absorbed a notable proportion of sound. That these features were noted by medieval authors is unsurprising: what is important is to understand the connotations that sound conveyed.

Moralists were not slow to invoke everyday sounds in support of their theology. The vanity of the cuckoo, crying its own name; the weeping of babies, a premonition of the miseries of mankind – these were sometimes, but rarely, matched from the pulpit by a positive delight in sound, for example in bird-song.[24] A Lollard sermon likened the glutton to the bittern. Sitting by the water, the bittern puts its bill to a reed, making a sound that can be heard all over the country; sitting on land, away from the water and lifting its bill up to Heaven, it then makes no noise. In the same way, gluttons in the tavern put their mouths into their bowls until they are noisily drunk and raucous, but sitting in church by their confessor, far from the tavern, they are dumb.[25] Other medieval lore praised the qualities of the song of certain birds, such as the swan, reputed for its sweet voice, the notes modulated by the bend in its neck.[26]

The solace provided by the music of bird-song was a popular literary motif, authors seemingly vying with each other for the longest list of birds they might introduce: the formulaic, late fifteenth-century author of 'The squire of low degree' enumerated no fewer than seventeen in thirteen lines.[27] Beyond the list, however, we are to understand that these were familiar sounds and that the different calls of birds were known to many. More exotic birds, distinguished further by their cries, were a common sight in aristocratic households, caged or, like peacocks, free to wander in the grounds.[28] Even an Oxford scholar might have in his room a caged blackbird, a vexation to his colleague, who cut out its tongue with his penknife – only for the bird's song (but not its tongue) to be miraculously restored by the intervention of St Richard.[29] The overwhelming nature of bird-song was remarked on by the peasantry, who complained at the noise of rooks and crows.[30] The sounds of birds and animals were sometimes referred to as 'language', but although philosophers debated the intentionality of the bark of the dog, it was clearly differentiated from human speech.[31]

Natural sounds could be immensely frightening. The sound of the sea so terrified a young horse bought in Scotland that it fell dead as it was about to be shipped back to England by its new owner, only reviving after the spiritual assistance of Simon de Montfort had been sought.[32] According to Isidore of Seville, thunder had that name because its sound was terrifying.[33] The sound had many explanations, centred on the clashing of clouds, the ignition and extinction of vapours there. Richard Lavenham's treatise on the weather, from the second half of the fourteenth century, followed an Aristotelian explanation: extinguishing the flash of the lightning in the dampness of the cloud made a sound much like submerging hot iron in water.[34] The delay between the flash of the lightning and the sound of the thunder had been well observed by Isidore, whose acuity was repeated by Bartholomew. It was like a woodcutter striking a tree – one witnessed the action before the sound was heard.[35] Thunderbolts, however, once they had fallen, were beneficial. An English lapidary from the first half of the fifteenth century, echoing several earlier works, held that the man who bore it 'clenely' – as a virtuous individual – would not be struck by lightning in his house (analogous to the proverb that lightning never strikes the same place twice); and its power would not allow the man to be overcome in debates with men or in battle, perhaps a recollection of the strength of the sound that had produced it.[36]

Many man-made sounds had significance in the daily round. The sound of bells, ringing out the hours for church services, was intentionally widespread and rivalled subsequently only by the temporal hours struck by mechanical clocks. William Drake of Little Marcle, 'a very simple man', whose son, John, had been miraculously revived by Thomas Cantilupe, recounted how, on the Tuesday before Ascension in 1304, while he was making a ditch between his field and the highway, he heard the hour of nones struck. He estimated that it was halfway between then and sunset that he found his wife Edith crying outside their house at the death of their son.[37] A primer in Anglo-Norman, of 1396, referred in a model conversation between a traveller and a lady to hours sounded by a mechanical clock. She informed him that the hour they had just heard struck must have been eleven o'clock, as she had earlier heard the clock strike ten.[38] Peter Idley, writing in the mid-fifteenth century, lamented the degenerate tendency for people to work on Saturday afternoon, after the point at which the bell had rung for nones.[39]

For those whose life was rigorously organised in religious devotion, particularly those who lived in silence, almost the entire day was regulated by a series of signals designed to punctuate or define the day in sound. Archbishop Lanfranc's constitutions for Christ Church, Canterbury, prepared probably in the 1070s, formed a model for other Benedictine establishments in England. He set out the round of the day, with its variations throughout the liturgical year, describing not only services themselves but other features. Throughout the month of October, the monks, who returned to bed after lauds, were to be awakened at daybreak by a small bell (*parvulum signum*) rung quietly by the warden of the church. The monks then went into the church to sing prime, then proceeded to the cloister. Later, another small bell, the *skilla*, described as 'the smallest sign' (*signum minimum*), was rung to announce the preparation for terce. This bell was separate from the larger one (*maius signum*) that announced terce itself; at this point initial prayers were said in the church; and then a lesser bell (*minus signum*) was used to indicate the beginning of the service. Once the service had ended, the abbot or prior ordered the smallest bell – probably the *skilla* – to be sounded and the monks went out of the church to chapter. At the conclusion of the reading of the Rule, the abbot or prior struck a board or flat piece of wood (*tabula*) with a mallet, like a gavel, announcing the point at which speech might begin in the cloister. The pattern was continued throughout the day, with the *skilla* used in order to give individuals time to prepare for a service or activity. In addition to these sounds, in the cloister, close to the refectory, there was a gong (*cimbalum*) that was rung a short while before mealtimes.[40]

Different monastic houses employed other patterns, but the principles were the same. The customs of Eynsham Abbey, composed by Brother John of Wood Eaton between 1229 and the early fourteenth century, were extensively based on those of Lanfranc and the Augustinian house of St Victor in Paris. The description of the refectory at mealtimes, although incomplete, gives further detail. At *prandium*, the main meal of the day, the community was hastened into the refectory by two strikes of the *tabula*. At the third strike, the meal was over: two novices were to get up from table and collect the crumbs. Food and drink that had been partly consumed was to be covered at the conclusion of the reading that accompanied the meal. Any food that remained untouched was to be cleared away, along with the cups and jugs. When the abbot or presiding monk made the signal with his knife for the remains to be collected, cups were to be wiped and dried, and put in their places. Once the crumbs had been collected, the presiding monk would make a sound to indicate to the reader that he should finish his task. Finally, the presiding monk would sound the *skilla*, handed to him by a novice, who kissed his hand. The monk would continue to ring the *skilla* until all were standing and had moved from the table. At the conclusion of the ringing, all would turn to the east to say a final verse. The reader, those serving the meal and the kitchener and refectorer would then eat if they had not previously done so.[41]

The sound of church bells had a further range of significance. Why were bells employed? Their use in the early Christian Church appears to have been connected first of all with funerals in the catacombs, and subsequently with hermits in the desert. In both cases the force of sound from the bell was used to keep demons away.

27 Hailes Abbey, a Cistercian house: the refectory, constructed between 1246 and 1251, viewed from the south.

This was one of the reasons why they were used in the later Middle Ages: bells that had been consecrated or had specific virtues, or prayers inscribed on them, conveyed these attributes by almost the loudest sound that man could make. Ringing the bells therefore created a positive environment: the operation of the holy, in church services, might continue unmolested and, more generally, the sound gave protection to the neighbourhood. Handbells used by itinerant preachers may have given a special character to the sweetness of the word of the Holy Spirit that would come with the service; bells were sometimes named to reflect this quality.[42] Ringing bells outside the usual pattern of canonical hours would have been recognised instantly as marking something distinctive. In 1229, when masses were said for the succour of the Holy Land, Bishop William de Blois ordered bells to be rung in the diocese of Worcester, so that those hearing them outside church might kneel down and say the Lord's Prayer in this cause.[43] Bells might announce sentences of excommunication[44] and a peal of the principal bells, sometimes accompanied by the singing of the Te Deum, marked the occurrence of miracles.[45] It was forbidden, however, to ring the bells in the presence of those who had been excommunicated.[46] When, late in the reign of King John, probably in 1213 or 1214, St Osmund cured a man suffering from paralysis, the celebration was muted. The country was under interdict: no Te Deum was sung – it was said, quietly (*sub silentio*), as Master Thomas of Chobham recollected.[47] It was noted at York, in 1520, that it was the custom to summon people to church using rattles rather than bells during the last three days in Holy Week: the corporation forbade children to participate in the practice, which had become over-exuberant.[48]

The use of bells – the passing bell for the dying, the death knell for the dead, and a subsequent bell for the funeral procession or translation of a body – created a protective aura of sound at a critical moment of transition, when it was of especial importance to reinforce the powers of the sacred. In 1421, the body of Thomas, Duke of Clarence, was brought back from France where he had been killed at the battle of Baugé. On landing at Sandwich, it was taken to Canterbury for burial,

28 A funeral: the sound of church bells was commonly employed to protect both the corpse – before its burial in hallowed ground – and the living against the forces of evil. From the Smithfield Decretals, written in Italy for French use, but illuminated in England, c. 1330–40.

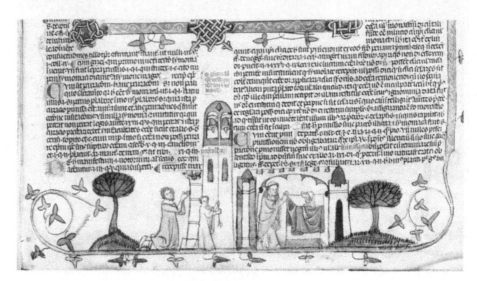

29 Church bell of c. 1320, inscribed + IHC: NAZARENUS: REX: IUDEORUM (Jesus of Nazareth, King of the Jews), from Hales in Norfolk.

where a special payment of 16s. 8d. was made to the bellringers of the Cathedral Priory and St Augustine's Abbey for their labours on the day.[49] When the body of Richard, Duke of York, was moved from Pontefract to Fotheringhay in July 1476, bellringers were again in evidence, at Doncaster, and possibly at other churches where the body rested.[50]

Handbells might accompany a funeral procession to the same end. The houseling bell was commonly used to accompany the host as it was taken to the sick, bringing powers of protection but also warning people of the host's presence, to stir them to devotion and to salute it.[51] In the early 1240s, Bishop Raleigh of Norwich forbade lepers and all others except priests carrying the sacrament to have handbells in the public streets, as they were the means of arousing devotion and reverence.[52] Handbells were further used at the elevation of the host during mass – like a small trumpet announcing the coming of the judge, the saviour, stirring the soul to rejoicing.[53] Archbishop Pecham's Constitutions of 1281 envisaged that the principal church bells would be rung at the point of elevation, so that the people, who would not normally be present at mass each day, might kneel down, whether they were in the fields or in their houses.[54] The principal bells were rung out of custom when the king or queen passed through the parish, or at episcopal visitation.[55]

30 The host carried in procession, preceded by a man with a candle and bell. From the *Omne bonum*, late 1330s and 1340s.

The civil authorities also used church bells. The curfew bell in the city of London marked the closure of the city gates and taverns: the watch patrols might then challenge those on the streets.[56] The watch, or waits, had signals of their own. The use of horns was common practice.[57] In winter, from the end of September through to Maundy Thursday, the wait in the household of Edward IV 'pipeth the wache within this court iiij [four] tymes, and in the somer ny3ghtes iij tymes'.[58] Others using horns included those attending animals, like the swineherd of Cuxham, Oxfordshire, who recovered his horn in the manorial court of 3 August 1315,[59] and huntsmen.[60] The two minstrels who preceded Edward IV when he went riding probably blew trumpets so that the remainder of the household could follow their progress.[61] Trumpets were associated with judgment, following the trump of doom in Revelation.[62]

Horns and brass instruments were popular for their music – Henry VII paid a total of £1 on two occasions, one near Rochester, in April 1498, to horn-players[63] – but had a variety of other uses. *In extremis*, they might be used to revive the senseless. Near Gloucester, in the 1170s, a horn, brought from a well, was used on a boy who was kicked by the horse of Robert de Buckler, who was on his way to Ireland.[64] In 1196, the monks of Eynsham Abbey, in an attempt to raise from his comatose state the monk whose vision was to be widely reported, vehemently blew a horn in his ear.[65]

Raising the hue and cry might be done by horn,[66] but the noise of the hue was distinctive in itself and horns were not necessary for creating the general clamour that signalled its urgency. Geoffrey de Smalmedewe was digging in the ditch near the house of Gilbert the Reeve of Little Marcle when he heard 'the clamour that it was accustomed to make in that place when anyone had been found dead by accident or killed'.[67] Lucy, the wife of John the Fisherman, whose son had drowned in the River Wye, was desperate for his revival not least so that the family would not be subject to the scandal caused by the clamour of the hue and cry.[68]

If the hue and cry bound the individual to participate in the communal cause, the act of hearing in other instances might be anti-social. Eavesdropping was regarded in law as a common nuisance – an offence against the community as a whole rather than a private matter – and it was for this reason that it appeared in the articles for presentment at the manorial court.[69] Given the construction of medieval houses, overhearing a conversation must have been an unexceptional event. Malice of purpose in the hearer, however, changed its nature. The ghost of Robert, the son of Robert de Boltebi of Kilburn, not far from Byland Abbey, when caught, confessed that it had made a habit of standing in the doorways of houses, by windows and against walls, as if listening, so that it could seize upon those who came out of the building.[70]

If the act of hearing could be offensive, some sounds in themselves carried this connotation directly or, indeed, were much more dangerous. There was a whole range of sounds, vocal and non-vocal, that might carry an awful message. Even bells – if they resounded from below the surface of the water, from a site that had been submerged – might indicate demonic presence or impiety. Bells reputedly heard in the forest near Carlisle, probably from Tarn Wadling in Inglewood Forest, referred to by Gervase of Tilbury as *laikibrais* (in Anglo-Norman, possibly 'the lake that cries'), may have had this character.[71] The consequences of these noises could be dire. Besides some aspects of speech considered in the next chapter, particular sounds were associated with the Devil, especially cacchination or cackling, mad and angry voices, mocking and bellowing.[72] Benedict, the Abbot of the Cistercian house of Stratford Langthorne, recounted a tale to Peter of Cornwall, who included it in his *Liber revelationum* around 1200. When Benedict was in the monastery at Revesby, the Devil tempted two novices to renounce their calling and to go back into the world. One did so and, the following night, the Devil appeared to the other, mocking him, cackling and uttering terrible threats.[73] Diabolical noise is a feature of other visions. Prominent in one, which may have taken place at Rievaulx, were further mocking, cackling demons.[74] All these and very much more were present in the vision of Thurkill, of 1206. Here angry and incandescent demons dismembered sinners, for another of their number to toss into a boiling stream of oil, the liquid sending out on each occasion a great hiss like boiling fat thrown into cold water.[75]

Another sound, a great crack, was heard on occasions when the living encountered spirits and demons. When William, the cellarer of Stratford Langthorne, was at the point of death, there was such a loud crack on the roof of the infirmary that the whole building shook. The sick man was able to tell the monks, who had rushed to see what the noise was, that it was caused by the arrival of all the souls for whom he had omitted to pray – on account of press of business or negligence – demanding that the deficiency be made up.[76] The cracking sound heard on the performance of some miracles may have a similar provenance, perhaps marking the departure of a demon that had caused paralysis or possession, although it sometimes accompanied the straightening of bones. A girl who went to Barking and prayed there to St Ethelburga and St Erkenwald was cured by the latter, in a dream, while the nuns there were singing lauds. At this point, there was a great crack, as if a dry hedge were being destroyed.[77] Close to this was the sound of thunder, which was also associated with the Devil. In July 1316, the verdict on John Sone of Woodadvent, in the Brendon Hills in Somerset, who died in his sheepfold at Empnetridge while preparing to drive his sheep to pasture, was that he had been struck by thunder and that his death was caused by the Devil.[78]

These noises feature in a moral soundscape at many points throughout the Middle Ages. The association of sibilants with evil found voice in the late eleventh-century life of St Rumwold, which decried the 'false hissing' of blasphemers who impugned the lives of saints.[79] Cacophony, pure noise, senseless words – nonsense – were associated with the Devil and form a prominent element in his characterisation in late medieval drama. In the late medieval play, *Mary Magdalen*, at the casting out of seven devils 'the Bad Angyll entyr into Hell wyth thondyr'.[80] The crack of breaking wind was a popular element in this burlesque.[81] In the N-town play of the Creation and the fall of Lucifer, the Devil descends to Hell with the words:

> Now to Helle þe wey I take,
> In endeles peyn þer to be pyht.
> For fere of a fyre a fart I crake !
> In Helle donjoon myn dene is dyth.

And, addressing God, in the second play of the sequence

> At þi byddyng fowle I falle,
> I krepe hom to my stynkyng stalle.
> Helle pyt and Hevyn halle
> Xul do þi byddyng bone.
> I falle down here a fowle freke;
> For þis falle I gynne to qweke.
> With a fart my brech I breke !
> My sorwe comyth ful sone.[82]

Breaking wind might be combined with other diabolical sounds as the stage directions for the *Castle of Perseverance* indicate:

> And he þat schal pley Belyal loke þat he have gunnepowdyr brennynge in pypys in hys handys and in hys erys and in hys ars whanne he gothe to batayl.[83]

From its association with the Devil, breaking wind gained a moral opprobrium. If it was taken up in courtesy books from the point of view of manners and polite behaviour, it was an action that was replete with diabolical connotations. A popular scatological tale of an adulterous wife, circulating in the mid-fifteenth century, concerned Genulphus, a Frenchman who had lived a holy life and in whose name miracles had been performed after his death. In utter disbelief at the news of these works, his wife had the misfortune to exclaim, 'It is als trew at Genulphys duse meracles, as it is at myne ars syngis', which was then her fate every Friday whenever she spoke.[84] In the same vein, courtesy books also outlawed laughing loudly, or behaving raucously, all of which must have seemed to some eyes at least very close to patterns of behaviour commonly ascribed to the Devil.[85]

The sounds of the mad marked them out as possessed by demons. They were commonly restrained physically, but it was not unusual to bring them to places where a beneficial influence might have an impact on their condition. They were to be found in the greater churches as much as at lesser places where the influence of holiness might prevail. In the mid-1280s, within a few years of the death of Thomas Cantilupe, Edith, the wife of Robert the Ironmonger of Hereford, was brought to Hereford Cathedral in her madness to obtain the assistance of the reputed saint. According to one witness, she spent fifteen days in chains in the cathedral, but made so much noise that, in anticipation of the services of Palm Sunday, she was moved to a pulpit, where she was bound between the two women, her neighbours, who had brought her there. Given the demoniacal nature of the illness, it was not surprising that her husband, under examination some twenty years later, was asked about the words she spoke – but he could not recollect whether she had blasphemed God and the saints.[86] John Bute, Cantilupe's baker, recalled another madwoman, from Ledbury, who had been brought to Hereford, bound, gnashing her teeth and shouting horribly. Beating her, even to the point of bloodshed, had not tamed or disciplined her. She spent fifteen days at the Bishop's tomb in desperation, to be led out of the cathedral when her shouting became too great; subsequently she recollected how she dreamt at this time that she was surrounded by many demons.[87] In 1486, a woman of Ashby St Ledgers, described as delirious and full of demons, came into the church there, shouting. The wiser and more honourable members of the community wanted her bound with ropes, but the vicar forbade any such cruel treatment – and she was cured by prayers to the Virgin and Henry VI.[88]

There were, on the other hand, sounds connected with the supernatural that were good and beneficial. Supreme among these was the sound or voice of God. On the first Whitsun, as the disciples were gathered in a house in Jerusalem, there came a sound from Heaven, translated in the late fourteenth century as 'a greet wynd comynge', which brought the disciples the Holy Spirit (literally, the breath of God) and the gift of tongues.[89] The conversion of St Paul, riding to Damascus, was effected by God's direct speech, asking why Paul persecuted the disciples. Reported by Paul's servants in the play *The Conversion of St Paul*, the sound they heard was 'of won spekyng wyth voyce delectable' and 'a swete dulcet voyce'. Both descriptions were medieval embellishments of the biblical text, as was the addition of a tempest to the blinding light that overcame their master.[90]

The sounds of Heaven were as familiar as the sounds of Hell: the choirs of saints and angels, the sweetness of melody. These were a commonplace of medieval visions and devotions. In the 1150s, Peter Peverell, formerly a knight but at this point a monk of Norwich, had a vision of a vast crowd of men dressed in refulgent clothing, entering the church to varied and harmonious sounds; and he was told that the Queen of Heaven had come to talk with the Cathedral Priory's young saint, William.[91] It was accordingly no surprise that in the N-town play of the Assumption of Mary choirs of angels sang 'Alleluia', the whole heavenly court sang, and cithara were played.[92] Celestial harmony may have been inaudible to human ears in the Boethian scheme of music, but in Christian thinking of the high and late Middle

Ages it was far from imperceptible. *Jacob's Well* told how a girl, unjustly hanged, was kept from death for two days by an angel: 'sche felte no peyne but sche herde swete melodye and song of aungellys.'[93] Practitioners of the more extreme forms of sensory devotion heard heavenly melodies in their physical ears.[94] The monk of Strata Florida Abbey, in Wales, who had a vision, *c.* 1202 – aided by an angel who placed a burning coal from the thurible into his open mouth – made a telling contrast between the band of monks he saw in choir singing psalms, censed and blessed by the angel, and those who performed their accustomed tasks with less rigour.[95]

Divine intervention might play a leading role in the healing of the deaf. The miracle brought a moral dimension to sound, indicating to the alert listener that good was conquering evil. The late eleventh-century life of St Birinus, whose cult was linked with Dorchester on Thames and also with Winchester, recorded how he had cured an elderly woman who was both blind and deaf. She prayed to the saint, but her speech was confused, mumbling and bellowing, and she was incapable of speaking distinctly, until her faculties were restored by Birinus.[96] Godiva of Stratford – 'created as a human, but without the ability to hear' – suffered from noises in her head, a great sound of thunder, although she could hear no sound with her ears. Her ears were washed with a little water from Becket's shrine, which she subsequently visited, and she was cured of her infirmity.[97] Two of the miracles of St Gilbert of Sempringham involve the positive force of sound, of bells. A woman of Anwick, near Sleaford, had become so deaf that for nine weeks she could not hear the sound of bells. The morning after a night of prayer at Gilbert's tomb, she heard first a bell, and then, when her ears had been washed with water from the saint (presumably used to wash his relics or blessed at the tomb), she heard men speaking and, as she left, the tramp of feet. A second miracle involved a deaf man from Braceby, near Grantham. In a dream, he heard St Gilbert preaching. He awoke to hear the sound of the church bell ringing for matins and told his amazed wife that the bell was ringing and that it was time to go to church.[98]

Conventional medical treatments for deafness and afflictions of the ear varied in their basis, from the simple prescriptions of an early fifteenth-century translation of Macer Floridus that the smell of burnt hyssop would drive out ringing of the ears, or that black hellebore should be used for those that were deaf as the result of illness,[99] to more sophisticated analyses based on imbalances of complexion. An earache, for example, without a swelling, was caused by heat and the remedy was the application of 'cold' medicines, such as oil of roses; if there were a swelling, however, and the cause was cold, then 'hot' medication, such as oil of laurel, might be applied. Ringing in the ears, caused by 'ventosite' or windiness, could be relieved by preparations that destroyed wind, such as anise, calamint and oregano.[100] An Anglo-Norman collection of medical recipes from the first half of the fourteenth century employed a little juice from mint, heated, against earache and 'worms' in the ears.

Other warmed liquids might be used, such as those derived from fennel seed or plantain; or a plaster might be made of juice from groundsel mixed with mutton fat. For deafness, there were plasters and mixtures to be placed in the ear, with Avicenna's caution that nothing was to be placed in the ear unless it was warm.[101] Obstructions placed in the ear were a common impediment to the hearing of children, but even in 1486 the case of Richard Dyonyse must have been exceptional, miraculously relieved by Henry VI of a bean that had been in his ear since childhood – for thirty-seven years.[102]

There was considerable emphasis on the appropriateness of sounds, and the moral connotations that went with them were strong. Religious weeping might legitimately encompass devotion at the tombs of saints,[103] or the practice of enthusiasts. Margery Kempe, taken to her inn at Canterbury after meeting the monks there – who had thought that she might be burned as a Lollard – cried so copiously and continuously that it was marvellous her eyes endured and her heart lasted, preserved by the ardour of her love for Christ.[104] The weeping and wailing of penance was as appropriate to the criminal in confession before execution as it was to the religious spirit on Good Friday, contemplating the passion.[105] These forms of penitence were close to mourning and the madness of grief – the mother of St William, on learning that her son had been found dead, ran through the streets of Norwich tearing her hair, repeatedly clapping her hands, crying and wailing [106] – but they were very different in their quality from sounds of disorder or of questionable moral virtue.

31 Insanity (*amencia*): tearing the hair was also a feature of the extremes of mourning. From the *Omne bonum*, late 1330s and 1340s.

An edifying tale, designed for inclusion in a sermon by a thirteenth-century English Franciscan in Ireland, focused on immoderate celebrations. He had been told by a friar from Dacia (the Franciscan province of this name covered Denmark, Sweden and Norway) how, when a woman was in childbirth, it was the custom for neighbouring women to come to her, dancing and singing in an immoderate way. The women then went on to fashion an effigy of straw, which they dressed, calling it 'Bovi', dancing and singing before it in a lascivious fashion and calling upon the effigy to sing. To which the Devil replied, 'I will sing' – that is, not the effigy, but the Devil that existed in it, making such an awful noise that some of the women fell down dead, others barely escaping with their lives. The moral, rather obviously, was of the dire consequences of this form of celebration.[107] Noise might be a form of sacrilege. In 1229, the synodal statutes of William de Blois for his diocese of Worcester again forbade the use of the cemetery for secular activities, law pleadings, fairs – unless in aid of the church – dancing and wrestling while services were being sung, as well as any form of Sunday trading anywhere. In addition, he forbade secular song in a house where there was a dead body, as well as dancing and wrestling in the house at this time. This extension of the prohibition again centres on the susceptibility of the corpse to attack by the forces of evil, some of which would have taken place through sound.[108]

These restrictions commonly circumscribed the activities of priests. It was not unusual to require their absence from taverns, public drinkings and places where lewd songs or tales might be heard; and if these occurred at events to which they were legitimately invited, they must at least conduct themselves in such a way as not to appear to listen to or encourage the activity.[109] Others in pursuit of religious benefit needed to act with circumspection. A Wycliffite sermon-writer of the late fourteenth century, addressing the text 'See brothers that you walk cautiously, not as fools, but as the wise', held that men filled with the Holy Spirit should speak to God, singing in their hearts; but the glad songs and whistling of pilgrims indicated their fear of spiritual enemies, much as the common saying held that a traveller without money 'syngiþ sure bifore þe þeef', whereas one with funds was in dread.[110] Archbishop Arundel, examining William Thorpe in 1407, defended the songs and music of pilgrims for the spirit they gave to the difficult enterprise of pilgrimage, maintaining the expedition through difficulties. But the Lollards saw things differently, decrying the noise of the singing, bagpipes and jingling of their Canterbury bells, not to mention dogs barking after them. They made more noise than if the King had passed that way with his trumpeters and minstrels. Worse still was the jangling that pilgrims made long after their journeys were complete.[111] A thirteenth-century confessional formula probably intended for monastic use outlined, among the sins that might be connected with hearing, the pleasure received from listening to sweet sounds and secular songs, rather than the tedium with which Scripture had been heard.[112] The sounds of sex might even intrude on the ecclesiastical court, as it did at Scredington, near Sleaford in January 1337. Margaret Walters, speaking in support of Hawise Pendere's contention that Robert Duraunt should marry her, described how Robert Duraunt had twice lain down on a bed with Hawise – noisily – before going out into the garden to continue their activities.[113] If other sounds, such as the babbling of infants, might have no moral connotation,[114] that of resuscitation clearly did. The revival of Gilbert Sporun, the London goldsmith's child, was marked by a noise in the infant's throat as the breath of life returned.[115]

Two further categories of sound merit close attention: music and speech. As part of his examination of mathematics, Isidore in his *Etymologies* discussed arithmetic, music, geometry and astronomy.[116] This wider connection was noted by Bartholomew, who held that music – harmony and melody – was necessary to know 'mistik menynge of holy wryt', as the world was made in harmonic proportions and the heavens moved in consonance and accord of sound.[117] He noted the impact that music might have on the individual, because the body was bound together by harmony. The trumpet in battle comforted warriors: the stronger the sound, the greater the strength and boldness of men to fight. Music might diminish the power of evil spirits over men, just as David, with his cithara, had relieved Saul of an evil spirit; and it might have an influence over animals, birds and dolphins.[118]

With its transcendental power, music inevitably had a close association with the liturgy of the Church and with the powers of speech discussed in the next chapter. In the early thirteenth century, Pope Innocent III explained that the prophet David had introduced cantors into ecclesiastical service in order to make the service of God more solemn. It was therefore necessary for the singers to sing together in consonance and sweet harmony, to stir listeners the better to the devotion of God.[119] But how might that be achieved? The injunction of the Rule of St Benedict was that monks should consider how they should behave in the presence of God and his angels, and that psalms should be sung in such a way that the mind was in harmony with the sound of the voice.[120] Much in the structure of medieval liturgical music depended on the close analysis of the text, the words of the psalms, of hymns and parts of the services that were to be chanted: one of the purposes of music was to clarify the rhetorical canvas of the words – sound was meaning and the meaning had to be conveyed correctly both to listener and participant. The intonation of prayers, or psalms, on a single note would be followed by simple melodic inflexion immediately before pauses in the text, the rhythm at this point following the *cursus*, a stress inherent in the words. The study of psalmody allowed the singer to identify those points at which melody might occur in order to make the text clear.[121]

Music, of all sounds, had a vivid secular sense and the question of appropriateness was acute in ecclesiastical use, in settings of the word of God. The application of musical theories to the liturgy had a distinctive impact in the monasteries. The twelfth-century Cistercian reform of the liturgy sought to employ a ten-tone scale

derived from Scripture, quite unlike traditional patterns of music.[122] The intention
was to create a plain, simple style of song, that would reinforce the spiritual message
of the text. The harmony of the choir was to be a reflection of the harmony of the
monastery – and a means by which one might move to the harmony of Heaven.[123]
A later Cistercian, Stephen of Salley (d. 1252), successively Abbot of that house,
Newminster and Fountains, wrote a treatise on psalmody for his colleagues. In it, he
explained from a practical point of view the ways in which the psalms might be said
or sung, starting with a literal sense and moving on to a mystical or anagogical
understanding as a basis for meditation. He noted that some stuck with the literal
interpretation, in order not to be distracted during the service, whereas others, the
more simple monks, used almost mathematical correspondences between the verses
of the psalms and the mysteries of the faith, to serve as subjects for meditation.[124]

The manner in which the texts were sung was also the subject of close regulation.
The duties of the precentor at the Augustinian house of Barnwell, in Cambridgeshire,
c. 1295–6, gave directions on how he was to lead the singing, how all were to follow
him in pitch and tempo, intoning the psalms with a soft and sweet voice.[125] Especial
dignity might be given to an occasion by variation in tempo: Lanfranc's constitutions asked that on All Saints' Day the office of vespers for the dead – the prelude to
the following day, the feast of All Souls – be sung at a slower speed than usual.[126] But
the very sensuality of music was also an issue, a tension that might only be resolved
for some by avoiding its use in the service of God. Gilbert of Sempringham forbade
the nuns of his order to sing: a silent, inner psalmody was required.[127] Two centuries
later a similar argument was employed by the Lollards: that it was a ghostly music
that was intended in the service of God, not a physically present and sensual
sound.[128] More typical of general experience at this point, however, was the explanation included in the sermon in *Mirk's Festial* for Advent Sunday, that as the
coming of Christ brought joy and bliss into the world, therefore the Church used
melodious songs, such as 'Alleluia', with other songs of thanks, the 'Te Deum
laudamus' and 'Gloria in excelsis', to mark the passing of Christ to judgment.[129] A
tale in the fifteenth-century English version of the *Gesta Romanorum* illustrated the
conflict of sound. While out hunting, the Emperor Theodosius (Christ) heard the
beautiful sound of a harp (Scripture). It was played by a fisherman (a preacher),
who explained that his melody brought the fish (sinners) into his hands and he was
thus able to provide for his family. But recently there had been competition, from a
'hisser or a siblatour' (the Devil), who hissed so sweetly that the fish were drawn
to him. Accordingly the fisherman asked for the Emperor's help and was provided
with a gold hook (the grace of God) to place on his fishing rod, to draw the fish in
when he played his harp. The distance between metaphor and practice was not
large.[130]

Some musical instruments were felt inappropriate for ecclesiastical use and some
musical chords, such as the tritone, were also avoided. Care might be needed to

obtain harmony, even within a single instrument. It was believed, following Pliny, that to fit a harp or fiddle with strings derived from the guts of a wolf, alongside those from sheep gut, would lead to the corruption of the latter by the former.[131]

Minstrelsy as a whole was considered an inappropriate occupation by the Church. In practice this position moderated in the thirteenth century and music was widespread in secular life.[132] Inventories listing musical instruments[133] and bone pipes or whistles from excavations[134] are testimony to its ubiquity. The accounts of the chamber of Henry VII list payments for secular and liturgical events side by side, for the routine, the ceremonial and for pleasure. In November 1495, payment was made to a woman who sang with a fiddle; in December 1495, for the wages of the trumpeters, sackbut players and string musicians; in February 1496, £7 was given for new organs; and in August 1496 there was a payment to a bagpipe player.[135] In 1505, there were payments to the children of the chapel for singing 'Gloria in excelsis' on Christmas Day,[136] to Dr Derley singing at Westminster for the King, to the man who played the organ in the gallery at the newly rebuilt Richmond, to the chantry priest who sang on the King's behalf at Walsingham, and for a lute for Princess Mary;[137] and, in 1508, to John Rede's sailors, who rowed up and down singing before the King's manor at Greenwich.[138] Civic drama recognised that music was essential to give grandeur to an occasion. In the York mystery plays, Christ's entry into Jerusalem was orchestrated with a procession, with branches, flowers, 'unisoune' and children singing 'myghtfull songes'.[139]

An audience might thus be provoked to a wide variety of responses, depending on context, ceremonial, liturgical, private or public. Sound in itself might constitute good or evil; it might cause fear or delight; and it might allow the soul to reach out to Heaven, to celestial harmony. Unstructured sound or pure noise might be the most dreadful of all. In the operation of this sense, there was much that was implicit that it is difficult if not impossible now to recover. There is more to be said, however, about hearing the spoken word or song – and the operation of speech itself.

Chapter 5

THE SENSES OF THE MOUTH: SPEECH

Bartholomew the Englishman summarised an accumulation of ideas about the mouth, its physical functions, its relationship to taste and speaking; he went further in considering the link between the mouth and the inner soul, how its operation was revelatory of much beyond the physical and how the voice might affect others. Drawing on Isidore, he noted that the mouth was the messenger of the soul, 'for we tellen out by þe mouþe what we conseyven raþer in soule and in þouȝt'.[1] He went on to cover the views of others, including Aristotle, Gregory the Great and Constantine the African. The tongue received the influence of the 'animal spirit': it formed speech and told the meaning of the thoughts of the soul.[2] Considering the qualities of the voice, Bartholomew believed that an even, clear, strong, binding and flexible voice, lying between heavy and sharp, was to be praised. The opposite was a quaking voice, hoarse and rough, feeble and discordant. Too heavy or sharp a voice was evil and blameworthy: a single discordant and disproportionate voice shattered the accord of many voices. On the other hand, a sweet voice, one that was well ordered, made people joyful and stirred them to love, showing the passions of the soul, its strength, virtue, pureness and goodness of disposition. It relieved labour; put away trouble and sorrow; it distinguished male from female; it won much praise; and changed the affection of listeners.[3]

Moral dimensions of speech were prominent in many discussions, not least those of oaths or anathemas meant and unmeant. A well-known tale, of the mother who cursed her daughter, was repeated from Robert Mannyng of Brunne's *Handlyng synne* of the early fourteenth century to Peter Idley's *Instructions to his son* of the mid-fifteenth century. The mother went to bathe in a stream and handed her clothes to her daughter. She called the daughter to bring her clothes, but the daughter did not come. In her anger, the mother then used the expression 'Þe Devyl com on þe' – which was exactly what happened, and the Devil took possession of the child.[4] The mother's speech had a literal effect, her passion reflecting that of her soul, speaking directly to the Devil. In the late eleventh century, Alexander, a monk of Canterbury, recorded the opposite effect, the prevention of speech. Under investigation at a council, a bishop, asked by the Pope to repeat openly the words *Gloria patri et filio et spiritui sancto* in order to clear himself of a charge of simony, despite repeated attempts and pronouncing other words clearly, remained unable to say the words *spiritui sancto*. This was clear indication of his guilt: the bishop's offence and poor moral state prevented him from saying holy words.[5]

This transcendental power was commonly understood in the Middle Ages. Speech, like other sounds, could effect direct changes in listener and speaker.[6] It was thus extremely powerful, nowhere more so than when dealing with the word of God or his agents, or with evil and the Devil. By the later Middle Ages, the Immaculate Conception was commonly understood as the direct result of the Annunciation, that Mary had conceived as she heard God's speech addressed to her by the Archangel Gabriel. *Jacob's Well* recounted that the Godhead entered the Virgin's

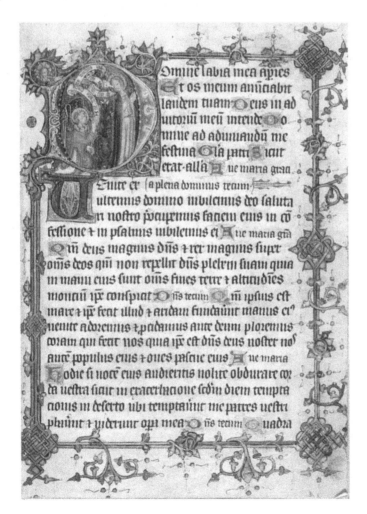

32 The Annunciation, from a book of hours and psalter of *c.* 1380–90: in the words of God to Gabriel in the Towneley play of the Annunciation, 'She shall concyf my derlyng / Thrugh thy word and hyr heryng.'

womb by way of her ear 'in þe heryng of Gabreyellys woordys'.[7] The power of speech had a place in even the more complex of theological arguments, such as Wycliffe's exposition of the process of transubstantiation. Although he did not believe the bread and wine changed in their material substance, the words of consecration added to them the spiritual being of Christ.[8] A sermon of John Mirk quoted a letter to Christ from King Abagarus, a leper, reporting that he had heard how Christ healed all manner of sick, without herbs or medicine, curing the blind and raising the dead 'wyth a word' – with the healing power that touched them through Christ's word.[9]

What the Holy Spirit might do in Christ's miracles, it might perform vicariously through the speech of saints and other holy individuals. Typically we find cures that were the results of words of blessing: in this way, Hugh of Lincoln cured the tumour of the son of Laurentia of Lincoln.[10] The process of excommunication was likewise effected by the word, by a formal anathema.[11] The power of speech was effective through those working on God's behalf, in sermons or in prophecy.[12] A Wycliffite sermon for the first Sunday in Advent, on John the Baptist in the wilderness, fulfilling the prophecy in Isaiah announcing the coming of Christ, noted that the

33 The angel gives John the book to eat, which is bitter in his belly but sweet on his tongue, a prelude to his work of prophecy (Revelation 10: 9–11), a theme commonly portrayed in depictions of the Apocalypse. From the Apocalypse sequence of tapestries at Angers, 1373–87.

Baptist said that he was no more than a voice, for he had no being but the word of God within; as a voice bore the sense of the word in the soul, so he bore the sense of God's word without error.[13] The same power of the Spirit might be invested not only in priests, but in anyone who had to carry out baptism. The formulas make it clear that it was the articulation of the words of baptism – in their exact form, in whichever language – that conveyed the benefit of the sacrament, in conjunction with the wetting of the child.[14]

Just as a sermon transferred the moral benefit of the word of God to the listener through speech, so did the recitation of holy words. The continuous round of the liturgy, the reading of the monastic rule or readings from spiritual texts were much more than practices of piety: they created a moral force through sound. The ritual surrounding the death of a Benedictine monk, outlined in the late eleventh century, illustrates how the protection of holy sound, in part the spoken or sung word, was maintained continuously. Once it was clear that a monk was close to death, he was never to be left by himself. Two monks were to read to him, day and night, from the Gospels and stories of the passion. When he lost consciousness, the monks were to switch to the psalter, which they were to recite continuously as long as their fellow remained alive. When death approached, the rest of the house were to be summoned and were to come, chanting the creed – they were then to pass to the seven penitential psalms and some were to continue with the psalter if death had not arrived. After death, there was a round of services for the commendation of the soul, a ritual for the washing and dressing of the body, the censing of the corpse and sprinkling it with holy water; it was then brought to the monastery's church. From this point on, the corpse was never to be left without the chanting of psalms, except during other services in the church. While the corpse remained unburied there was to be complete silence in the cloister: there was to be nothing that might compete with or counteract the sound of devotions. As the body was committed to the ground, the words of the final absolution were read over the monk – and the written text of that absolution was placed on his chest.[15]

Disturbances during services upset the environment of devotion and corrupted the liturgical message. Clarity in diction, correct pronunciation, due pauses, and so on, are frequently mentioned, from works on the liturgy, monastic injunctions and customaries,[16] to synodal statutes.[17] The development of punctuation in biblical and liturgical texts was designed to ensure this clarity; some monastic orders, such as the Cistercians and Carthusians, maintained uniform systems of punctuation, the latter as late as the fifteenth century.[18] Arrangements in the diocese of Worcester in 1240 set out the importance of making sure that the texts of the service books were correct, so that no mistakes in reading and singing were introduced.[19] This was essential because the power of the word was significant in much more than its grammatical sense: its ethical and moral qualities had to be assured. One popular tale told of a clerk who, encountering in church a devil with a sack, asked him what it contained: 'I bere in my sacche sylablys & woordys, overskyppyd and synkopyd, & verse & psalmys þe whiche þese clerkys han stolyn in þe queere, & have fayled in here servyse.'[20] Singing also had to be carried out in an appropriate way. Another devil with a sack was reported to have been seen in church gathering up the voices of clerks that 'wer syngand & makand a grete noyse'.[21]

Almost all reading at this date was reading aloud, perhaps *sotto voce*, but nonetheless with the intention that the words be pronounced. In monastic terms, there was little that separated reading from meditation, an activity requiring body and soul to participate[22] – and the moral force that went with reading spiritual texts was of particular consequence. The fifteenth-century translation of *The doctrine of the hert* emphasised the benefit of reading devotional works not only with the lips of the mouth, but with the lips of the soul. 'Hertly redyng is a gracyous mene to gostly feeling.' Internally, the word of God might convey an especial sweetness. God had arrayed the listener with silver, for just as silver gave a sound that was sweeter than all other metals, so the words of the Lord sounded more sweetly in the ear of a devout believer than the words of any other creature, whether in the reading or hearing of devotional treatises, or in listening to sermons.[23]

Readings at mealtimes in monasteries served to divert the attention of the devout from the delights of food, creating a positive force that might overcome a sensual pleasure. When Hugh of Avalon ate in the refectory on feast days, we are told he had eyes fixed on the table, his hands on the dish, his ears on the book that was being read and his heart fixed on God.[24] The scholars of Winchester College were to listen diligently to the Bible, the lives of the Church fathers, the sayings of the doctors of the Church or other holy writing, while eating in silence.[25] Although the readings prescribed in September 1473 for the two-year-old Prince Edward (the future Edward V) included such noble stories as it behoved a prince to understand, of 'vertu, honor, cunyinge, wisdom, and deedes of worshippe, and of nothing that should move or styrre him to vyces', the benefit to the Prince must have come from the nobility conveyed by the sound of these stories as much as from his comprehension of their content.[26]

The creation of a continuous aura of sound was also the aim of certain religious practices. Hugh of Avalon, in contemplation and even while he was asleep, was reported to have said 'Amen' repeatedly, concluding the prayers of his inner soul.[27] Some types of meditative text were designed, through their repetitive, rhythmical nature and their patterns of cadence, to produce an effect of incantation, for example the English prose works of Richard Rolle – an *oratio perpetua*.[28] In the aftermath of the Black Death, a group of more than 120 men, mainly from Zealand and Holland, came to London. This group, known as the Flagellants, performed public penance, beating each other and chanting the whole time.[29]

There is in addition a suggestion that the sound of English prose and verse at this period was much more continuous than our modern language.[30] These effects

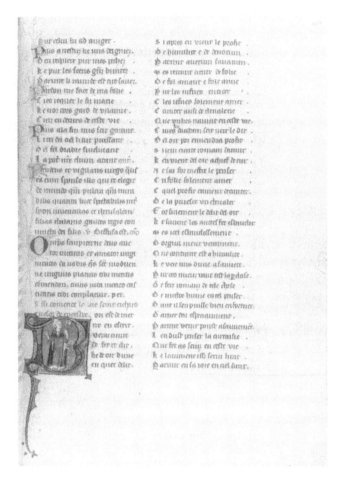

34 A volume of the lives of saints, including those of Richard of Chichester, Thomas Becket, Edward the Confessor, Edmund of Abingdon, Osith, Modwenna, Faith and Catherine, which was bequeathed to Campsey Priory in Suffolk, where it was to be read at mealtimes, as a fourteenth-century note on its final page indicates. The leaf shown is the start of the life of St Richard, late thirteenth century.

might be especially apparent in oral performance, such as the *carmina ritmatica* that were recited by John le Barber of Swansea for Queen Philippa in 1331.[31] While these were undoubtedly secular (although we do not know what the poems or songs were – and they may have been in Welsh or French), the aura created by the sound of this verse would have conveyed meaning additional to the sense of the words, much as the *cursus* generated this in Latin.[32] Poetry and rhythmic material were also popular in the household of Henry VII, where, between November 1495 and September 1497, more than £36 was paid to poets: to Hampton of Worcester for making ballads; to an Italian poet, possibly Master Peter of Florence; to Master Bernard, the blind poet; and on two occasions to 'a Walshe rymer'.[33] It is no longer clear, however, what some forms of punctuation were intended to convey in terms of the pace and rhythm of speech. The gemi-punctus, for example, the honorific double point – .. – placed in front of names of distinction, may have indicated a pause to emphasise the names that followed.

The force of the speech of many together, particularly the noise of crowds, was more influential than that of the individual, just as the strength of touch or vision was amplified in similar circumstances, but the force was not always a beneficial one. In 1378, at the time of the Wycliffite controversy, a knight from the royal household went to Oxford. That night a group of students stood outside the inn at which he was staying, singing rhythmically in English words in part derogatory to the King, as well as shooting arrows at the windows of the inn, an episode that led to the resignation of the chancellor of the university and the imprisonment of the vice-chancellor.[34] The woman taken in adultery, in the N-town play, asked to be killed as she was judged, rather than having her crime published before the people:

I pray ȝow kylle me here in þis place
And lete not þe pepyl upon me crye.
If I be sclaundryd opynly,
To all my frendys it xul be shame.
I pray ȝow, kylle me here in þis place
Lete not þe pepyl know my defame.[35]

The sound of the uncontrollable mobs of peasants, described in the accounts of the Revolt of 1381, was like the bleating of sheep or baying of wolves, terms that indicated their low birth and their lack of claim on the legitimate order.[36]

A more positive association of speech was with the power of prayer. Prayer was addressed not with a bodily voice, but with the speech of the heart; for the deity 'to hear' a prayer was an act of compassion and depended greatly on the moral state of the individual making the prayer or the one for whom it might be said.[37] *Jacob's Well* reiterated this message, drawing on a well-known account of a man who paid no attention in church, constantly chattering and keeping others from hearing God's word. When the man died, his body was brought to church for the funeral, but when the clerks said the Placebo and Dirige, the figure of Christ on the bier took its hands from the cross and placed them in its ears.[38] We cannot consider these ideas in isolation from other trends, such as reservations about the practice of purchasing or ordaining prayers for one's soul which appear not least among the followers of Wycliffe in the late fourteenth century. In theory, prayer, said by a holy man or priest, was particularly effective, but it depended further on the spiritual state of the man saying the prayer – speech was the expression of the soul. If the priest was damned, his prayer would be of no profit. 'Preyer of lippis bigiliþ many, and specialy whanne lippis ben pollut . . .'[39]

Good words countered bad. Immoral speech, such as lying, risked instant exposure through the challenge of good words. Walter Daniel tells of an occasion when Aelred of Rievaulx was subject to a verbal attack by a knight at court, in the presence of the King. The language was horrific, consonant with a prostitute rather than a knight, and it was countered by Aelred's meek reply. This took place before Aelred became a monk, when he was living in the world, where the voice of the dove was not heard, rather the persistent hiss of the serpent.[40] A tale from a thirteenth-century sermon collection reported how a clerk, in Warwickshire, going at night to visit his concubine in a neighbouring village, was confronted by a large dog which spoke to him, demanding his sword. To which he replied, 'You lie, by the death of Christ'; and the dog – the Devil, who had appeared in this form – thereupon disappeared, unable to sustain the force of good conjured by the mention of Christ's virtue, even when pronounced by a sinner.[41]

The moral force of bad words might contaminate the hearer. It was this that spread heresy: merely to be present when it was preached was sufficient to transfer the contagion.[42] In the mid-1160s a group of heretics, perhaps no more than thirty in number, came to England. They were arrested and in 1166 were brought before Henry II and the clerical council. In addition to the punishment meted out to them, the Assize of Clarendon prohibited anyone in England from receiving them. If that happened, the house where they had been received was to be taken out of the village or town and burned – the building that had been polluted by the words of heresy was to be destroyed. This was no symbolic act, but the only effective way of destroying the lasting contamination of this evil: heresy was transferred both by the spoken word and by physical proximity to a heretic.[43] It was, in the words of abjuration of a group of Coventry Lollards in 1486, 'evyll sonynge in the eerys of wel disposyd Cristyn men'.[44] One sentence employed against heretics was that they should hear mass, that good sound might counteract their evil.[45]

Similar corruption might result from hearing non-Christian doctrines and liturgy, or again, by physical proximity to non-Christians. By royal statute of January 1253, all services conducted by the Jews were to be said in a low voice (*submissa voce*) so that they could not be heard by Christians.[46] The synodal statutes for the diocese of Exeter of 1287, reiterating the injunctions of the Fourth Lateran Council of 1215, ordered all Jews to keep their doors and windows closed on Good Friday because they had been accustomed to mock Christians that day.[47]

Words might have a special force in other ways. Oaths, for example, were effective if uttered according to the correct formula; ineffective if they were mistaken in construction. The requirement for exactitude foreshadowed the precision required for pleading in legal proceedings: both plaintiff and defendant had to make their case orally, without deviation from the prescribed form of words. The Statute of Wales of 1284 made an exception to the custom of 'qui cadit a sillaba cadit a tota causa' – he who drops a syllable loses the whole case – as it was not possible for the defendant to know the plaintiff's demands from the form of the writ, as there were very many potential causes for a suit. Harsh the custom may have been, as the statute recognised, but it is impossible not to note its resonances with the requirement for exact diction in ecclesiastical contexts.[48] On the other hand, the effectiveness of nuncupative wills – like prayer – required no such formality: uttering the will was sufficient for it to be effective.[49] Gesture might also have an impact on the effectiveness of speech. A thirteenth-century manual on confession prepared for the Dominicans asked that the person confessing should bow his head – for that was how Christ on the cross had given up the spirit.[50]

Words might be employed as charms, or as curses when uttered with malicious purpose. The use of charms may have been hard to distinguish from that of prayers.[51] In March 1520 Henry Lillyngstone of Broughton was brought before a church court on the grounds that he used magic to cure people. He confessed the truth of the allegation, that he used the words 'Jhesus that savid both you and me from all maner deseasses I aske for seynt cherite Our Lord iff it be your wille', as well as various herbal mixtures which were claimed to have cured many. He told the court that he was not literate and had no medical training, that he had this know-ledge only by God's grace.[52] Analysis of collections of charms shows them to fall into different categories, ranging from words to be said while collecting herbs, to mystical expressions repeated over a patient or applied as an amulet, or when trans-ferring disease to an animal or object.[53] Their use was clearly both widespread and deeply rooted in tradition: for example, the recitation of the first words of St John's Gospel, or their use as a talisman or amulet, was commonly held to bring general benefit.[54] Synodal councils regarded these manifestations as evil, potentially corrupting – they did not doubt their power – and they therefore legislated against the use of incantations and charms.[55] Confessional manuals, such as Grosseteste's *Deus est*, asked whether God, the saints or the Devil had been invoked to assist in good or evil.[56] One might defeat a charm by not hearing it, as, it was alleged, was done by the asp: when that 'adder supposyth þat he schulde be charmyd he stoppyth his owyn ere with his tayl and his oþer ere with þe ground so þat he may noȝt here þe charmys ne þe charmer.'[57]

Curses were sometimes formulated from a perversion of psalms and prayers, such as saying prayers backwards, particularly the paternoster,[58] or the imprecatory psalms back to front. In the mid-eleventh-century life of St Kenelm, his sister, who had him murdered, attempted to do this with Psalm 108, to defeat the triumphal procession she saw celebrating her brother's glory at Winchcombe; she was stopped just before reaching the critical verses.[59] Impersonating a saint, speaking in parody and blasphemy, had severe consequences. The late eleventh-century *Miracles of St Swithun* include an account of a citizen of Winchester, who had been to an impressive wake and was manifestly drunk. He claimed to be St Swithun, offering – if suitably placated with gifts and prayers – to intercede. He was instantly struck down.[60] In another miracle, possibly based on this formulation in St Swithun's miracles and probably written in the 1140s, Eustace, a silversmith, drunkenly and inadvisedly tried out the new sepulchre of St Erkenwald then under construction in St Paul's. He called out raucously that he was the saint, demanding gifts and a sepulchre of silver: on uttering the words he was struck with a severe pain and died shortly after.[61] Liturgical parody was a feature of demonic forces, particularly in evidence in the late medieval mystery plays.[62] Talking in church, jangling and japing before God, frequently prosecuted in ecclesiastical courts, were offences similar in nature and inspiration.[63]

Both the sound and content of speech might indicate the holiness of a man, or otherwise, as with the raucousness of the speech of Eustace the silversmith. The hoarseness or lack of speech that accompanied leprosy,[64] the noises of the mad and the possessed, or even speech impediments, might be indicative of these qualities. For a horse to be well reined, it should have a short bridle:

> So most þou do ȝif þou wilt speke without defaut or stomblyng. Put in þi mouth a bridel of sad spekyng, ellis but þou can governe þe þus þou art a feble religious woman.[65]

The speech of ghosts, typically of those who had died unshriven or in the hands of the Devil, often revealed their parlous moral state. The ghost, recorded by a monk of Byland in the late fourteenth century, that appeared to the tailor, Snowball, spoke 'by an interior mouth, forming his words in his intestines and not speaking with a tongue', a literal ventriloquism. In another case, the same monk recorded that a

35 Gossip was not to take place between clerics at meals or between women in church. From the *Omne bonum*, late 1330s and 1340s.

ghost spoke in his viscera, not with a tongue, but as if in an empty barrel.[66] The study of physiognomy, determining moral qualities from physical characteristics, had much to say on the sound of the voice. John Metham, *c.* 1448–9, pointed to vocal features that delineated foolishness, gluttony, deceit, wanton desire, a manful heart and a great understanding. Those who had small voices and spoke shrilly, like a bird, were lecherous, and so on – a moderate voice was the most desirable.[67]

36 Byland Abbey: the monastic church looking west down the nave. In the late fourteenth century, a monk here recorded a number of stories relating to ghosts appearing in the neighbourhood.

Speech, as well as being a force for good or evil, was therefore also a signifier. Hagiographic literature attributed remarkable powers to the speech of individual saints. According to the life, from the second half of the eleventh century, of the infant St Rumwold, a seventh-century Mercian who lived no more than three days, the saint was able, miraculously, to speak, demand the sacraments and preach.[68] In *La passiun de Seint Edmund*, written about 1200, the severed head of the saint, the King of East Anglia killed by the Danes in 869, cried out to be reunited with the rest of his body.[69]

37 The head of St Edmund calls out and is discovered. The story was included by John Lydgate in his Lives of St Edmund and St Fremund, written to commemorate Henry VI's stay at the abbey of Bury St Edmunds, in his manuscript presented to the King between 1434 and 1439.

The twelfth and thirteenth centuries brought new opportunities for speaking in public and a rise in professionals whose task this was, often using specialised procedures, for example lawyers making their pleadings in precise forms of French, or academics constructing arguments for their disputations. Previously public speaking had been confined to bishops, abbots, monarchs and a few members of the aristocracy, but this facility was now necessary in law courts, parliaments and universities and, especially from the time of the Fourth Lateran Council onwards, in the pulpit, with sermons in the vernacular directed to the laity. A new interest in the arts of memory and rhetoric, in how public speaking might be cultivated, came with this,[70] as well as ways of making it more effective through accompanying ceremony or dignity. It is perhaps not surprising that both Caesar Augustus and Herod in the Towneley plays expected silence when they were speaking,[71] although there may be more here than characterisation. The autocratic nature of Richard II's kingship may be encapsulated in his claim, openly stated, 'that his lawes were in his mouthe . . . And that he allone myht chaunge the lawes off his rewme and make newe'; but we should take into account as well the possibility that a special power attached to the speech of an anointed monarch.[72]

Quality of speech delineated the standing of the man. High standards of speech were required of professionals, such as physicians and surgeons.[73] They were advised not to dress like minstrels, but like clerks, so that they would not be out of place at a gentleman's table. They were to have the appearance of gentility, with clean hands and nails; and, further, they were to be courteous in speech, to hear many things but say few, and those to be 'faire and resonable and without sweryng. Be war that ther be never founded double worde [duplicity] in his mouthe, for ʒif he be founden trew in his wordes fewe or noon shal doute in his dedeʒ.'[74] The Benedictine monk was not to give himself to idle chatter;[75] and, at Barnwell Priory, at the end of the thirteenth century, the customary advised the elderly Augustinian canons in the infirmary that they were not to engage in foolish reminiscence with a delight and jocularity that was inappropriate but typical of the elderly in secular life.[76] Another, with experience of the cloister, noted the unfortunate tendency of some who were subject to prolonged periods of silence to indulge in bickering and strife as soon as speech was permitted.[77] Provided it was not adulterated with pride – as monks were warned by a model for confessors[78] – spiritual benefits came from temperance of speech. The fifteenth-century author of *Jacob's Well* enjoined cleanness in tongue, citing Proverbs, 'Pure speech is the most beautiful' and advising the listener to have the discretion to examine his speech as it passed out of his mouth.[79] *The doctrine of the hert*, in its instruction of devout or religious women, argued that there were five things to be considered if one were to speak well: what one said, when it was said, where it was said, to whom it was spoken and how it was said. The last had three aspects: the sound of the words – mild and easy, without 'crying' (probably shouting); accompanying gesture – demure and well-mannered, without waving the arms or pointing; and the moral quality of the words, that is, their truthfulness, without duplicity or sophistry.[80]

Speech was in stark contrast to the virtues of silence. Voluntary silence was synonymous with holiness. Silence in the cloister was rigorously imposed, with the aim of creating interior silence; not only was speech excluded but also, and just as importantly, extraneous noise. The Rule of St Benedict forbade unnecessary speech: it avoided the sins of the tongue and it was the disciple's duty to be silent, to listen to the master. This regime was enforced even more closely after compline and at night.[81] The pattern was confirmed in Lanfranc's monastic constitutions. Here, the guestmaster controlled external visitors, taking those who had been given permission to speak to a guest out of the cloister in order so to do. When taking visitors into the cloister – only at times when no monks were present – they should nonetheless be attired in a way not to cause disturbance, without spurs or riding boots.[82] The whole community, wherever they were within the monastery, in its offices or courtyard, with the sole exception of those of the sick in the infirmary whose suffering might not permit it, was enjoined to keep silent during all services except compline.[83] The initial siting of the monastic community was crucial in

38 Byland Abbey: low stone benches flank the walls of the parlour. Outside the chapter house, this was the only room within the cloister where the monks might speak, and then only when absolutely necessary.

attempts to banish noise. Hugh of Avalon, arriving at Witham, set out to establish the physical structure of the monastery on a permanent basis. He therefore arranged to exchange land or to grant freedom from villeinage to those whose properties might impinge, through their noise or frequent comings and goings, on the extreme silence that was required by the monks.[84] At Barnwell, the canons might break silence in four cases – in the presence of thieves, sickness, fire and workmen – and, additionally, to reply to a king, prince, bishop or archbishop.[85] A further consequence of silence was that monks wrote because they could not speak, frequently adopting a highly literate form of rhetoric – like most written texts, designed to be read aloud.[86]

Silence or the absence of response by an accused felon in court, however, was a serious omission. Under the Statute of Westminster of 1275, possibly re-enacting an older provision, a notorious felon who refused to plead – that is, to answer the charge that had been uttered against him – was to suffer imprisonment 'forte e dure'. Shortly afterwards, the legal text, Fleta, indicated that until he made an oral response (for the court could not plead on his behalf), the alleged felon was to be kept in prison, wearing but a single garment, without shoes, and was to have no more food and drink than a farthing barley loaf and water every other day. The penalty was interpreted more severely by the fifteenth century: unless a response was made, the accused was pressed to death.[87]

Language as an indication of evil was commonly recognised, especially speech made in anger, cursing, mocking and bad language (*maledictiones*). A nun, chaste in body but unable to restrain her tongue from bad language, died and was buried in church. The night after her burial, the keeper of the church saw her body brought before an altar, cut in two, one half burned and the other replaced in the grave, reflecting the dual nature of her life.[88] Features like these are apparent in the characterisation of late medieval drama: devils speak in alliterative verse, as do evil characters in general; arrogance, impertinence, power and bombast are all signified in this way. In *Mary Magdalen*, probably of *c.* 1515–25, the pagan King of Marseilles speaks in alliteration until his conversion, when most of this feature disappears. Distinctive stanzaic forms also mark the parts of these characters.[89] In the N-town plays, King Herod speaks in a thirteen-line form used by no other character, larded heavily with alliteration.[90] In the play of the Last Supper in the same cycle, Mary Magdalen, from whom Jesus is subsequently to exorcise seven devils, enters with alliterative speech, resounding proof of her possession:

> As a cursyd creature closyd all in care,
> And as a wyckyd wrecche all wrappyd in wo
> Of blysse was nevyr no berde so bare
> As I mysylf þat here now go.[91]

The characterisation might be reinforced by scatalogical language, bad grammar, broken Latin, the nature of the musical accompaniment, or pure noise, the opposite of ordered speech, as well as by colours of clothing and styles of costume.[92] In *Mankind*, probably written *c.* 1465–70, Mercy warns Mankind to beware of New Guise, Nowadays and Nought: 'Nyse in þer aray, in language þei be large' – an understatement of the linguistic depravity that follows in the attempt to distract Mankind from his virtuous labours digging in the field.[93]

In spiritual terms, these forms of speech were associated primarily with what were known as the sins of the tongue. These were enumerated and catalogued in pastoral works from the thirteenth century onwards, in sermons and tracts on vices and virtues.[94] From idle words, boasting, flattery, lying, backbiting, forswearing, striving, murmuring, grouching and rebellion to blasphemy and lisping, these descriptions were strong on the moral consequences of speech. Backbiters were likened to mermaids, who, singing so sweetly that they made mariners fall asleep, then slayed and devoured them. 'Missayers' bit like adders, by treason: the venom killed three at one stroke, he who spoke, he who heard and he of whom it was spoken. Foul and violent speech betokened filth and villainy at heart.[95] It was not without reason that gossiping women, depicted in wall paintings in churches, had devils seated on their shoulders.

The second commandment – not to take the name of God in vain – and the eighth – not to bear false witness – charged speech with a high moral currency. The employment of oaths, curses or excommunication, all calling on God, was therefore carefully regulated by the Church. The context in which these utterances were made was particularly important. To swear by God, or another holy person or thing, that something was true or false attracted a great deal of clerical comment; to many it was blasphemy. These topics were addressed in cycles of sermons for the laity in the later Middle Ages.[96] 'The mouth that lies slays the soul' was one early fifteenth-century comment from the pulpit, taking up the words of Solomon;[97] and a Lollard sermon addressed the question of whether it was lawful to lie, deploying authorities back to St Augustine.[98]

Speech was powerful in creating contracts and bonds, from oaths for loyal service to feudal bonds and promises of marriage – and public performance had an effect on reputation. When the Lollard, William Thorpe, declined to swear on the Gospels, holding that a man's soul might not be seen or touched by any perceptible thing, his argument was that nothing additional was brought to the already moral quality of his speech by touching the Gospels as he made his oath.[99] Oaths of compurgation on the one hand and cases of defamation on the other constitute a useful touchstone of the importance of these forms of speech. Their association with indicators

39 A redrawing by E. Clive Rouse of a wall painting of cursing, Corby, Lincolnshire, early fifteenth century. The Devil inflicts on each man the punishment associated with the words of the curse.

of sanctity or holiness persisted well beyond the Middle Ages; but some types fell out of use, or were modified for practical reasons. Oath-helping, or compurgation, for example, was in decline in English manorial courts by *c.* 1300, in favour of juries: that is, the cumulative effect of swearing to benefit an individual was no longer held to match the rigour of a verdict delivered by a jury. While this reflected a change in procedure, it continued to be underpinned by oaths and speech, by a jury that was sworn to deliver the truth in its verdict. Compurgation survived much longer, perhaps as one would expect, in ecclesiastical courts.[100]

The history of defamation, a verbal attack on reputation or *fama*, the moral aura that attached to the individual, has a similar trajectory to compurgation. From its origins in England in a constitution of the Council of Oxford (1222), it was primarily a moral crime, tried in the ecclesiastical courts, unless financial damages were sought. In the fourteenth century, the ecclesiastical courts conceded to the secular all cases where anything other than a spiritual offence was at issue, where there was material damage. Two fifteenth-century cases are typical. At Wisbech in 1460, it was alleged that John Freman defamed Margaret the wife of John Digby publicly, in church, at vespers, calling her 'stronge hoore' – which was the occasion of great uproar. He was not alone: Alexander Frauncess and Joan his wife had previously called Margaret 'strong hoore' and 'stronge thefe' publicly in the town (it is in the very common expressions like this that the term 'strong language' must originate). The case was complicated: a fortnight before his alleged offence, John Freman had been caught in the middle of the night, dressed in women's clothing, in John Digby's chamber; and while it counted against his reputation, it was not known why he had behaved in this way.[101] In 1492, a case of defamation featuring two clerics, Thomas Couley, a chaplain, and Nicholas Barton, the rector of part of Waddesdon, came to court. Witnesses said that Nicholas, departing in considerable anger from Thomas, used the words 'Avaunte chorle and I wolde prove the a chorle of condicione' or called him 'chorle and boundeman', saying that he would prove his status.[102] In both cases the impact of speech and its moral force was the central issue. Defamation increased in business in the ecclesiastical courts of the fifteenth and sixteenth centuries.[103] Curiously, however, sexual swearwords did not feature strongly in this development and their use in, for example, street names suggests that they were not offensive in the way that we currently regard them.[104]

It has been argued that there is a secular parallel to the rise of the sins of the tongue in the ways that developed of controlling deviant speech, especially its use to attack authority. Analyses of the business of manorial and other courts have pointed to the prosecutions for false use of the hue and cry, superseded after the Black Death by the growth of scolding as a crime.[105] The suggestion is that speech was controlled in two ways: for its moral qualities and for its material consequences.

Late medieval England was multilingual. Did this have any impact on the understanding of the sense of speech? There is little doubt that the use of some languages

– Latin or words from Hebrew and Greek – gave a special force to some forms of speech. Isidore of Seville described these as sacred languages: those written on the cross.[106] 'Alleluia' – the quite literal injunction to praise God, given in Hebrew, God's language – was particularly significant. Gregory the Great recalled that the British had made the transition from Old English – a language that was barbarous, like the gnashing of teeth – to sing divine praises, that is, Alleluia, in Hebrew.[107] The maintenance of Greek words in the mass had a similar impact. Choirs of angels sang in Latin[108] and the Church's determination to maintain both the liturgy and the Bible itself in that language set the sacred word in a form that, while unintelligible to most, brought with it an aura of sanctity so special and distinctive that it was heretical to say or read the Bible and the liturgy in the vernacular. Further, it was argued by speculative grammarians in the thirteenth century that there were parallels between the grammatical structure of Latin, the expression of thought and ultimate reality.[109] Some urged that Latin was a better language for theological debate,[110] but to many the knowledge of the 'neck verse' (the opening words of Psalm 50), which an accused had to read to claim benefit of clergy, was a stronger argument for the transformational quality of the language.[111]

The position of Anglo-Norman is of interest. Its use was intimately associated with power, prestige and hierarchy, the language of the conquerors and a continuing means of subordinating the population. By about 1200, however, it had lost its connection with continental French as a living language, and at this point its speakers became more closely involved in linking their past with that of the Anglo-Saxons, in a burgeoning literature recalling the lives of English saints and, for example, the myths of Bevis of Hamtoun, Haveloc, Waldef and Tristan.[112] In the thirteenth century, beyond literature and the households of the upper classes, Anglo-Norman was the day-to-day language of the law and of royal and seigneurial administration. On the other hand, the language had to be learned: most French speakers had English as their mother tongue. It was strongly associated in cultural terms with the women of the nobility: literary works were dedicated to them, and sometimes houses of nuns, as at Aconbury Priory and St Michael's, Stamford, even had their accounts written in Anglo-Norman rather than Latin.[113] By the fourteenth century, English was well established as a literary language, although it was not until the advent of the Lancastrian dynasty that it became the pre-eminent language of power and government, superseding Anglo-Norman.[114] In the course of three centuries, the trajectory of Anglo-Norman French saw a close association with power, with literature and with the mechanics of government. Even if it was only ever a language understood by a minority – and all those who spoke it knew English and, in the case of clerics, Latin as well – it was remarkably robust and persistent, and marked its adherents with the distinction of an élite.

But one's position in that group might be marked by diction: languages can be spoken well, or badly. This was of extra significance when there was a moral dimension involved: French might be spoken after the school of Stratford at Bow;[115] and Chaucer's Friar was stigmatised for his pronunciation of the vernacular:

> Somwhat he lipsed, for his wantownesse,
> To make his Englissh sweete upon his tonge . . .[116]

Walter Map's satire of the resignation from the see of Lincoln of Geoffrey, the illegitimate son of Henry II, which took place at Marlborough in 1182, played on the supposed spring there, of which it was reported that whoever tasted it spoke bad French; hence when anyone spoke French badly (*viciose*) he was said to speak Marlborough French.[117] Henry of Lancaster was conscious, in writing the *Livre de seyntz medicines*, that French was not his first language: he was English and begged to be excused for his infelicities.[118]

Although some were proficient in several languages, the majority of the population was not. Middle English was striking for its diversity: dialects were recognisable and not always mutually comprehensible.[119] Abbot Samson of Bury was fluent in French and Latin, but when preaching in English did so in his native Norfolk dialect.[120] In addition, pejoratives easily attached to those who did not speak correctly. In the late fifteenth century, Caxton, bemoaning the difficulties of finding a standard English language for his translations, outlined difficulties of comprehension for visitors to London. A merchant visiting the capital asked for 'eggys', but the lady of the house answered him that she spoke no French, which perplexed and annoyed the merchant as he could not speak it either. Fortunately another party explained that the merchant meant to ask for 'eyren'. Caxton concluded that his reworking of the *Aeneid* was not for 'a rude uplondyssh man', but for one that understood chivalry and feats of arms, and that he should aim for a mean between the two, turning it 'into our Englysshe, not over rude ne curyous . . . for this booke is not for every rude and unconnynge man to see but to clerkys and very gentylmen that undertstande gentylnes and scyence'.[121]

Languages and dialects had their own pattern of sound. A late eleventh-century account of the miracle of a deaf and dumb boy cured by St Swithun shows some perception of this. The boy went on to repeat exactly what was said to him, in English, Latin and French. He spoke the words of each language in exactly the same way: he did not speak any one either more distinctly or more openly than another, as he had not before become used to speaking one of them. From this we can infer that the monks of Winchester expected a different pattern for each language, for some to speak some languages more fluently than others, or with inflections brought from their mother tongues; and that these variations were the normal sound of speech – the author of the miracles, but not the child, expected to make a clear distinction between individuals speaking the different languages. Nor did the boy make the customary inflexions of speech, replying to query with query, without grammatical variation of case or persons.[122] That the boy did not make distinctions between the three languages, while perhaps self-evident, may have been a part of the miracle, which in some ways parallels the exegesis of the Pentecostal speaking in tongues. There were a number of different approaches to this question. Did all the disciples speak all languages; or did each speak a different language; or did all understand all languages spoken to them, the miracle being that the foreigner understood the reply when the disciple answered him in Hebrew? The last became a favoured explanation, that contact with the Holy Spirit allowed the Spirit to be understood in any language.[123] It was the Holy Spirit that helped the deaf and dumb child to speak in a way that was beyond language, crossing, unknowing, between the tongues.

Other accounts of curing the dumb show the miraculous transition from incomprehensible sound to ordered speech. The mid-eleventh-century life of St Kenelm contains an account of the cure of the huntsman of Aelfric, a nobleman. The man had trained the hounds to hunt according to the sounds he made, even though he uttered no comprehensible speech. He had a vision of the saint, who brought a lighted candle to him and placed it in his open mouth; from which point he could speak in words.[124] If speech might be returned miraculously, it might also be lost through malice. At a great feast in the second year after the battle of Evesham, the vituperation of Robert the Deacon against Simon de Montfort was such that his host, William de la Horst of Bolney, had to ask him to desist. At which point Robert lost the power of speech, as well as the power to move his limbs. After promising that he would never again say anything against Earl Simon, he managed to escape danger.[125] While most discussions of the senses of the mouth indicate that speech represented the sense giving out sound and the characteristics of the body, with taste acting in the contrary fashion, as a receptor, there were occasions when speech might act in that guise. In the *Alphabet of tales*, a clergyman in Burgundy was reported as scorning a passage in the Gospels: he had made himself meek before his enemies, but had not received the profits of the church that he desired. 'And forthwith a levynnyng [lightning] like a swerd went in at þe mouthe of hym þat spak, as he was spekand, & onone it killyd hym.'[126]

Medical conditions might also prevent speech. Loss of speech resulting from a fracture to the skull might be mitigated by a mixture of violets in wine: crushed violet had to be applied to the left foot if the right side of the head were injured, and vice versa if it were the left. The juice of onions mixed with water was held to be good for those who had temporarily lost their speech, probably through soreness of the throat.[127] Others looked to the suffocation of the womb as the cause of loss of speech.[128] Speech might be encouraged in the infant by placing butter or honey on the tongue.[129] Other aids to speaking might be found in the virtues of certain stones.[130]

Speech was enumerated as a sense, or presented in lists of bodily faculties in ways that suggest that it was commonly interpreted as a sense, particularly at a popular level. This gave speech a special power, conveying not only the sense of words in terms of grammar and vocabulary; but also, beyond that, the moral or spiritual qualities that might be associated with the speaker or with the literal sense of the words. As well as the sense of words, other qualities would be conveyed to the listener, entering his body and spirit through his senses. Words might be used to create an environment for good or for evil, or to convey positive or negative suggestions. The aspirational chant of the contemporary football supporter might be no match for the benefits that flowed from Henry V's chapel royal and its continuous round of prayer, stationed in the rear at Agincourt along with the baggage.[131] The spoken word might heal or infect, condemn or persuade, save and preserve.

The operation of speech needs to be understood in plural ways that were in currency simultaneously. Speech was used to create a range of effects: it might contribute to characterisation in drama, drawing on stock connections; its sound might mark sanctity or evil; but these features might equally be discounted from analysis, depending on circumstances. To the post-Enlightenment view that it conveyed nothing of the essence or nature of the individual, some would have been sympathetic. Speech, however, was only one of the senses of the mouth, and we must now turn to taste and the faculties it brought to the individual.

Chapter 6

THE SENSES OF THE MOUTH: TASTE

If the sense of taste is more familiar to us than a sense of speech, a closer investigation of medieval perception of taste reveals much that is less commonplace. The range of tastes was different; and while the power of the mouth to give out sensory information was much more clearly associated with speech than taste, that power should not be excluded entirely when considering taste. The operation of taste was explained in physiological terms by both philosophers and academic medicine. In the first part of the thirteenth century, Bartholomew the Englishman argued that taste operated by the 'animal spirit', which was brought to the tongue by nerves running down the middle of the tongue and branching out to the sides. As the substance to be tasted touched the pores of the tongue, the animal spirit took a likeness of the taste, presenting it to judgement of the soul. The sense of taste was most like the element water. The tongue itself was held to have no flavour and its effectiveness might therefore be impeded by illness which gave a preponderance of a particular humour within the organ. Red choler would make all things seem bitter; salty phlegm would impart saltiness to all flavours. What was tasted might affect the qualities of the sense: the bitterness of aloes, for example, impeded subsequent taste.[1] The mid-fifteenth-century English translation of Guy de Chauliac's *Cyrurgie* reported that the tongue was made of flesh, was soft and spongy, that its principal purpose was taste, although it helped shape speech and controlled food in the mouth. It was directed by two pairs of nerves, to convey tastes to the brain and to govern movement.[2] Unlike Antiquity, in the medieval period man's ability to taste was reputed to be inferior to that of some animals, notably monkeys and deer, but also bears, pigs, crows and the ostrich, which was renowned for its ability to eat anything.[3]

However closely smell might be associated with taste – and it was held by some that part of the action of smelling was a form of internal 'tasting' – it was possible to define a wider range of tastes than smells.[4] If we now recognise four tastes – sweet, sour, bitter and salty – common opinion in medieval Europe, largely following Constantine the African and his translations of Arabic medical works, was

40 Longthorpe Tower: the monkey or ape represents the sense of taste.

that there were eight, or nine if one included a wearish savour or 'tastelessness' (*insipidus*). These were sweet (*dulcis*), greasy (*unctuosus*), bitter (*amarus*), salty (*salsus*), sharp (*acutus*), harsh or styptic (*stipticus*), salty like the sea (*ponticus*) and vinegary (*acetosus*). Two things made taste: complexion (make-up in terms of the humours) and substance (the textural quality or consistency of the substance in which they were found), and the tastes were further grouped in this way. Those tastes that were sweet, greasy, salty, bitter and sharp were held to have a hot complexion; the others were cold, although there were sometimes degrees of warmth in between. The substances were divided into three consistencies: thick (bitter, sweet and salty like the sea), interim (salty, harsh or styptic, and tasteless) and thin (sharp, greasy and vinegary).[5] Further distinctions might be made. Sweetness, for example, was sometimes divided into four, a pure sweetness as in sugar, a viscous sweetness as in dates, a bitter or biting sweetness as in honey, and an insipid sweetness.[6]

These analyses of flavour became the touchstones of medieval dietary theory, the principal aim of which was to find foods that matched one's own complexion, that either complemented it or mitigated adverse effects. Isaac Israeli's *Liber dietarum universalium*, summarised by Bartholomew, noted the preference of the body for sweet things. The qualities of sweetness – heat and moisture – were most close to mankind's general complexion and nourished it best. Conversely, things that were bitter might be considered harmful and needed to be consumed with care.[7] The bitter and bad-tasting medicines, recollected by Henry of Lancaster in 1354, were doubtless carefully considered remedies, part of the effect of which was to balance the humours of the patient.[8] Guy de Chauliac argued that bad breath arose from corruption in the stomach and was to be mitigated by further doses of things that were of the same humour as that which caused the disease. To this end, the patient was to have sour things, such as pomegranates, oranges (only bitter, Seville oranges were available in Europe before the sixteenth century) and vinegar, dry things such as partridge and small birds, and those digestives, taken after the meal, that suppressed wind, such as quinces or pears.[9] Another fifteenth-century translation, of Lanfranc's *Science of cirurgie*, demonstrated how two men, both the same age, wounded at the same time and place with a sword or knife, had to be treated differently in terms of their diet because one of the men had a complexion that was hot and moist, the other cold and dry. The first was to be forbidden wine, milk, eggs and fish, and fed a strict diet of gruel of oatmeal or barley meal with almonds. The diet of the second was hardly moderated: he was to be fed wine and meat as normal.[10]

Given this impact on man's physical constitution, there was exceptional interest in the transmission of flavours, first of all from the natural environment. This was evident in the flavours of honey, where the blossom and flowers supplying the nectar imparted different qualities.[11] Bartholomew, following Constantine, argued that fish from standing water – particularly lakes or fishponds that had not been well managed – were unwholesome, gaining a poor taste from the inferior quality of the water in which they lived; late medieval palates, however, promoted freshwater fish, often pond-fish, to rank among the most sought-after items of upper-class diet.[12] Favourable qualities, on the other hand, were transmitted to fruit and sweet-smelling things by the clean air of hilltops and mountains, and sheep, pastured in the hills, where the grazing assumed these qualities, were considered better than those that had been pastured in valley bottoms or on the plains.[13]

Secondly, flavour might be transmitted by artificial means, by mixing flavouring with things of insipid taste, or strong flavourings with lesser ones. It was in this way that salt might be employed: many held, following Isidore, that nearly all foodstuffs were flavourless without it.[14] In practical terms, it was quickly observed that goods had to be kept apart in storage, to avoid mixing of flavours as well as other contaminants. The York civic ordinances of 1301 required vendors to keep bread separate from oil, butter, fat and other goods that might affect it; bread was to be placed by itself, cheese by itself, and so on, to avoid cross-contamination.[15] Spices were similarly kept apart. Inventories show spice boxes with separate compartments, or containers that were dedicated to a single spice, such as pepper boxes.[16]

The biggest influences on taste and flavour were cooking and practices of consumption. Across late medieval Europe, among the upper classes, there was a common pattern of cuisine that favoured highly spiced food, along with acidic, but thin, sauces, frequently based on wine, vinegar or verjuice, rather than fat or oil.[17] Salt was a prominent flavour. Methods of food preservation employed a good deal of salt, sometimes in conjunction with other preservation techniques, such as drying, smoking or using spices – and these were extensively employed for meat and marine fish. Although foods preserved in this way required preparation to remove some of the salinity, that was sometimes achieved through the use of salt water: it was believed that salt water was better than fresh for drawing the salt out of meat, probably because of the likeness of the two substances.[18]

It is interesting that medieval cuisine should exhibit these characteristics, given the common division of flavours into eight. The gradations of the sour, bitter and salty tastes are finer than we would employ and suggest a more acute perception of these tastes. These categories were largely taken over from Antiquity, which had an equally wide-ranging, if different, pattern of cuisine, and they created distinctive elements in upper-class diet. Henry of Lancaster, examining his conscience against the sin of gluttony, had experienced this at its best. Foods of extraordinary cost had passed his mouth both morning and night – and they were made as delicious as man might, well spiced, that is, with spices that were neither too hot nor too cold but exactly right, and with mordant sauces.[19] The fifteenth-century *Boke of nurture* held that sauces provoked a fine appetite: mustard for beef or salted mutton, verjuice for capon, veal, chicken or pork; pepper, garlic and vinegar for roast beef and goose; ginger sauce for lamb, kid, pig or young deer; sauce cameline – made with cloves, ginger, cinnamon and currants, mixed with vinegar – for young herons, egrets, cranes, bustard, bittern, spoonbill, plover, and so on.[20] There was a similar range of sauces for fish, with prominence given to mustard and to green sauce, a combination of herbs, pepper, ginger and green herbs, with vinegar.[21] Other sharp flavours came from citrus fruits, a commodity available to a privileged few. Among the goods of Henry VIII at his death in 1547 were strainers for orange juice.[22]

We know most about these dishes from cookbooks, but these, in England, principally reflect the activities of the households of the higher aristocracy. Some further detail is available in domestic accounts, which show that distinctions between the cuisine of the aristocracy and the gentry lay not so much in the range of spices as in the quantities and regularity with which they were consumed.[23] Records of three gentry families in the 1330s and 1340s show the range of these flavourings. In 1336–7, Dame Katherine de Norwich bought almonds, ginger, galingale, cinnamon, cloves, maces, saffron, pepper and two types of sugar. Besides these there was a series of sweet confections, either preserved fruits or sweet powders, including boxes of *gingebradz* – not at this date a sweet biscuit, but a confection derived from preserved ginger – pinionade, a similar confection made from pine-nuts, and festucade, made from pistachio nuts.[24] The household of Thomas de Courtenay, at South Pool in Devon, in 1341–2, made a large purchase of salt for preserving meat – 8½ quarters –

in November 1341. The household brewed its own ale, as well as drinking wine and cider, and would therefore have had sour ale or vinegar for sauces. There is little detail of other flavourings beyond the eighty onions bought for 2d., the garlic for 3d., and 8 gallons of honey.[25] The household of John de Multon of Frampton in Lincolnshire, making its purchases at Lynn, Boston and Lincoln, bought a typical range of spices: almonds, ginger, sugar, cloves, mace, cinnamon, cubebs and fenugreek. Between Michaelmas 1342 and Epiphany 1343, the establishment spent 3s. on ready-made sauces: ginger sauce, mustard and galantine. A further 9d. was invested in *salsamentis*, or sauces generally, in November 1347. On the Sunday before the start of Lent 1348, purchases were made against a feast probably associated with a tournament. For this saffron, pepper, mustard, ginger, galantine, vinegar and sugar were all required: 2s. 3d. was spent on flavourings, about 10 per cent of the cost of the occasion. Other flavours came from the manorial gardens and were not closely recorded in the accounts. In 1342–3, the 4s. spent on onions and garlic for planting at Frampton, along with 2s. for garlic at Ingleby, and a further 2s. invested in leek seed, indicate that these were grown in considerable quantities.[26]

This pattern of cuisine permeated down the social scale, but with an increasing element of garden and wild produce to provide flavourings. Winchester gardens in the early fifteenth century grew leeks, parsley, sage and hyssop as well as unspecified vegetables;[27] fruit – apples, pears, plums, damsons, cherries – and nuts;[28] as well as hosting beehives.[29] In September 1305, Thomas, son of William of Donnington was on his way to collect crab apples in the woods of Radnor for making sauce – probably a form of verjuice – when he found the body of William, son of John le Lorimer, in the Somergeld stream.[30] In 1492, William Sugory of Burnham in Buckinghamshire was alleged to have collected crab apples and acorns on Michaelmas Day, Sundays and other feast days, at the same season and probably in part to the same end as Thomas nearly two centuries earlier.[31] The availability of other spices changed, making them accessible more widely. In the twelfth century, pepper was a commodity whose value might mark it out as an exceptional item: William de Beauchamp received from Master Alvaric the Cook an annual rent of a pound of pepper for land in Stoulton.[32] But the fixing of rents at a peppercorn – a trifle – in the later medieval period indicates how far the value of the commodity had changed and how it could now be much more widely consumed.

Aside from speech, we would regard taste as the sense that was most clearly open to cultural influence and individual choice. Was a child taught about tastes or was it naturally endowed? It was believed that babies initially required to have their taste stimulated. This might be done with honey, which being both bitter and sharp, was of particular merit if rubbed inside the mouth with a finger.[33] There was also control of texture and tastes: when it was first weaned, nurses and mothers were expected to chew a child's food in their own mouths first before passing it to the child.[34] In the household of Henry VII, while the nurse of the royal child was breast-feeding her charge it was expected that her food would be subject to assay, that is, it would be tested for poison lest anything transfer to the infant either through her milk or, once she had weaned the child, through anything else she gave it from her food.[35]

If an infant gained its knowledge of food directly from its mother and nurse, little further can be said about the teaching of taste in terms of gastronomy and cuisine, at least in England. We have almost no knowledge of how cooks were trained. Although we have cookbooks, the purpose of these works does not seem to have been directly didactic; they may have served more as *aide-mémoires*, or for the use of others interested in household management, with works on which they are some-times found. Most training in cookery must have taken place on the job, to learn the effects of sauces or spicing. We are better informed about the operations of the kitchens of the élite on the Continent, for example those of the Duke of Burgundy. Here the role of the physician can be seen, advising the lord on what to eat as the foods came to table. Medical advice was crucial and it was more than coincidental that it supported the use of spices. For the late medieval period at least, it was dietetics that drove the pattern of consumption. The absorption of qualities, first by taste and then by digestion, had a significant impact on the body. Hot spices – hot in terms of their humoral composition – and other flavourings, such as garlic, were believed to help the circulation by thinning the blood and assisting its distribution. Their addition to heavy foodstuffs, such as meat, might, from a medical point of view, assist not in cooking and flavouring, but by their effects on the body once the food had been consumed. Both before and after consumption spices of this nature were held to be a valuable force against the corruption of the food and the body that had consumed it.[36]

At an élite level, cuisine was about transformation, about changing the nature of substances, a form of alchemy but also a form of conjuring, creating dishes that might transcend taste. In terms of alchemy, the consumption of gold-coloured foods, *endorred* with egg, coloured by saffron, or even by the application of fine gold leaf itself, might allow the body to assume some of the incorruptibility of that substance. The quire of gold leaf bought by Elizabeth Berkeley, Countess of Warwick in 1420–1, listed among the purchases of spices, was almost certainly acquired for this purpose.[37] Recipes indicate that the ingredients were typically reduced to a formless consistency, through grating, grinding, pounding and fine chopping. They might then be sculpted into new shapes, some quite fantastic, or might be reintegrated with skins and plumage, or those of other creatures. In this way, English cooks might create a dish called *poumes* (apples): these were meatballs, made of veal, cut small, bound with egg and spiced with pepper, cinnamon, cloves, dates, saffron and currants. The finishing touch was to roast them on a spit while basting them with a green batter, coloured with parsley, mallows or young shoots of wheat.[38] A similar creation, based on pork and chicken, imitated pomegranates, coloured at the conclusion with two batters, one green and one yellow.[39] Ersatz eggs appeared in Lent (when the normal commodity was not served) made of almond milk, part coloured yellow with saffron.[40] The cockatrice was made from a capon and pig, each cut in half, and the half of one sewn to the other, stuffed with a well-spiced mixture of bread, suet and egg, then roasted.[41] These visual effects were as important as any manipulation of taste. In fact, the availability of standard forms of spices, such as blanchpowder, analogous to modern curry powder, suggests that there may have been a common taste to at least some categories of food.

As well as changing the body's physical composition or complexion, some foods were held to affect the individual's moral or spiritual state. A translation and reworking of the *Secreta secretorum*, by Geoffrey of Waterford, an Irish Dominican, working with a Walloon, *c.* 1300, argued that this was especially the case with wine. Drawing also on the work of Isaac Israeli, they argued that wine could change the moral character of an individual, reforming cruelty to pity, meanness into largess, and pride into humility. Wine was like medicine, however: the right dose was beneficial, but too much was injurious, causing drunkenness and upsetting the five senses.[42] Henry of Lancaster confessed that he had enjoyed both white and red wine, particularly asking for strong wine so that he and his companions might lose their senses in drunkenness.[43]

Cooking and the alteration of flavours were sometimes considered acts against both nature and God. John of Salisbury, writing in the mid-twelfth century, complained about culinary changes that induced one food to surrender its flavour to another. Bernard of Clairvaux wrote against cooks and the sauces and spices they employed to take away the natural taste of foods.[44] A further argument, that spices excited the palate and were therefore to be eschewed, became part of the moral crusade against sensuality and costly luxury. It was not only spices that fell into this category. The effects of eating meat were of great concern to canonists and moralists alike: avoiding carnality in all its forms was essential to Christian life. This led to the adoption of patterns of abstinence from meat that had a major impact on late medieval diet. The rise in fish consumption from the mid-eleventh century may be associated with this development: while fish-eating carried a connotation of virtue, its growth may also be connected with wider economic developments, especially trade with northern Europe and the provisioning of towns. Other aspects of abstinence, such as drinking water rather than ale or wine, distinguished individuals in a similar way. Choice in diet, however, required both means and the availability of alternative foodstuffs. It was probably not until after the Black Death and the endowment of the food supply with proportionately much more meat per person that these possibilities became significant for the population beyond the upper classes and the inhabitants of the major cities.[45]

The very monotony of some diets was a mark of their virtue. The diet offered to the poor by Dame Katherine de Norwich in 1336–7 was one of almost perpetual abstinence, of herring and bread made from a poorer corn than that eaten by her household.[46] The plain and unvaried food at Witham was no bar to those former monks and canons who wished to acquire a richer spiritual fare under the leadership of Hugh of Avalon.[47] In ascetic diet, the avoidance of stimulus from taste might be taken even further. St Godric was an extreme example. His food consumption was severely restricted: at times he ate almost nothing, although he chewed the roots of herbs, leaves of trees and flowers; and at other times he used unpleasant-smelling pastilles of vegetation, bread made with ash mixed in the dough, and he mainly ate at night. Gifts of ordinary food – bread, cheese, butter – from admirers, placed him in difficulties as he did not wish to offend by the refusal of these goods.[48]

Beyond carnality, the general war against sin and sensuality did not spare taste. Contrasting the humble location and trappings of the birth of the King of Heaven with the daily round of the luxurious great household, a late fourteenth-century sermon writer asked:

> Wher weren thoo kny3tis and squieris to brynge service to this ladi of noble metes, costeli arayes, with hoote spices and deutevous drynkes of diverse swete wynes?[49]

The sin of gluttony featured in every confession. In a manual written for the Dominicans towards the end of the thirteenth century, the friar was to enquire particularly with regard to taste whether unnecessary sauces had been consumed – which would identify a tendency to luxuriousness – or unwatered wine or other drink, and whether fasts had been broken. Aspects of the sin of gluttony were considered. Among these was the case of those who could not wait until mealtime before eating: this featured the example of Jonathan, who, not knowing that his father Saul had enjoined on the Israelites a fast during daylight hours, ate honey that was in the woods.[50] It was probably not coincidental that one of the leading conspirators in the Croxton *Play of the sacrament*, composed soon after 1461, was a Jewish spice merchant called Jonathan, a character marked by name and trade as much as by religion.[51]

That taste might have bad moral and spiritual effects was known to all from the paradigm of the tree of knowledge in Eden.[52] The act of eating might itself distinguish good from bad. In many accounts of the life of Edward the Confessor, the death of Earl Godwin, choking at table, proved his guilt in the matter of the death of Edward's brother Alfred. The metrical Anglo-Norman life by Matthew Paris, written *c.* 1236–45, portrayed this incident as a form of ordeal: Edward blessed the fragment of food Godwin was about to eat, thereby constituting God judge of the case.[53] Unnecessary tastes and tastings had to be avoided. Nuns were forbidden, by the Council of London of 1268, from entering the refectory at any time other than the appointed hour, lest the flesh be provoked by unnecessary eating.[54]

41 Earl Godwin choking at table, from a series of images of the life of St Edward the Confessor, late fourteenth century.

Good tastes could indicate merit. The words of Hugh of Avalon, preaching at Grenoble, were seasoned with a marvellous, honeyed sweetness.[55] A story in *Jacob's Well* argued that good works were more meritorious than the outward signs of religion, such as fasting. A sinful man went to a hospital in order to help the sick. He washed the feet of a leper, an act that disgusted him, but he reflected that his response was a bad one and accordingly drank the water that had been used for washing. This, eventually, tasted as sweet 'as it hadde be made with spyces': his sin was forgiven.[56]

This way of thinking was close to a whole series of metaphors linking the sweetness or pureness of water to the pureness of the Christian heart. The fourteenth-century *Book of vices and virtues* spoke of the well of love in the heart, cleansed and purged of the love of the world and so sweet-tasting and flavoursome that those who drank of it forgot all other sweetness. The less the well smelled of earth, the better and more wholesome it was to drink.[57] A further dimension was achieved by the miraculous transformation of pure water into other substances, especially wine (echoing the biblical marriage at Cana), by the virtuous. In her Anglo-Norman life of *c.* 1230, St Modwenna visited a rich abbey in Ireland. Because of the large number of visitors, there were not enough bowls and, on Modwenna's advice, the bowls that were normally reserved for phlebotomy were also used. When the bowls were filled from a spring they were found to contain a strong wine, as indicated by its colour, smell and taste.[58] Later in the poem, a spring where Modwenna used to bathe miraculously produced ale for a visiting bishop, a brew he recognised as exceptionally distinguished.[59]

Heaven itself was marked by its taste, by its spices as much as its jewels and flowers. In the fifteenth century, Peter Idley's vision had much of the conventional about it: a meadow, of fresh colour, 'With all maner frutis and of divers spicerie / Delicious in taast and of suche odoure'.[60] In contrast, the flesh of the sinner who was guilty of gluttony was beyond all preservation that might come from herbs and spices.[61] Purgatory provided another sensory environment, less awful than Hell, but unpleasant. In the vision of Thurkill of 1206, the souls that were crossing the fire of Purgatory were immersed in a great pond of extremely salty and incomparably cold water.[62]

These metaphors for the delights of Heaven or the punishments of Hell were closely linked to forms of devotion focused on the senses that allowed these pleasures to be anticipated physically.[63] A key text was Psalm 33: 9, 'Taste and see that the Lord is sweet',[64] and taste manifested itself markedly in this form of spirituality. Aelred of Rievaulx had accordingly tasted and seen the sweetness of the Lord.[65] *The doctrine of the hert* explained that any sweetness that came from God in this world prefigured the sweetness that would come in heavenly bliss.[66] The salutations in the play of *Mary Magdalen*, probably late fifteenth or early sixteenth century, are indicative of similar thinking: Martha and Mary hail Christ as 'ower melleflueus swettnesse'; the priest greeting Mary in the wilderness describes her as 'swetter þan sugur or cypresse' (galingale, or cypress root).[67] As the translator of the account of the miracles of Henry VI explained in his dedicatory letter to John Morgan, the Bishop of St David's (and formerly Dean of Windsor), the work literally allowed a foretaste of the sweetness of holiness.[68]

Spiritual joy, however, was not all nectar. To the author of *Jacob's Well*, wisdom was a sweet, spiced wine, but those spices – myrrh, cloves, cinnamon, and sugar or honey – were tinged with bitterness to mark suffering. It brought 'þis savery kunnyng' to the heart of the devout: with myrrh, the bitterness that Christ tasted on the cross; with cloves (*clowys*), the nails (modern French, *clous*) that held Christ to the cross; cinnamon, the bark of the 'sweet mind' of the cross; and sugar or honey, the sweet mercy of God that he showed to man by his death on the cross.[69] *The doctrine of the hert* described an aromatic sauce that made all penance sweet; but, following the Song of Songs, noting that myrrh was harvested with the aromatic spices, it interpreted myrrh as the temptations and tribulations of religion, and the aromatic spices as the hope of endless reward.[70] A Wycliffite sermon of the late fourteenth century noted that the things of the world tasted as bitter as wormwood.[71] It is sometimes difficult to distinguish absolutely between taste and smell in accounts of sensory devotion – and our impulse to do so may be anachronistic. Margery Kempe, visiting the church of the White Friars in Lynn, felt a savour so wonderfully sweet and heavenly 'þat hir thowt sche myth a levyd þerby wyth-owtyn mete or drynke ȝyf it wolde a contynuyd'; which, it was explained to her by Christ, was an indication that there would shortly be a new prior at Lynn.[72]

There were other ways of tasting truth. The virtues of some stones, placed under the tongue, could be productive of insight. A late fifteenth-century lapidary from Peterborough, a compilation entirely recycling the dicta of other texts, records three types of stone that could have this effect. In the first case, a man's mouth had to be washed out before the stone was placed under the tongue; but this would then allow him to divine the future, depending on the state of the moon. The second, placed on the tongue, allowed a man to know what others were thinking about him. The third again required the mouth to be washed – so that there was nothing to impede the virtue of the stone – also allowing the man to predict the future.[73]

Taste could be a useful indicator for the physician. An abnormal sensation in the mouth, with those things that were sweet tasting bitter, was evidence of a corrupt humour – in this case choler – in the patient.[74] In addition, the physician might tell from the taste of the patient's blood if the humours were out of balance, bitterness again indicating the presence of a superabundance of choler.[75] The tastes of some foods, such as onions, might transfer to the body.[76] Guy de Chauliac's *Cyrurgie* noted a number of other causes of bad breath, pointing to diseases or ulceration of parts of the mouth, including the teeth or the nose, but also sometimes a less proximate cause, in the brain, the stomach or the chest.[77] Others concentrated on oral remedies, to freshen the mouth. An early fourteenth-century *Physique rimee* advocated drinking a concoction of mint and pennyroyal to clean out the mouth. If there were no rotten teeth, then it would be worth taking spices, particularly cloves and nutmeg, which would overcome malodorous breath.[78] A version of the *De ornatu mulierum* (On women's cosmetics) from the Trotula ensemble, copied in England around 1300, advocated washing the mouth after meals with wine and cleaning teeth with an abrasive powder, drying them first with new linen cloth, then with a woollen one. After that, one ought to chew the seed of fenugreek, lovage or parsley, which would reform the breath, comfort the gums and heal the teeth. Another recipe recommended keeping some laurel leaf and musk under the tongue.[79] The quality of breath was important: here was an exhalation, effectively from the sense of taste, that might indicate not only disease but also the poor moral or spiritual state of an individual.

Taste was a sense of practical use to the physician, especially in the later medieval period, used in diagnosis in much the same way as smell. In linking this information from observation to disease, the use of the senses had advanced beyond the earlier medieval framework, which focused on the connection between the moral or spiritual and the physical. This view of taste continued, however, particularly when associated with speech as 'the mouth'. Tastes might be analysed within a more extensive framework than that to which we are now accustomed, and the process of tasting itself – eating and drinking – brought with it both virtue and vice. In elaborating this, in connection with medieval humoral and dietary theory, we get a glimpse of how the nuances of this sense might operate. The operation of taste was seen as having close connections to smell: in some ways, they may have been indistinguishable in medieval analysis, but smell, as we shall see, had other connotations.

Chapter 7

SMELL

Smell is unlike other senses in that we lack a specific vocabulary to describe many of the sensations that can be perceived through this faculty. Some can only be put into words by borrowing terms descriptive of other senses, particularly taste, for example sweet, bitter or acrid; or, by analogy, things 'smell of something', such as roses or decay. This phenomenon was equally true of sensory language in the late medieval period. Like sound, before the latter part of the nineteenth century, or taste, there is no way of storing smell for long periods of time, which means that we are entirely dependent on written descriptions for our information about the sense and its operation; our understanding is further constrained by the inherent need to transfer descriptive terms to olfaction.

The sense of smell has a special emotional or psychological impact: it often provides a crucial dimension to a multi-sensory experience, even if it is not in the foreground of our perception. The smell of individuals, of objects or of environments frequently makes a lasting impression that we can recollect or understand without words – and we should not underestimate the immediacy or importance of these associations in the past.[1] The range of smells and their impact are therefore of peculiar significance in understanding medieval sensory perception and its cultural impact.

Why did things smell? The medieval answer was not straightforward. In Salernitan problem literature, a series of questions address olfaction. Why do smells seem good to some but awful to others? How many things are necessary to make a good smell? Why can a dog perceive the smell of a far-off deer and no other nearby animal? Why does musk placed in a fetid place smell greatly, but when with other odoriferous spices very little? Why do fruits of some plants smell good and the plant itself bad? Why can we see further than we can hear or smell?[2] The answers drew especially on humoral analysis. Similarity of make-up was an indication of those smells that would be well received by an individual; dissimilarity indicated what would be noisome. So men with a predominance of melancholy or of evil complexion avoided what smelled good and associated with places that smelled bad

– 'loveþ stynkeng place' – as did those who could not smell or who smelled bad themselves, such as lepers, who fitted both categories.[3]

Smell, like all other aspects of sensory perception in the Middle Ages, was charged with moral and spiritual dimensions. The association between good smells, holiness and the divine was of very great antiquity. Moses was given instructions for the preparation of aromatic ointments, based on a mixture of myrrh, cinnamon, sweet calamus, cassia bark and olive oil, for the tabernacle, the ark and the sacred vessels; and of perfumes, based on myrrh, galbanum and incense.[4] There is then a long history of divine manifestations, or of individuals of exceptional sanctity, both in Judaism and Christianity, associated with good smells.[5] These events were well established in clerical writing long before the twelfth century.

The appearance of the odour of sanctity coincided not uncommonly with the death of a saintly individual. On the demise of St Erkenwald, at Barking, his late eleventh- or early twelfth-century life recorded that an amazing fragrance filled the whole monastery.[6] Reginald of Durham, in his account of the burial of St Godric (d. *c.* 1172), noted how:

> At the same time, just as the foremost citizens [of Durham] were made more joyful at the ascension of such a man, their rejoicing was made more abundant by the marvellous effusion of sweetness of the air of Heaven. Indeed the air of Heaven in all the surrounding woods became honeyed, because the trees and branches, wood and leaves, fruits and green places had appeared to be filled and moistened by a mellifluous dew from Heaven, which taste, look, savour, touch, substance and odoriferous vapour they retained for two whole months. Many, marvelling at this, cut down branches and leaves and took them back to the town and there were accustomed to extract the liquid honey with their fingers and to taste it. This all happened through the merits of the man of God [Godric], by God's gift; because while his sweet spirit was going to the Heavens through the air, by chance the mellifluous odour which came from him filled the space he left, and which, falling as dew to the earth, his holy soul was as a mellifluous sacrifice to God, which was openly shown to all on earth.[7]

The martyrdom of Becket was likened to the breaking of a perfume box, suddenly filling Christ Church, Canterbury, with the fragrance of ointment;[8] and the account of the opening of the tomb of St Wulfstan in 1198 used almost exactly the same words to describe the fragrance, drawing on the descriptions of Mary Magdalen anointing the feet of Christ in Bethany.[9]

There were many similar occurrences on the discovery of a saint's body,[10] or at the opening of tombs,[11] accompanying miracles,[12] and more generally during the life of a saint or in association with his relics. St Hugh of Lincoln visited Meulan and the shrine of St Nicasius (Bishop of Rouen, martyred *c.* 340) on his way to Paris. He failed to extract one of the teeth from the skull of Nicasius, but he placed his fingers in the nasal cavity, where the saint had always breathed the 'good odour of Christ',[13] and broke off part of the orbital bone.[14] Matthew Paris, in his metrical life of St Edmund, Archbishop of Canterbury – written probably between 1255 and 1259 – recounted an incident in the saint's youth. Although he went along with his school friends to bathe in the Thames, somewhere in the meadows between Abingdon and Oxford, he did not join in. He saw a bush unseasonably in flower, from which came an amazing smell – and Edmund then had a vision of Christ. Christ told him not to be afraid of these extraordinary signs – the bush, an accompanying bright light, the flowers and the smell – as Christ had caused these to happen.[15]

Where we have descriptions of the odour of sanctity, it was often categorised as similar to honey, flowers, or as a most sweet smell (*suavissimus*). The reputation of St Birinus spread as an odour wafting from a garden of lilies.[16] In 1144, Henry de Sprowston, burying the body of St William of Norwich, noted a smell like that from a great mass of odoriferous herbs and flowers. When St William's body was subsequently exhumed for reburial, no bad smell was sensed, but rather the fragrance of spring flowers and herbs, even though there were none around.[17] The body of Aelred of Rievaulx was redolent of incense[18] and, after a disastrous fire in York Minster, the area around the tomb of St William of York (whose body and silk clothes remained unscathed, although part of the structure fell on his tomb) was fragrant with incense or the most precious ointment.[19] In a sermon on the nativity of the Virgin Mary in his *Festial*, Mirk likened her to a spicer's shop, 'for as a spycers schoppe smelleþe swete of dyverse spices, soo scho for þe presens of þe Holy Gost þat was yn hur and þe abundance of vertues þat scho smellyth swettyr þen any wordly spycery'.[20] These attributions represent a commonplace in medieval perception of odour.

Visions of Paradise were similarly populated by sweet smells and a multitude of flowers.[21] In the closing years of the twelfth century, Roger, a Cistercian lay brother at Stratford Langthorne, recounted an extensive dialogue he had in a vision of his friend, Alexander, a monk who had died about a year earlier. Alexander told how he was in Paradise and that there they lived off smell, which at the start of day descended from Heaven, satisfying and refreshing each according to his merits by a differential sweetness.[22]

Incense was symbolic of prayer, for prayer made in charity ascended to Heaven and smelled sweet to all round about, just like incense, thrown on the coals.[23] Censing formed part both of ecclesiastical ritual and of monastic life, following the monk to his grave and filling it with divine odour.[24] Sermons on the Epiphany explained that the purpose of incense as one of the offerings of the three kings was to put away the stench of the stable,[25] to employ a good smell to counter an offensive one, an offering repeated by the fourteenth-century Kings of England on this day.[26]

42 The Ramsey Abbey censer, *c.* 1325, perhaps an idealised chapter house, and incense boat, *c.* 1350.

Fragrant flowers and odours were generally accepted as signs of virtue and grace, sometimes considered a foretaste of Paradise. Twelfth- and thirteenth-century descriptions of Cistercian monasteries, for example Meaux and Rievaulx, frequently equated them with Paradise, redolent with the fragrances of flowers, trees and spices.[27] The locations of others imply similar imagery, the foundation at Strata Florida literally located in the 'floral way', and Dore Abbey in the Golden Valley.[28]

The use of flowers was a tangible mark of honour and grace. Hugh of Avalon, arriving at Grenoble on his way to Grande Chartreuse, was greeted outside the city by the Bishop of the place and brought to the cathedral through streets decked with flowers and silk cloths.[29] It was in a similar spirit that the N-town play had children strew the way into Jerusalem with flowers as Christ approached.[30] Flowers were required for the decoration of altars and other focuses of religion, such as images. Grown in monastery gardens,[31] or supplied by the devout, like the poor woman in a thirteenth-century tale who frequently adorned an image of the Virgin with roses and lilies and whose merits led the Virgin to restore the woman's son to health,[32] they were a regular part of ecclesiastical life.

43 The monastic church of Rievaulx seen from the cloister.

The odour of sanctity might be maintained in other ways. Some gold and silver images in the possession of Henry VIII, including those of the Virgin, St Peter and several of kings, were constructed on an underlying framework ('stuffed') of wood and cinnamon; and the warmth of adjacent candles would have brought out the fragrance of the spice.[33] The associations of this odour would have been widely understood. A fourteenth-century poem on fortune-telling by casting dice made clear that the thrower of three sixes would be without comparison, 'as synomome that ys of odour sote [sweet]'.[34]

Just as these pleasant odours were redolent of good things, so the correspondence between bad smells and evil, Hell and the Devil, was equally ancient and widespread. Hell was frequently characterised as sulphureous or reeking of corruption.[35] Two accounts of the cure of Elias, a monk of Reading who suffered from leprosy, suggest either that he visited the thermal springs at Bath, but was not cured, or that he proposed or pretended to visit; he subsequently went to Canterbury and was cured by the water of Thomas Becket there. The waters of Bath were not only of no merit in his case, but their designation as 'sulphureous' tainted them with undesirable, other-worldly characteristics that were no match for St Thomas and his water.[36] The places of torment seen by the monk of Eynsham in his vision encompassed baths of pitch, sulphur and other liquids with a revolting stench.[37] If the Savoyard origins of the Bishop of Hereford, Peter of Aigueblanche, were not enough

44 The translation of St Edmund: the body reaches Stapleford. From the Lives of St Edmund and St Fremund, written by John Lydgate and presented to Henry VI between 1434 and 1439: Lydgate records how the way was strewn with fresh flowers, walls were hung with coloured cloths and the road made clean. The cart carrying the shrine rests on a carpet, a fifteenth-century recognition of its honour.

to earn him opprobrium, the ingenious idea he had, about 1255, for raising funds for the Crown made sure that he merited it. By getting the English bishops and major religious houses to seal blank charters of obligation – which he took to Rome, where the charters were completed with promises to pay large sums of money and the documents handed over to the bankers of Florence – he ensured himself an ill-starred place in monastic chronicles. Matthew Paris, in his account of this *démarche*, had no doubt that Peter was, so to speak, taking a diabolical liberty: 'The Bishop, whose memory exhales a sulphureous and very foul stench, went to the King. . .'[38]

There was a strong association between anal imagery and evil, doubly diabolical in that breaking wind involved both a bad smell and a reprehensible sound. St Godric had a vision of two demons and he confessed the details to Roger, Prior of Durham (1137–49). The following night, one of the same demons returned to the half-sleeping Prior and addressed him: 'Because you have not refrained from turning that stinking Godric against me and my sister, you will receive such a reward from us.' He then broke wind – a most disgusting stench, which did not completely clear from the Prior's nostrils for three days.[39] In Chaucer's 'Summoner's Tale', Friar John, groping in search of lucre under the clothing of Thomas, a sick churl whom he had been canvassing for funds for his friary, received instead a fart, louder than that of any cart-horse. The lord and lady, whose confessor the friar was, did not hesitate to suggest that the fart was the work of the Devil and that 'a cherl hath doon a cherles dede'. The lord then posed the remarkable problem, readily solved by his carver, of how this savoury contribution to the convent might be divided into twelve – parodying the division of the Pentecostal breath among the apostles.[40]

A bad smell, 'stinking', might be both a characteristic of those whose activities were nefarious or hypocritical and a term of abuse. In the N-town plays, the nature of characters is illustrated by the use of language. Cain, showing his true mould, calls his brother Abel a 'stynkyng losel' who has God's love when he does not; and Lameth, in turn, who has killed Cain believing him to be a beast, turns on the accomplice who has induced him to shoot, a 'stynkynge lurdeyn', words indicative of the evil intent that led to this death.[41] At Leverington, near Wisbech, in 1467–9, William Freng was brought before the ecclesiastical court for calling John Sweyn a 'stynkyng horysson', defamation that had a strong moral odour.[42]

45 Jesus explains that those who praise God and boast of it are to be considered hypocrites and stink; they resemble the tomb, which is beautiful and rich on the exterior and contains stinking carrion within, as the figure inside the church demonstrates by holding his nose. From the Holkham Bible Picture Book, c. 1320–30.

Evil smells marked out bad things or those destined for Hell: the stench of bad breath or of those who had eaten onions, leeks and garlic – with its sulphureous odour – was a sign that could not be ignored.[43] In 1307 Brother Robert of St Martin reported that for a period of three weeks he had not been able to raise his arm, or celebrate mass, and that he had been cured by Thomas Cantilupe's intervention, but not before he had tried, fruitlessly – and not perhaps unsurprisingly, given the moral implications – a remedy that involved rubbing crushed garlic on the affected area.[44] Mankind, confessing his sins in the *Castle of Perseverance*, noted that:

We have etyn garlek everychone [everyone].
Þou I schulde to Helle go,
I wot wel I schal not gon alone,
Trewly I tell þe. . .[45]

Those that sinned might acquire a stench. The fifteenth-century sermon cycle, *Jacob's Well*, likened the rotting of an apple to the corruption of the body by sin: 'Also as rotynhed doth awey þe swete smel & þe good odour of an appyll so dooth synne awey the smel of swetnesse of vertuys out of þi lyif and makyth þi lyvyng to stynke in þe syȝt of God.'[46] A bad stench, however, could have merits. The *Alphabet of tales* noted the case of a labourer, Arsenius, who changed his clothes just once a year. He then told himself that it was necessary for him to feel their stench to balance the good smells he had encountered.[47] The soul could recover from the effects of the bad odour of sin. Cleansing the taint of original sin took place at baptism and further sins might be washed away through contrition.[48] Medieval authors likened baptism to bathing. In a metrical life, written *c.* 1300, the conversion of St Katherine was completed when she had bathed: in a medieval, aristocratic bath, richly curtained, which would have been replete with spices and good odour.[49]

The idea that one virtue, or a smell, might counteract another was widely held. A saint's goodness could protect him from the excesses of bad odour. In May 1198, St Hugh of Lincoln officiated at Bermondsey at the burial of Simon, Abbot of Pershore. Hugh's biographer recorded a very unpleasant occasion. The swollen and rotting body produced an overpowering stench, and some made use of strong spices or incense to lessen the ill effect, but Hugh needed none of this, even though he was peculiarly sensitive to smell and often objected to bad smells, a sensory predilection that echoes the powers of a number of saints. Indeed, he was completely unaware of the problem: the sweet odour of Christ protected him from the stench of death.[50]

Good smells might be used to counteract bad, not just as one might use a perfumed disinfectant, but also to create or return something of the odour of sanctity. Wimarc, a woman taken as a hostage at Gainsborough during the reign of Stephen, had conspired with her fellow inmates to poison the gaoler's drink; but he had forced them to drink it first. She was the only one to survive, although very unwell. She visited doctors and shrines, but then came to the cathedral at Norwich, where she was cured at the tomb of St William. The cure was marked by her vomiting the poisonous discharge over the pavement of the cathedral: the sacrists hastened to clean up the area and strew it with fragrant herbs.[51] The complete absence of smell might also be counteracted: a most sweet smell filled the nostrils of Eilward of Tenham on his visit to the tomb of Becket, 'that odoriferous flower of England', curing the anosmia from which he had suffered for some years.[52] Bad odour might be used to combat a false, good scent. When, in 1440, the Lollard Richard Wyche was burned at the stake in London and a makeshift shrine sprang up at his place of execution, the vicar of All Hallows, Barking, mixed spices with the ashes there in an attempt to promote Wyche's sanctity. The authorities responded by countering smell with smell and, in a deliberate act of pollution, established a dunghill on the spot.[53] A bad smell, such as garlic, could mitigate other, foul smells. Encyclopaedists noted that the stench of garlic would defeat the odour of the dunghill.[54]

These ideas about smell endured through the medieval period, particularly at a popular and uncomplicated level, for example in preaching, in visions and descriptions of Heaven and Hell, of saints and demons, and in descriptions of bodily corruption. They might also transfer to support individual versions of sanctity. The Lollard William Thorpe justified his self-serving account of his interviews with Archbishop Arundel in 1407 in this way:

> For truþe haþ þis condicioun: whereevere it is enpugned, þer comeþ þerof odour of good smel, and þe more violentli þat enemyes enforsen hem to oppressen and to wiþstoonde þe truþe of Goddis word, þe ferþir þe swete smel þerof strecchiþ. And no doute, whanne þis hevenli smel is moved, it wol not as smoke passe awei wiþ þe wynde; but it wol descende and reste in summe clene soule þirstinge þeraftir.[55]

The association of smell with moral qualities was not favoured in some circles, especially amongst canon lawyers. The papal commissioners who came to England in 1307 to enquire into the miracles and sanctity of Thomas Cantilupe, as a prelude to the canonisation process, asked each witness to a miracle how it had happened. Although there are references to smell, there are no instances of a divine smell accompanying the presence of the saint or the performance of a miracle or the translation of the body, or of diabolical smell associated with demons. The closest a witness came to mentioning the association, in the commissioners' account, was in the testimony of Cantilupe's successor as Bishop of Hereford, Richard de Swinfield, who reported that his predecessor had lived honestly, seriously and in the odour of good repute and without any infamy.[56] On the other hand, the mysticism of sensory devotion, preached by Richard Rolle among others, and followed by Margery Kempe, maintained and uplifted the divine link with smell. An extreme personal delight in the senses was probably what characterised both the devotion of Rolle and Kempe, but on the Continent, for example in the case of St Bridget, the sense of smell was used for establishing the character of individuals.[57]

A further, widely accepted dimension to smell was to be found in carnal pleasure. Monks were urged to confess both their delight in smells, such as the good-smelling herbs they might spread in places where they wanted to sleep, or the smell of spices, sauces and foods, such as meat cooking, which might provoke a desire for a more luxurious diet; and, at the same time, their unreasonable turning away from bad smells, those of disease or the sickness of one of their colleagues.[58] A formula used by the Dominicans for secular confession from the mid-thirteenth century enquired whether musk or other perfume had been worn by women, with a view to its odour attracting men.[59] In 1354, Henry of Lancaster recorded how he had sinned through the nose, delighting in the smells of flowers, herbs, fruits, women and scarlet cloth.[60]

There was an enduring belief in the regenerative or debilitating effects of odours. If smell was a substance, a vapour, and if it was drawn directly into the brain – as we have seen, considered by some to be the sense organ in this case – it could have a direct effect on the body by cooling or heating it. There was therefore potential to use smells as a means of therapy and also a danger from misapplied smells. The effect of the smell varied with the humoral composition of both the odour and the person perceiving it: cold smells (such as roses) would benefit those whose nature was hot; warm or hot smells (for example musk) those whose temperament was cooler.[61] Smell was held to have a powerful effect on the spirits: aromatics were a first line of therapy to bring the patient's humours back into balance, or as a prophylactic to keep the body well adjusted.[62] The monk of Eynsham, lapsing into an unrousable state while he had his vision, caused great consternation. Potions of various spices and herbs were forced into his mouth by his fellow monks in an attempt to revive him, but all ran straight out.[63] In the late thirteenth century, the infirmarer at Barnwell Priory was rarely without ginger, cinnamon, peony and similar substances, so that he could readily assist if someone were suddenly taken ill.[64] Taking this idea one stage further, the smell of food was believed to be nourishing, rather than just appetising.[65]

As smell passed directly into the body, it was widely believed it was the channel both for benefit and for the transmission of disease. The Middle Ages inherited the gynaecological theories of Hippocrates and many believed that the uterus might move around the body, causing cardiac and pulmonary disorders as well as those associated with the reproductive system. The use of odour, applied as ointments or as fumigants, was a standard element in the treatment. For 'suffocation' of the uterus, strong and unpleasant smells, such as those of burnt cloth and feathers, were placed near the woman's nose, to drive the uterus downwards; and a 'suffumigacion' of good-smelling spices and herbs was placed close to her genitalia, to attract the uterus. For 'precipitation' (prolapse) of the uterus, the opposite course was employed, with one late-medieval treatise placing 'all wel savird thynges to hire nose and benethe all ill savyrd thynges'.[66]

The subject need not be immediately present to feel the benefits or detriment of an odour. Fumigations were an essential part of the practices of both astral magic and necromancy.[67] Their presence in mainstream religion, manifest in the wide-spread use of incense, conveyed a supernatural or spiritual benefit directly to partic-ipants, much as the fragrances of unction – at coronations, or at the end of one's life – marked the touch and presence of divinity and passed on benevolence. Fumigants counteracted disease: good smells counteracted the bad.

Disease might come from close contact with infected individuals or, more gener-ally, through bad air.[68] The miasma of lingering odours suggested the transmission of particles, by a smoke or vapour, and characteristics from one body to another. Eleanor of Provence was exercising reasonable caution when she wrote to her husband advising him not to take their son, the future Edward I, to the north: 'When we were there we could not avoid being ill, because of the bad air, so we beg you to arrange some place of sojourn for him in the south, where the air is good and temperate.'[69]

Bad smells, however, were indicative of more than disease: at the start of the thirteenth century they were considered as revealing the nature of the individual. Remedies might treat the symptoms, but they could not change the underlying state of the person. Physiognomy emphasised the link between these manifestations and the moral character of the individual. Unpleasant body odours displayed one's character as a person. There was a real concern about bad breath. Treatises on cosmetics therefore gave a high priority to ways of concealing these failings.[70] Malodorous breath could have severe consequences for the medical profession. John Arderne, in his treatise on clysters, noted how it was believed that air changed the body and, in consequence, the wound. He could demonstrate from experience that the breath of a menstruating woman irritated a wound if she came close, or the breath of a physician if he had recently slept with a menstruating woman, or if he had eaten garlic or onions.[71] From the end of the thirteenth century, however, physi-cians began to seek the underlying causes of bad odours, using smell itself as an indication of physical disease and the sense of smell increasingly as a tool of discrimination in their diagnoses.[72] This change in emphasis, however, was slow to make an impact on popular beliefs.

To control infection, the malodorous were set apart. Lepers were isolated, outcast by the Third Lateran Council of 1179 and stripped of their property, their dreadful stench marking them as spiritually deficient as well as physically afflicted. Many of the leper hospitals adopted strict regimes of mortification of the flesh, prayer and abstinence to treat the spiritual causes of the disease.[73] The foods that were suitable for them also had a stench: when butchers in York were convicted of selling rotten, or measled, meat, that meat was then to be given to the lepers.[74] In a tale reputedly told to Henry I of England, the odour of sweet spices that greeted Theobald, Count of Champagne, sent by Louis the Fat to visit a leper who was on the point of death, was a poignant way of demonstrating that his act of mercy was rewarded.[75]

According to Isidore of Seville, the sense of smell, beyond its spiritual abilities, was the judge between the sweet and the stinking. Neither was lacking in the terrestrial realm. The pleasure taken in the smell of flowers and the countryside was widespread and is particularly evident in English poetry from the fourteenth century onwards.[76] Herbs and flowers were a ready means of creating perfumes: gathered fresh, they would be hung around the house or in clothes, and flower-gathering expeditions must have been common.[77] Purchases of flowers for the royal household at the end of the Middle Ages are indicative of a small-scale trade; and flowers were among the many commodities considered appropriate gifts for a queen.[78]

The construction of enclosed gardens, possibly modelled on those of the Middle East or Muslim Spain, an epitome of Paradise from Eden to the Apocalypse, was a feature of upper-class horticulture in the Middle Ages. Here plants were grown for their smell as much as for their colour.[79] The gardens of Henry VII show something of expectations at the end of the fifteenth and the start of the sixteenth centuries. The King had gardens and gardeners at the Tower and at Sheen. After the fire at the latter in 1497, there was extensive investment in the new palace at the same site, Richmond. This included the creation of new garden features, alleys and galleries (some of which had to be further rebuilt on account of poor carpentry), as well as the planting of vines in the orchard. The King created gardens at newly acquired properties at Hanworth, where he planted apple and other trees in 1508–9, at Wanstead and at Woking, spending £16 9s. 9d. on the last between October 1506 and March 1507.[80] These were elegant gardens where the court might wander and linger. In the poem 'Why I can't be a nun', the prospective nun goes into a garden on a May morning, enters an arbour and falls asleep among the fresh and fine herbs, on a bench of camomile.[81] Paradise was a garden, redolent with scents.[82]

In the natural world, unpleasant smells indicated danger, corruption and even death. Standing water, in puddles, became 'corrupte and stynkynge', just as a man would unless he emulated the sea, continually in movement in a constant quest to stir for good works.[83] Bad smells char-acterised industrial areas, slaughterhouses, tanneries and the premises of fullers.[84] The atmosphere underground could be noxious. On 13 September 1337, John de Maldone, who had been hired to clean a well in

London, descended into it and was overcome and killed there by foul air.[85] In Nottingham, in 1551, a man who went down into a cellar by a rope was suddenly overcome by a suffocating damp that caused him to fall to his death.[86]

Kitchens, prisons, latrines and dunghills featured prominently in the litany of evil-smelling places. As bad air was considered responsible for the transmission of disease, much attention was paid to mitigating its effects. Kitchens were moved or rebuilt. Innovations were made in the construction of garderobes and rubbish pits were lined with stone. Regimes of latrine- and street-cleaning were the responsibility of conscientious municipal authorities.[87] Individuals might adopt their own remedies, combating bad smells with good. That musk might be placed in latrines is envisaged by one, at least, of the answers to the Salernitan prose questions; and John Russell's *Boke of nurture* enjoined the aristocratic chamberlain to keep the privy 'fayre, soote [sweet], & clene'.[88]

Those engaged in occupations associated with dirt and bad smell were stigmatised and stereotyped, from the cook and scullion to the dyer and tanner. A demon, questioned by a master of the Dominicans, was asked why it delighted so greatly in the stench of sin. It replied noting that in towns there were those whose job it was to clean the latrines: they reeked of the most awful smell, yet they did their work freely because of the considerable financial rewards. Demons similarly rejoiced in the gain they made, as the soul of the sinner was their meat and drink.[89] In 1339, William Wombe, whose job it was to clean the public latrines in London, was found dead in the Thames. He had gone into the river about the time of curfew to wash himself – understandably – and had drowned.[90] These were degrading professions, to be avoided for the sake both of honour and of physical contamination. Moreover, spiritual impurity was close to physical contamination and pollution.[91]

From this stereotyping, it was a small step to condemning whole groups of society on the basis of odour, moral as well as physical. There is an extensive literature about smell and gender, particularly in the use of smell to elevate or demean women. There were standard repertoires of bad smells to associate with prostitutes, and old and ugly women.[92] Sensory epithets for the Virgin Mary indicated her place as Queen of Heaven. In the play *Mary Magdalen*, probably *c.* 1515–25, Christ, on the Heavenly Stage, praises his mother, her good odour countering a series of bodily ailments.[93] Smell, however, might be a false indicator. Chaucer's host incorrectly suggests that odour confirms his notion that the parson, who has rebuked him for swearing by God's bones, is of suspect orthodoxy:

Oure Host answerde, 'O Jankin, be ye there ?
I smelle a Lollere in the wynd,' quod he.[94]

The poorly sighted Isaac, calling his son Esau to him, in the Towneley plays, expects him to smell like a field of flowers or honeycomb – and fails to realise that it is his son Jacob he is about to kiss.[95]

There is good anthropological evidence that different cultures regard each other as smelling unpleasant; there is not necessaily any pejorative connotation beyond the physical observation, but in some cases that is a concomitant. The notion that the Jews had a distinctive smell does not appear in English sources before the expulsion of 1290, but it was current on the Continent. While the archetype for the blood libel (the allegation that Jews sacrificed Christian children in order to employ their blood in Passover rites, to rid themselves of this odour) is English – the murder of St William of Norwich in 1144 – neither this example nor subsequent English cases involved the use of blood for ritual associated with odour or anything of a sacrificial nature.[96] In fact, some held that any odour associated with the Jews was good, because of their devotion. A late fourteenth-century Wycliffite sermon argued that the prayers of many were not heard because their prelate was 'so stynkyng afor God'. This exclusivity could work in reverse. The devout who made offerings smelled well to God, but badly to men: David desired that his prayer pass to God as incense, but stink unto men, 'as Godus children stonkon to Pharao'.[97]

Although its purpose was to attain virtue, fasting could be a cause of bad breath. As a practice that was widely prevalent in medieval England, in differing degrees, it could nonetheless have inspired the opprobrium associated with bad odour.[98] The odour of the poor, with sickness, from which, to his shame, Henry of Lancaster resiled,[99] was matched in the *Castle of Perseverance* by Mankind's determination, in consultation with Avarice, not to feed or clothe a beggar: instead, he should starve and stink.[100]

The odour of decay represented a striking example of the transitory nature of the flesh: Lazarus and the stench of the tomb were common literary motifs.[101] The embalming of the bodies of the upper classes, both lay and clerical, may have been designed not only to preserve them from putrefaction until burial, but also to imbue them with an odour that might reflect the virtue of their lives. The body of St Hugh of Lincoln was treated with sweet spices before its journey from London, where he had died, to Lincoln, where he was to be buried.[102] Typically the viscera were removed – in the thirteenth century, sometimes for separate burial – followed by the application of spices and ointments. The body was then wrapped in waxed linen cloth. In the late fifteenth-century romance 'The squire of low degree', the body of Sir Maradose, the steward, mistaken by the King's daughter for her lover, the squire, was eviscerated by the daughter who 'sered that body with specery, wyth wyrgin waxe and commendry' [either cumin or prayers].[103]

A Lollard sermon likened the stench of the flesh – lechery – to the smell of a boar. Not only did the animal rest gladly in fowl sloughs or muddy places, but it left a 'foule stynkynge savur' wherever it went. Another sermon in the same sequence likened men who did not speak the word of God, or who did not speak virtuously, to pigs 'þat raþer rennen to toordis þen to flouris'.[104] The smell of some creatures, resulting from their behaviour or the aptitudes of their senses, was carefully observed, marking them out as metaphors for spiritual impurity or burlesque. As a model for covetousness, the otter was believed to take fish back to its den and keep them piled up until they rotted, for it was never sure of having enough – the stench then making it sick and causing it to die.[105] Foxes – the sins described by Henry of Lancaster – were remarkable for their strong scent, which allowed the hounds to seek them out and destroy them, just as Henry's five senses should track down and eliminate his sins.[106] In the Towneley Second Shepherd's Play, the three shepherds come to Mak's house in search of the missing wether, which Mak's wife has disguised as a baby in the cradle: a wether beyond compare, no animal domestic or wild smelled 'as lowde as he smylde'.[107] A tale of the ingratitude of a steward to a poor man who had saved his life recounted how the steward instructed a forester to make a hundred pits and to cover them with sweet-smelling herbs, in order to attract animals. The steward returned later to be hoist by his own petard and to be joined in the pitfall trap by a hungry lion, an ape and a serpent.[108] The vulture, reputed from Antiquity to have an excellent sense of smell, featured as representative of that sense at Longthorpe. Notable among other animals given a special

46 Longthorpe Tower: the vulture represents the sense of smell.

significance for their odour or their sense of smell were the panther, which gave off a marvellous odour (and which gained a place in bestiaries as a symbol of Christ and his good odour), the elephant (for its proboscis) and the bear.[109]

The link between odour and morality, goodness, even authority, on the one hand, and iniquity on the other, has a distinctive bearing on the creation and manipulation of smell. If sanctity was accompanied by a good smell, were those that smelled good saints or otherwise good people? Did groups of individuals therefore set out to be identified by smell, to be raised up by it, or condemned by it? By looking at how odours were created or controlled, the regimes of washing and bathing, the sources of perfumes and scents, who had access to them, and when, we can provide some answers. The odour of the individual is only one aspect of this: the wider environment might also be perfumed – in the first instance goods, such as bed linen, the contents of wardrobes and chests, then whole rooms and buildings. This last was least effective in certain conditions, for example where there was already a dominant odour of smoke in a building, and the investment was unlikely to be worthwhile until buildings and lifestyles had reached a certain pattern.[110] In England, buildings were increasingly divided into separate chambers, rather than communal, smoky rooms, especially from the fourteenth century. In the fifteenth century it was common for substantial buildings to have ranges of chambers, each with its own fireplace and chimney, for individuals or small groups. These arrangements had obtained earlier in other settings, particularly ecclesiastical buildings. Before this, most perfumery focused on the individual and personal objects.

Evidence for the odour of individuals is oblique. There are practices that tended to deodorisation, through washing; and some record of the opposite process, the use of odour in washing, through the employment of herbs, spices, waters and perfumes. The object of washing was not cleanliness for its own sake; washing was a mark of gentility, and dirt, conversely, a mark of denigration. Beyond that, there were a few personal objects associated with smell and preparations for looking after clothing and linen that would have been used by the individual. In all cases the evidence is better for the upper classes, but some common practices can be discerned elsewhere.

In terms of basic washing arrangements, both men and women bathed in streams. This was principally a summer activity: tubs in houses were used at other times of year.[111] Bathing the whole body was less common than washing the exposed parts, overwhelmingly the face and hands. Washing, as opposed to bathing, was an activity done at specific times, especially at the start of the day. It was the custom of St Hugh of Lincoln, while at Witham, to wash his hands at dawn without an attendant to serve him and without a towel.[112] The general pattern is also apparent in the records of coroners: at sunrise on 10 August 1375, Roger Alwey, of Loughton near Milton Keynes, went to wash his hands and fell into a well and was drowned.[113] Washing at dawn was probably more widely current in northern

Europe. A treatise on healthy living, composed in 1286 by a Savoyard probably for the royal foundation of Benedictine nuns of Maubuisson, near Pontoise, gave advice for summer: the nuns were to rise and comb their hair, then wash their face and hands with good, cold water from a spring.[114] A late fourteenth-century English sermon noted that Christ and his disciples washed their hands when they ate and always when they were dirty, as Christ was a 'most honest man'.[115]

The ewers, bowls, jugs and aquamaniles employed in washing are common in inventories and in the archaeological record.[116] Aquamaniles – shaped vessels for pouring water over the hands – appeared in the twelfth century in metal, and were copied in the thirteenth century in pottery. The elaborate forms found across Europe suggest that they were for use on occasions where they could be seen, for example in the washing of hands before eating that was a part of dining ritual both among the upper classes and more widely.[117] There are also references to brass bowls, which seem to have been used while washing the head. One, worth 2d., was among the goods of Alan de Hacford, a London chaplain, seized in 1345 after he killed a rival for his mistress. Even among the upper classes, bowls for this purpose might be of brass.[118] Moralists condemned the vanity of both sexes in attending to their hair 'as in wasschyng, in kembynge, in tresyng, and lokyng in þe myrrour, wherfor God wraþþeþ hym ofte and many tyme'.[119] Around 1200, a woman of Lynn, who washed her hair one Saturday afternoon at a time when all work was commonly prohibited because of the sabbath, immediately went mad and remained

47 The octagonal lavabo in the cloister at Much Wenlock, a design that was much more common on the Continent than in England. Sixteen monks might wash here simultaneously. There were carved stone panels around its base, two of which survive (although that shown is a replica), depicting the calling of Peter and Andrew, and John and another apostle.

48 Rievaulx: the entrance to the refectory from the cloister, bordered on either side by lavers where the monks washed prior to eating.

so for eleven weeks until she was cured at Sempringham.[120] By the sixteenth century, washing the hair, as much as other parts of bathing, required spiced preparations. In September 1501, the third Duke of Buckingham expected to use cinnamon, liquorice and cumin to this end.[121]

Washing the hands and feet was understood as preparation for a journey, although the ablutions undertaken by Thurkill in October 1206 may have been as much about removing the mire after a wet day digging drains in the field he had just sown as about purification before his celestial excursion with St Julian. In either case, it annoyed his wife, who believed he should not conduct himself in that way on a Friday.[122] The members of Winchester College who had been allocated rooms other than on the ground floor were forbidden by William Wykeham to wash their head, hands or feet, or anything else in these chambers, lest any water or any other liquid – wine, ale or other substance – passed through the floor to the detriment of those quartered below.[123] The routine outlined in 1396 by a treatise teaching the Anglo-Norman language covered the dressing of the lord, after which he was to go to one side of the chamber near the windows to wash and dry his hands. His newly-wed wife then combed his hair. He also did not go to table until he had washed his hands.[124]

References to washing other parts of the body are less common. St Hugh was assigned to look after Peter, the Archbishop of Tarentaise, when he was at

Chartreuse. He used to wash his feet and legs, but Peter, who believed his frailty received much benefit from this treatment, had to persuade him to wash further up his body.[125] Ranulph Higden, *c.* 1340, classed the daily washing of genitalia as a Jewish custom that, along with the practices of praying to the west (like the Jews), the south (like the Saracens) or kneeling before trees, was not to be tolerated.[126]

Aside from cleanliness, there were two reasons closely connected with odour for taking baths. Bathing formed a part of certain purification rituals from which it was intended that a good odour – and moral benefit – would result. It was a part of the ritual undertaken by catechumens. From the twelfth century it might be included in the ceremony of knighthood and the Order of the Bath preserves this important association.[127] At Darlington in 1292, the barber of Edward I prepared a bath for four new knights at a cost of 20s., with a further 3s. for a cover for the baths.[128] Bathing may also have featured in the ritual of preparation for weddings. Eleanor, sister of Edward III, married Reynald II, Count of Guelderland, in May 1332 and she bathed probably on the eve of her wedding: a valet of her chamber was paid 18d. for herbs and other aids in making a bath for her at Nijmegen.[129]

Secondly, odour gave the bath a therapeutic character. There are some references to spas, but most medicinal baths made use of herbs, with the tub cocooned in sheets in a tent-like arrangement.[130] There are occasional references to public baths in England, but little to parallel those on the Continent. In England, the association of the stews was with prostitution, although the continental experience appears to have been more sensuous and involved bathing.[131] A version of the Pseudo-Aristotelian *Secreta secretorum*, from soon after 1400, set out some of the expectations of the aristocratic bath, 'on of þe merveueylles of þys world'. It was to be prepared with attention to the four seasons of the year and to make use of the odours – preparations of spices and herbs – most appropriate to the bather in terms of his humoral make-up. In addition, the seat in the bath was to be wetted with rose-water. Once he had taken the initial bath, the bather was to pass into other parts of the bathing complex. If he was overcome with heat – presumably from the steam – his head was to be combed and he was to use an ointment: in spring and summer, 'Saracen ointment', in autumn and winter an ointment of myrrh and the juice of blites (wild or garden spinach). Cooling, scented waters were then to be cast on his head and he was to wash and rub himself with these. His body was then to be anointed with ointments fitting for the season. If he was thirsty, he was to have a syrup of roses and an electuary with musk. After eating and drinking a little wine – possibly in the bath, as was the custom on the Continent – he was censed, before going to rest.[132]

Peripatetic bathtubs feature in royal accounts, but some establishments had permanent bath-houses, as at Easthampstead, Eltham and King's Langley in the mid-fourteenth century.[133] Where there were permanent bathrooms, they might be tiled, as at Leeds Castle in 1291–2 and at Sheen under Richard II,[134] and there might be hot and cold running water, as there was at Westminster under Edward III.[135] The

sunken bath adjacent to Henry VIII's lodgings at Whitehall Palace indicates something of the character of the event. The presence of stove tiles here suggests that hot
water and steam – perhaps like a Turkish bath – were significant.[136] The expectation
of warm or hot water was common, even if it were not 'on tap'. An ampulla of
St Wulfstan's water resuscitated a boy who had drowned in a bath while his nurse
had gone out for wood to heat additional water.[137] In her metrical life of *c.* 1230,
St Modwenna instructed Bishop Kevin to bathe: he used a spring that miraculously
produced warm water.[138]

There is one ingredient we might expect that is missing from these descriptions
of bathing: soap. Although there are references in accounts to soap, many were
probably to soft soap, which was based on the potassium-rich wood ash common
in northern Europe. Wood ash from Mediterranean countries, where there was
more salt in the soil, had a higher sodium content and produced a hard soap.[139]
Mentions of 'Spanish soap' refer to this product and it may have been perfumed: in
1273–4, 2 lb. cost 8d.[140] In June 1289, four loaves of soap – that is, a hard soap –
bought for Queen Eleanor of Castile while she was in France cost 10s. in money of
Tours or about 2s. 3d. sterling.[141] The 122 lb. of white soap bought for Elizabeth
Berkeley, Countess of Warwick, and her three daughters, Margaret, Eleanor and
Elizabeth, in 1420–1, was mainly purchased in London along with the household's
spices and ranged between 1½d. and 3d. per pound. Again, this was probably a fine
product,[142] and one that implied a much higher level of consumption than that of
an elderly widow: Isabella, Lady Morley, managed in 1463–4 with 6 lb. of black soap
and 1½ lb. of white.[143] By the fifteenth century, elegant soaps were in much greater
use in the washing routines of the upper classes.

For medieval England, the principal source of spices, ointments, perfumes,
scented oils and waters, as well as exotic soaps – in fact, anything beyond commonly
found herbs and the produce of gardens – was the Mediterranean and what trade
routes brought there from the East. At the same time, Middle Eastern and Arabic
customs and literature influenced the way in which these materials were used,
particularly in areas where there were points of common experience for Muslims,
Jews and Christians, for example in Spain. Here, from the eleventh to the thirteenth
centuries, there was considerable knowledge of the methods of perfume-making:
obtaining essential oils by washing, sublimation, alembic (by sweating), and by
burning, to produce a smoke (*per fumum*). Odours were based on a range of
substances, including musk, ambergris and yellow amber, and camphor. The most
desirable ointments were designated by their dominant ingredient: violet, jasmine,
cloves, narcissus, rose, iris and camomile. In addition mousses of various compositions, soap, and washes composed of wood lye, honey and syrup of figs were among
the procedures described.[144]

As with cosmetics, from which perfumes are sometimes indistinguishable, there
were two significant changes in the later medieval period. The first, from the late

fourteenth to the late fifteenth centuries, was the wider availability of perfumes and an increase in the number of people using them in the countries bordering the northern shores of the western Mediterranean. For example, a spicer's shop in Perugia, inventoried in 1431, contained an extensive selection of ointments, waters and sugars, leading on to syrups, cosmetics and electuaries, powders and scented oils, besides flasks, vessels and equipment for making these products.[145] The second change, between the late fifteenth century and about 1560, saw a development in the technology of perfumery and an expanded knowledge of its products.

Perfume-making in the Middle Ages was markedly different from contemporary practice – and the most significant change happened in the sixteenth century. Today perfume largely has a base of alcohol. There were some parallels in Middle Ages, such as naphtha-water, a volatile mix of water and extract of bitumen, but most medieval perfumery was based on oils perfumed by absorption (for example, using almond oil as a base), waters, powders, pastes and some rare essences made by the distillation of lavender, rosemary or resinous woods. The principal techniques of manufacture were to prepare powders, using a pestle and mortar, which could be used for filling sachets and cushions. With the addition of a resin or gum like acanthus or ladanum, these formed pastes, which could be shaped and hardened, for example to create round balls – *pommes* or pomanders. Richly encrusted with gold and silver, these objects might be held in the hands, the heat of which would help release the odour, or the paste might be made into paternosters and chaplets. The powders were sometimes a source of perfumes for burning, although they needed the addition of charcoal. Sugar might also be used as a medium for carrying odour, as with mouth pastilles (*muscadins*).

The second main method of preparation was maceration, with glass phials of the raw ingredients, such as rose petals, placed in the sun, the gentle heat accelerating the process. These might then be used to produce scented oils and waters, and could also be used in conjunction with a third technique, distillation. Some of these techniques were described by authors whose works were written or circulated in England, including those of Bartholomew the Englishman, Henry of Lancaster and John Arderne: they described the preparation of rose-water and essential oils of other plants, but such methods were not widely applied in England to produce fine perfumes.[146]

Advances were limited by the difficulties of distillation, although that did not prevent many in England producing infusions from herbs and plants, olfactory waters of medicinal value. At Norwich Cathedral Priory, for example, the infirmarer had a still, fuelled by peat, in which he worked with the products of the infirmary garden.[147] In the late fifteenth century, however, there were improvements in the process of distillation and in the isolation of new, essential oils, and it was at this point that the first treatises on distillation, perfume manufacture and the use of filtration were published and widely disseminated.[148] There was a strong aristocratic market for these materials, although it was restrained in certain cases by religious practice. Throughout Europe, the fifteenth century saw an enormous broadening of the materials available: by the later years of the century there were specialist perfumiers, not just apothecaries or pharmacists with an interest in this trade.

The evidence for the use of high-quality perfumes and cosmetics is much less ample in England, although there were 'home-made' preparations for upper-class women from at least the thirteenth century. The few recipes for cosmetics from medieval England mainly focus on colouring, especially the face and hands, and dyeing hair. Smell was a secondary consideration, although some odour must have been associated with them. An Anglo-Norman treatise of the second half of the thirteenth century has a recipe to stop bad breath by inhaling a fumigant mixture of roses, a type of mint and other odoriferous herbs, burning on charcoal.[149] In the same collection, an ointment for the head involved the root of patience, cooked in olive oil and boiled with wax after the oil had taken on the substance of the herb; it was then strained through a cloth and, when it had cooled a little, mixed with quicksilver and put in a box.[150] Another thirteenth-century treatise describes powders composed of nutmeg, cloves and incense, to be used in a preparation for colouring the face.[151] A further recipe described a face-wash made of various herbs, including fennel, powder of roses and white incense. Once these had all been cooked together, the face was washed in a solution with a little of the powder, both night and morning, in order to produce a white, clear and healthy appearance.[152] These were expensive preparations and it is difficult to tell how widely they were used, but that may have been an essential part of the effect: the perfumed spices of the bath, or the cosmetic preparation, set the gentle lady apart. Herbal recipes may have been in use at other levels of society, on the analogy of the late fourteenth-century Languedoc, where peasant women used a wide range of herbs, plants and vegetables, including onion juice.[153]

If the purpose of the cosmetic was not primarily smell, some objects, like pomanders with musk and ambergris balls, were for scenting the person. Margaret and Eleanor, daughters of Humphrey de Bohun, Earl of Hereford, had each among their goods in 1322 highly prized objects of this sort. Margaret had a pomander (*poume de aumbre*) – probably of ambergris – set in three clasps of silver; and Eleanor had what was either a nutmeg or a musk-ball (a hollow ball to hold the perfume) to perform a similar function, set in silver with small stones and pearls.[154] The goods of Richard II included a *pomme de musk* garnished with silver gilt and a pearl worth 40s.[155] Those of Henry V, inventoried in 1423, included two musk-balls, one of gold, weighing 11 oz., the other of silver gilt; his brother, John, Duke of Bedford, also had two among his goods at his death, the gold in them weighing 15¾ oz.[156] In December 1507, Henry VII acquired from John van Utricke five gold pomanders and a pair of rosaries of gold beads, for £13 10s.[157] For commodities used by aristocrats and monarchs, the value of these goods was not exceptional and the numbers listed were remarkably few. What they portray is a gradual change in habit, in personal preference, perhaps as these items became more commonly known from continental connections.[158]

Other odour-producing items were scarce even in the fourteenth-century royal household. Among the goods in Edward III's privy wardrobe received by John de Flete on taking office were a basket with various phials of balsam, treacle and oil of St Katherine – some doubtless with a medicinal purpose, but some also with a strong odour; three phials of balsam were recorded among the goods used for the almonry and chapel.[159] These references are again unusual at this point in the fourteenth century. When, in 1354, Henry of Lancaster appreciated the smell of women, by which he almost certainly meant the spiced and perfumed women of the aristocracy, there must have been few of them.[160] There is a little more evidence for the fifteenth century. The inventory of the goods of Henry V prepared for his executors is the first for a king of England to include significant evidence for the use of cosmetics and perfumes. Among the plate and jewels is a group of items listed in a way which suggests that together they may have been the contents of the King's dressing table: three mirrors, one with jewels valued at £13 10s., the other two much less significant affairs, valued at 7s. 9d. and 15s. 6d.; and a series of pots, two of glass, one garnished with gold and valued at 70s., the other with silver and worth a mere 6s. 8d., and three of beryl, one of which was quite small, and one with a knop like a hawthorn, possibly a link to its contents.[161]

The use of olfactory waters as perfumes as much as medicines is probably to be understood from the small – and elaborate – bottles that were used for such items as rose-water. The inventory of Richard II's goods, taken after his deposition, lists two small, silver gilt bottles for rose-water, together weighing 23 oz. and appraised at 67s. 6d.[162] When Margaret, Duchess of Clarence, travelled to Normandy in November 1419, she took with her two gardeviands for holding rose-water and other things. There were three large phials of rose-water and twelve small ones, as well as a range of sweet powders that are likely to have been for cosmetic use.[163] Olfactory waters, particularly rose-water, were at this stage – and probably for much of the first half of the fifteenth century – high-quality imports from the Mediterranean, possibly even from the Middle East.[164]

Among the vain refinements reported of a duchess, in the fifteenth-century English translation of *The alphabet of tales*, was that she would not wash with ordinary water like others, but insisted that her ladies gather the dew on summer mornings to wash her.[165] Later in the century, that water might have been perfumed in other ways, as there is much greater evidence for the use of scented materials and in much greater quantities, the product of home-grown stills. When Elizabeth of York washed the feet of the poor at Richmond on Maundy Thursday, 1502, there is a possibility that the water was scented as well as hot, and even if it was not, there were flowers.[166] By the sixteenth century an English aristocrat, like his continental peers of the preceding century, would have expected nothing less. In 1511–12, the Earl of Northumberland had instructions given for a range of waters that were to be distilled for him each year. The still was to produce 120 bottles of water, each containing half a gallon – in other words, the Earl would have used a little over a gallon of scented water a week. He also expected an extensive range of waters: not just rose-water, but more than twenty-five other varieties.[167] In 1517–18, Edward Stafford, third Duke of Buckingham, twice had new stoppers – one, with a screw thread – made for a silver bottle for rose-water.[168]

An analysis of some of the earliest printed editions of treatises on perfume manufacture, mainly French and Italian, has shown upwards of fifty ingredients in perfumes, although only about ten were used regularly. Musk was the most often used, perhaps in half the recipes, followed by rose-water, aloes, dried red roses, ladanum, ambergris, civet and lavender. Those odours most often present – rose-water, water of orange flowers – were used as a base for others.[169] The use of manufactured perfumes travelled north in the fifteenth century. Inventories from fifteenth-century Dijon, one of the centres of the Burgundian court, contain few mentions of beauty products. Apart from an apothecary, however, there were five houses that had ovens or stills for making rose-water: one belonged to the Duke of Burgundy's cook, and two others were among the professional equipment of barbers.[170] Rose-water as part of the regimen for shaving had arrived in England by the sixteenth century. The barbers' coffer listed in the inventory of Henry VIII's goods made on his death contained two silver gilt bottles for rose-water, along with a casting bottle enamelled with roses, either the dynastic emblem or an indication of its likely contents.[171] Evidence for shaving arrangements from monasteries is much plainer:[172] at the Augustinian house of Barnwell, c. 1295–6, the chamberlain was required to supply hot water, as well as soap, for washing the heads of the canons and for bathing, if necessary; it was still an occupation in which some care, if not pleasure, might be taken.[173]

If moralists held that perfuming the person could lead to sin and was worthy of denunciation, it is perhaps reassuring that some individuals did intend odour to be an aphrodisiac or an encouragement in that direction. A group of recipes, circulating in England from the late thirteenth century with the *De ornatu mulierum* text of the Trotula group, advised a woman on spices to employ when she was intending to sleep with a man, as well as those for washing genitalia and armpits.[174] Gluttony, in the *Castle of Perseverance*, advises men to consume 'spycys of goode odoure', which will make them sexually attractive to women.[175] Camphor, on the other hand, had a completely debilitating effect: 'camphora per nares castrat odore mares' (smelling camphor castrates men) was the tag debated in Salernitan literature.[176]

Odour might be employed not only on the person, but on clothes, bedding and more widely within the domestic or other environment. Arrangements for laundry in both town and country show that a considerable investment was made in cleaning clothing and other fabrics. London was not alone in having a special jetty on the river for the laundresses, 'La Lavenderebrigge'.[177] At Eaton Socon, in 1271, there was a washhouse on the banks of the River Ouse, adjacent to the ford.[178] Others washed clothing in ditches and streams.[179] Medieval preachers were attracted to the example of the laundress – leaving clothes to soak in lye, taking them out, beating them, rinsing them and then hanging them to dry – as a metaphor for the bitterness of sin or the pain of Hell, that might be washed away with the water of fresh tears.[180] Washing household linen and sheets, as well as linen undergarments and shirts, was frequently recorded in royal accounts.[181] More delicate work was performed on silks, and in pressing clothes and sleeves and generally cleaning the textiles of furnishings.[182] Bran might be employed to scour armour.[183]

While these cares for general presentation, cleanliness and maintenance would all have carried an odour, perfuming clothing in addition was probably practised throughout the later Middle Ages: it was considered a sin worthy of confession before the mid-thirteenth century.[184] On one level this might be achieved by wearing flowers: roses and other flowers were bought for Henry, son of Edward I, to wear at his father's coronation in 1274.[185] Perfuming of clothing and bed linen was commonplace in aristocratic and royal households. In part, this had a practical foundation: the smell of bitter things, such as laurel, cedar or cypress, was a defence against moths and their caterpillars.[186] In the household of Edward IV fumigations were made within the office of the wardrobe of the robes by the yeoman apothecary;[187] the grooms and pages of Edward IV's wardrobe of the robes and wardrobe of the beds were to gather sweet flowers, herbs and roots to make the goods in their charge 'brethe most holesomly and delectable'.[188] In Henry VIII's household ordinances of 1526, Anne Harris, the King's laundress, was to 'provide as much sweet powder, sweet herbes and other sweet things as shall be necessary to be occupied for the sweet keeping of the said stuff'.[189] The connotations of odour, however, were wider: scented clothing and bedding were both a mark of the distinction of the lord and pleasant to all around. Smell was an essential part of the test of the cleanliness of the lord's clothing for John Russell's chamberlain:

Than to youre sovereynes chambur walke ye in hast;
all þe cloþes of þe bed, them aside ye cast;
þe fethur bed ye bete / without hurt, so no feddurs ye wast,
Fustian and shetis clene by sight and sans ye tast.[190]

Among the goods of Henry V, listed for his executors in 1423, were twelve pillows with lavender;[191] and the duchess in the *Alphabet of tales* had a bed that was so 'redolent savurand with spice þat it was a mervayle to tell off'.[192] Perfuming clothes and wearing scented objects were in addition wise precautions adopted by medical practitioners, to counter the bad odours of disease.[193]

The smell of the wider living environment was controlled by a variety of measures aimed at cleanliness generally: sweeping, cleaning, dusting and washing, changing floor coverings of straw, grass and rushes, and probably the use of herbs. On a May afternoon in 1304, Edith, the wife of William Drake of Little Marcle, was engaged in cleaning the house, throwing out the dirt or dung (*fimum*) and chaff, and generally cleaning up, believing that her one-year-old son, John, was playing with other children – he had only been gone about the length of time it took to walk a mile – when it was reported to her that he had been found drowned in a nearby ditch.[194] Similar vignettes of domestic cleanliness in peasant houses occur in sermons and devotional literature. Cleaning was a never-ending task, with domestic fowl scratching at the floor, scattering straw and worse. One must confess like a woman cleaning her house: she takes a besom and sweeps together all the uncleanness of the household; she sprinkles the place with water, to keep down the dust; and when all the dirt is gathered, she throws it out of the door, with great violence.[195] According to *The doctrine of the hert*, 'Knowleche of þe synnes by þe mouth in confession is noþing ellis but puttyng out filthes of þe hous of our hert by þe dore of þe mouth with þe brome of þe tonge.' Floors might then be strewn again with straw or rushes, but one had to avoid hypocrisy: it was of no value to confess in the manner of a lazy servant who, when she should sweep out the house and put out the filth, 'sche castith grene rusches above and hide þe filth'.[196]

Similar routines were present in the monastery. Lanfranc's monastic constitutions required Benedictine houses, at the five principal feasts of the year – Christmas, Easter, Pentecost, the Assumption of the Virgin and the main, patronal feast of the house – to clean the monastic offices and the cloister, and to strew rushes on the floor.[197] At Barnwell, the customs of *c.* 1295–6 required the subsacrist to assist the sacrist in sweeping the church and keeping it clean. The fraterer was to sweep the refectory, providing baskets for leftover food. He was also to provide mats and to strew with rushes the floor of the refectory and those parts of the cloister adjacent to the refectory door. These he was to renew. In summer he was to cast about flowers, mint and fennel, to create a good odour, as well as to provide fly-swatters. He was also to keep clean the washing-place, providing the canons with clean water for their hands and mouths; and to make sure that the windows of the kitchen were clean, so that nothing would fall into the dishes set under them.[198]

The sermon for Easter in Mirk's *Festial* urged that men should clean out their souls at annual confession at this time, just like the insides of their houses – putting down green rushes and strewing the floor with herbs and flowers. This spring-clean came at the traditional end of winter, after which the great household – and, presumably, many lesser establishments, to whose owners the sermon was addressed – ceased fuel allowances for its members, the fire no longer smoking in the hall.[199] Fresh rushes were bought for major occasions: the household of William Wykeham renewed them on the two occasions it was visited by Richard II, in July and September 1393;[200] and the room offered to Peter for the Last Supper, in the Towneley plays, was 'fare strewed'.[201] The odour of herbs and flowers might do much more than mask the smell of uncleanliness. An incubus which had been harassing a woman from the diocese of Exeter was kept at bay, on the advice of Hugh of Lincoln, when she strewed St John's wort around her house, a herb the demon found revolting.[202]

Without attention, the state to which floors might be reduced, the rushes covered in worms, was reminiscent of the worst place of torment seen by the monk of Eynsham in his vision at the end of the twelfth century.[203] The dampness of bad air was believed to be responsible for the spontaneous generation of flies and spiders, anathema to cleanliness and a well-known hazard of dining areas.[204] The refectory at Notre-Dame de l'Espinar at Barjols was investigated by Gervase of Tilbury to test claims that no fly could stay there. He even went as far as collecting flies and placing them on milk, honey and fat in the building, but he was personally unable to disprove the claim.[205] It was not surprising, therefore, that when seeking a room at an inn it was prudent to establish whether it had been decently cleaned of fleas, lice, dust and dirt. There may be doubt whether reassurance was to be had in the confirmation, in a model Anglo-Norman conversation of the late fourteenth century, that the room was free of these hazards, given that there were many rats and mice, which the innkeeper, priding himself on his traps, ventured would be of no trouble.[206]

Smoke might be the source of infection and its overweening presence must, at least, have been unpleasant.[207] Along with changes in room size came changes in methods of heating, the control of smoke through chimneys and the use of smokeless ceramic tile stoves, a fashion that gained ground in upper-class circles in the late fifteenth and early sixteenth centuries.[208] This improvement in the environment changed the potential of flowers and aromatics to perfume rooms. The gardens attached to manor houses and palaces had long supplied flowers, but the effectiveness of their odour might now be improved. They were present right from the start of spring – Henry VII received a present of flowers in the week starting 10 February 1499[209] – and gifts continued through the summer: his queen was given a present of roses in June 1502.[210]

Burning aromatics to perfume rooms became domestic practice in England in the late fifteenth century, although it is possible that it happened earlier, given that it was commonplace in France and elsewhere. By the sixteenth century there was a range of ceramic fuming-pots, a down-market equivalent of the precious metal perfume-pans that found a place in early Tudor palaces.[211] In the late fifteenth-century romance, 'The squire of low degree', the King attempted to cheer his daughter out of mourning for her love, the (allegedly) dead squire, by conjuring a sumptuous environment for her chamber that would inspire all her senses:

Whan you are layde in bedde so softe,
A cage of golde shall hange alofte,
With longe peper fayre burnning,
And cloves that be swete smellyng,
Frankensence and olibanum [an aromatic resin]
That whan ye slepe the taste may come.[212]

Perfumes were available at feasts on the Continent, for example in the elaborate subtleties that appeared at tables. Olivier de la Marche, in his *Mémoires*, describes one at a feast at Lille in February 1454: a castle like Lusignan, which had two towers from which orange-water flowed into the moat.[213] Whether they were used in this way in England is not clear, but flowers and perfumes were associated with pageants and other performances, such as the formal entry into London made by Richard II in 1392, and the entries of Henry VI into Paris in 1431 and into London the following year.[214] At Norwich, processions for Corpus Christi are first recorded in

49 Fuming-pot with a pierced and decorated top, a green-glazed London redware vessel of the sixteenth century.

the late fifteenth century, and plays from 1527. The Grocers' Play, performed from this year until the middle years of the sixteenth century, typically on a pageant cart, had a griffin which was supplied each year with fumigations or perfumes. The play centred on the Creation and the expulsion of Adam and Eve from Paradise: much of the fruit and flowers purchased for the occasion would have represented the bounty of Eden, some explicitly acquired for the tree; but it is unfortunately not clear how the griffin featured.[215]

Fumigation of rooms had a medicinal aspect, not just to counter bad smells, but as a prophylactic and a control against plague. Smoke from an open fire controlled pests living in thatch, and the fumigation of rooms, particularly after the construction of chimneys, was necessary to provide this and a range of other benefits. The fumigants might include herbs, such as bay leaves.[216] It was probably for this reason that many of the fumigants were employed in royal palaces. Henry VII paid one John Baptist, 'maker of fumycacions', a reward of £5 in the early summer of 1498, with a further £2 to his brother in the week before Easter the following year.[217] In the 1540s, the citizens of Norwich burned perfumes whenever the Duke of Norfolk came to their Guildhall, almost certainly as a defence against any mishap that might arise from bad air.[218] Other fumigants of special virtue were made from ground stones. The odour of a fire burning powdered jet was believed to drive away serpents, to aid in difficult conditions of the womb and to undo witchcraft.[219] Smoke from magnetite, dispersed around the building, would cause the inhabitants to flee, fearing the collapse of the structure, and was thus useful to thieves.[220]

With smell, as with the other senses, it was a commonplace that moral or spiritual qualities might be attached to sensory perception, that it might detect the celestial and the diabolical as well as the terrestrial. The medieval use of perfume and smell played on these associations and dynamics: if the monarch or aristocracy smelled of cinnamon, of incense or other perfumes, odours that appeared in religious contexts, in images, or arose from paternosters, was this not an indication of divine favour or powers, powers which might pass through proximity or touch? If there was bad air, or an evil smell, might these characteristics not pass to whoever smelled it? Good smells overcame bad ones, and good smelling individuals might thus overcome evil and noisome ones. These connotations were readily extended throughout the natural and supernatural worlds.

At the same time, there was an evolution in the practical use of odour and perfume, with more sophisticated preparations coming into wider use. Drawing on continental practice, perfumes and olfactory waters were created for the aristocracy in the thirteenth century by their apothecaries or from the ingredients they supplied. In the fourteenth century, high-quality perfumes were imported, again for a very limited market; but during the following century they became much more widely used and in much greater quantities. Aromatic fumigants also featured to a greater extent at this period, expanding their use into the sixteenth century.

Inventories of apothecaries' shops from the sixteenth century show how widely available these products had become. Much less is known of the use of odour among the lower classes, although the traditional lore that applied to garden herbs and wild plants may have extended to a basic perfumery. Certainly herbs, many of which must have been widely available, were used indoors by all groups within society. The use of plants to counteract the noxious effects of bad air was widespread.

Chapter 8

VISION

A case for the pre-eminence of the sense of sight in Western culture is no modern innovation, but one that was well rooted in Antiquity and early Christian tradition. The association of the sense with the element of fire was proof that it was more worthy than the other senses: of its own strength – and unlike the other senses – it was able to perceive things at great distance. Isidore had argued that it was quicker and more vigorous than the other senses: it was closer to the brain and in speech expressions were used such as 'see how it sounds' or 'see how it tastes', indicating its involvement in the discrimination of other senses.[1] With this pedigree, there is more that could be said about sight than about all the other senses. In common with the previous chapters describing individual senses, this one encompasses medieval understanding of the physiology and neurological processes that were attached to the sense; it then addresses the values that were attached to the process of seeing, the 'visuality' of the culture. The latter centres on the perception of light, colour, signs and gesture, on the gaze, both individual and communal, and on sight and medita-tion. I have chosen to write little about dimension – size, shape and number – or the perception of movement, even though the kinaesthetic sense is now considered a special element in perception, and these were without doubt important for some aspects of medieval sensation.[2]

Contemporary knowledge of the physiology of the eye was summarised in the discussions of Bartholomew the Englishman and the fourteenth-century French surgeon Guy de Chauliac, whose work was available in English translation in the fifteenth century. The eyeballs were each composed of seven enveloping membranes (tunicles) that enclosed and separated one from another three or four humours, viscous liquids. The third of the membranes was the retina, connected to the brain by the optical nerve, along which passed the spirit of sight. The humours were described as crystalline, glassy and white, of which the crystalline was the most important and it was here, in the middle humour, that sight was formed. The crys-talline humour – the 'middle eye' or 'black of the eye' – was clear and bright so that it might change to different colours and take their likeness without interference; it might also distinguish the shape of all things. It derived from the parts of the brain which were the highest and clean, pure, thin and bright; it did not let light pass, showing images as in a mirror. The brain itself was white so that it might receive the likeness of colour.[3]

Light was necessary for the eyes to function and was, itself, a divine quality. Central to the theories of extramission and intromission was the belief that objects emitted light – rather than our notion that most objects are illuminated by the light of the sun, either directly or by reflection, or by artificial means – a point to which we shall return. Equally prominent in both theory and popular understanding was the notion that there was direct contact, effectively touch, between the seen and the seer. It was this that led to a wide variety of beliefs about the force of sight.

Closely bound to these theories was a further dimension: that sight had moral or spiritual consequences. On one level, sight might convey good or evil between the parties. The 'evil eye' might operate in this way, as well as the awesome powers attributed to the legendary basilisk or cockatrice, and other creatures. The *Gesta Romanorum* recorded how Alexander the Great, while besieging a city in Egypt, lost a considerable part of his army to the power of a cockatrice concealed in its walls: the men were killed 'thorowe the venyme that passithe from her syght'. He was advised to place a mirror between his army and the creature, to reflect its gaze back upon itself, killing it.[4] Although the basilisk might kill through its smile or sibilant breath, it was its sight that was especially feared.[5] In an echo of this power and its antidote, in July 1329 Edward III gave Philippa, his queen, a mirror in the shape of a basilisk, with a belly of shell, bearing an image of an armed esquire with a helmet on his head and spear in his hand.[6] Even carrying a picture of a basilisk might give one strength: two early thirteenth-century lapidaries described cameos, which included depictions of the beast, that gave one the power to eat the flesh of poisonous creatures or to defeat wild animals.[7]

Unlike beasts, whose eyes naturally turned to the ground, the eyes of men were set high in their heads so that they might look up to Heaven.[8] Aside from the appropriateness of this aspect of human physiology, it was a signal that sight was a form of perception by the soul. For some, following classical precedent, that was its primary purpose.[9] Augustine had divided sight, beyond the corporeal, between a form of vision that enabled one to see in dreams or the imagination and a form of vision within the mind that allowed one to perceive divine truths.[10] The suggestion that sight was a form of meditation was not out of the ordinary to twelfth-century philosophers.[11] That sight might penetrate beyond the visible was a notion that had been widely current among the Christians of late Antiquity: the eye of faith enabled one to see beyond a place or object to the eternal realities that were associated with it biblically or hagiographically, which then became physically present.[12] To some fourteenth-century theologians, to look upon the pope, the vicar of Christ, was physically to look upon Christ himself.[13] These transcendent powers could also be invested in individuals or conveyed to their advantage in particular ways. St Mary Magdalen – the first person to see the risen Christ – was accorded special powers of vision which might assist those of the devout. In this way, in 1144, Legarda, a virtuous widow and nun living in Norwich close to the church of St Mary Magdalen, was among those who saw a bright light in the sky. The following day, that divine light shone into her mind, guiding her to the body of St William.[14]

The eyes were the tokens of the soul: they were the sense organ closest to it and the eyes showed its trouble and delight.[15] Not only did the sense of sight enable individuals to see the physical and the spiritual, but looking at the eye allowed one to see the state and character of the individual: this sense not only received images, real and spiritual, and other virtues, but also gave out information about the individual and projected his or her virtues. In an early twelfth-century account of her miraculous recovery from leprosy, the physical transformation of the eyes of a girl called Morvid, from Wyke near Much Wenlock, relieved them of their mistiness and brought to them a new brightness, a reflection of her new spiritual state.[16] Works of physiognomy paid special attention to the eyes, the 'most evydent schewerrys ['showers' or indicators] off man and womannys dysposycion', and what their physical characteristics might indicate. Eyes that were clean, bright and shining might indicate liberality and kindness, especially if this tendency was confirmed by other features in the face; small eyes, like those of apes and foxes, might augur deceit; the red-eyed were likely to be lecherous and untruthful; those with squints were malicious and wicked; and those with moist eyes were much more virtuous than those whose eyes were dry.[17]

If it might perceive the ineffable, however, the acuity of man's sight of the corporeal was more limited. A long tradition recognised the higher abilities of the eagle and lynx; and at Longthorpe the sense was represented by the cockerel.[18] Nowhere was this difference more apparent than with night vision. Darkness brought obscurity and uncertainty to human sight, a transition that was defined by twilight, commonly known as 'between dog and wolf' (*inter canem et lupum*).[19] Night – *nox* – was so called, according to Isidore, because it was harmful to the eyes (*eo quod oculis noceat*).[20] The creatures of the night had special characteristics, sometimes described as 'hating light'.[21] This brought to them associations with evil, frequently reinforced by their dark coloration and their eerie cries; their activities were often considered to focus on disreputable or foul matter and their attention to be of ill omen.[22] Walter Map numbered here the light-shirking owls, vultures and night-crows (*nycticorax* or *noctua*), the last probably also owls, whose call was a portent of evil. Map noted too their sense of smell and interest in carrion.[23] A thirteenth-century compilation of questions about the natural world, that was among the medical collections of Peterhouse, Cambridge by 1418, argued that creatures that could see at night could not see in daylight.[24] The nocturnal activities of the cat, a

50 Longthorpe Tower:
the cockerel.

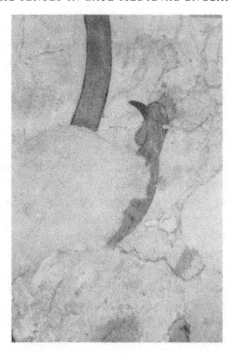

creature often associated with the Devil, were made possible by the light that shone
from its eyes.[25]

Darkness was a phenomenon that was shared only by night and Hell. The
remainder of the universe was constantly illuminated by the sun[26] or by the divine
ray of God.[27] What humans saw was light – but what was to be understood by light?
Light, from its creation at the start of Genesis, had a significance that touched everything on earth, an element of the divine. Late medieval authorities might debate
whether light was a material substance or an 'accident', a quality attached to material substances and indicating part of their nature or form, such as colour, but they
were not in any doubt that it provided not only physical illumination but spiritual
enlightenment. The study of optics had the potential to reveal divine truth as well
as the processes of cognition. Discussions about light were therefore an essential
component of the work of philosophers and theologians, notably in the thirteenth
and fourteenth centuries.[28] To be touched by light, like any substance of virtue,
conveyed a benefit, and unlike the potential corruptibility of many virtues, nothing
could defile light.[29]

While we believe that objects that shine reflect light, medieval belief was that
these objects were themselves a source of light, and that, because of the divine
quality of light, made them objects of virtue in their own right. Here lay the origin
of the medieval interest in lustre, often held to be of more significance than the hue
of colour. Particular regard was paid to jewels, stones and precious metals. The

amethyst, according to Bartholomew, cast out gleaming beams of light.[30] The quality of emitting light and the association between light and truth made gemstones, gold and silver fortunate components of reliquaries and shrines.[31] This further manifested itself in many ways, from the investment in the *retonsio* or second, close shearing of high-quality cloth, which gave it a shiny appearance,[32] the use of calenders or slickstones to a similar end,[33] to the poetic interest in the gleam of armour and weapons.[34]

At the same time, there was a general interest in optical phenomena. Mirrors were present in most great households[35] and the study of reflection held a wide fascination. Philosophers discussing sensory error had a close interest in the reflecting surfaces of different planes.[36] A poem on the art of love, dating from the late thirteenth or early fourteenth century, included among the weapons of the enemies of love a siege engine, representing Jealousy, that was armed with a giant mirror.[37] Those without mirrors gazed at their reflections in water, with wells proving a disastrous attraction to small children seeking their own image.[38] A late fourteenth-century Wycliffite sermon on the words of St James, urging men to be doers of the word rather than considering the face of their youth in the mirror, had an audience that could follow the metaphor into the observation of optical effects and the three modes of perspectivist vision. There were three kinds of sight: direct sight; sight reflected back immediately, as in a mirror; and sight that was 'reflexud', seen indirectly, as one might see the moon as a penny in a dish of water – and their exegesis placed an individual differently in terms of spiritual benefit.[39]

A blinding, bright light was a manifestation of the divine. The midwives sought by Joseph in the N-town play of the Nativity, arriving tardily, after the birth of Christ, were afraid to enter the stable on account of the light, brighter than the sun by day or the moon by night.[40] A popular tale, recorded in Mirk's *Festial*, of the painter sent by King Abagarus to paint the picture of Christ, noted that he could not do so directly: 'But when þys paynter lokyt on Christ, hys vysage schon so bryght þat he mayght noþyng se of hym.' The image was captured by Christ placing a cloth on his face, which the painter was then able to copy and bring back to the King.[41] No mortal might look upon the divine: that was reserved for the elect in Heaven. According to a late fourteenth-century sermon on the opening chapter of St John's Gospel, 'God no man saw evere.'[42] The blinding light was reflected in illustration and costume: medieval illuminators commonly gilded or silvered the face and hands of God and other heavenly beings, a practice also found in the costumes, especially the masks, used in medieval drama.[43]

Many other holy objects radiated light, or light was present to radiate holiness. An early fifteenth-century Lollard sermon on the the book of life in the Apocalypse noted that it was never to be closed up: the book itself was so bright that no lantern or lamp was ever needed to read it.[44] The use of lights in religious contexts was extensive and conveyed especial significance and benefit to all those who saw them.

Ecclesiastical statutes and monastic regulations paid careful attention to lighting arrangements. Candles or cressets (oil lamps, burning in mortars) accompanied the sacrament at all times, in church or when it was taken to the sick or dying, the statutes, probably of Bishop Grosseteste, of 1239, for the diocese of Lincoln, citing the Book of Wisdom as witness to the practice, 'for it had the radiance of eternal light'.[45] The synodal statutes of Bishop William Raleigh of Winchester, c. 1247, were not unusual in ordering each parish church to have two large wax candles, which would be held to the right and left of the priest by two clerks at the corners of the altar at a given part of the mass ('Te igitur'), until the priest had taken communion.[46] The lights were commonly extinguished on Maundy Thursday, with the exception of a single one on the altar, in anticipation of the extinguishing of the Light of the World on the cross on Good Friday.[47] The sole light left burning in front of the altar at Wellingore, a village south of Lincoln, on Good Friday 1336, fell into the straw on the floor and was the cause of a fire that night, destroying the goods around the altar.[48] The light of divine presence was an essential part of rites of excommunication[49] and exorcism.[50]

Rather than a direct liturgical purpose, some uses probably had a strong element of custom about them; but to all the candle brought the benefit of light. At the tomb of St Swithun at Winchester, in the eleventh century, those seeking cures attended with lighted candles in hand.[51] In the diocese of Salisbury in the mid-thirteenth century, women, on their marriage and nursing after birth, were to come to church holding lighted candles.[52] *Jacob's Well* explained the practice of putting a lighted candle in the hand of an infant at baptism as signifying that the child should give light, by his own good example in good works.[53] Another practice placed a lighted candle in the hand of the dying, as they breathed their last.[54] Setting lighted candles around a corpse maintained this protection: the Benedictine monk in late medieval England was to have a lighted candle at his head and at his foot all the time his body rested in church, until it was taken for burial.[55] Lights burning in church might also create an effect far beyond those who saw them. Henry VII, as well as maintaining a priest at Walsingham to sing before the image of the Virgin, had a substantial candle burning there for his benefit: in July 1505 a wax candle weighing 52 lb. was purchased to this end.[56]

Among the many categories of miracle, the divine rekindling of candles was prominent.[57] In the 1210s, a widow, at the tomb of Bishop Osmund, reported seeing the lights around his sepulchre go out and then re-ignite, and that there was no one near who could have relit them.[58] Sacred qualities of lights were not to be abused: candles that had been blessed at Candlemas (2 February, the Feast of the Purification) were not to be used for non-religious purposes, unless they had first been melted down.[59] There was also, at the end of the fourteenth century, a belief that it was dangerous to look directly at fire or light after an encounter with a ghost, perhaps to be explained in terms of the competing strengths of the moral forces that were attached to each of them.[60]

The division into light and dark, moral as well as physical, had an impact on day-to-day life. Distinctions between activities that might take place during the day and those of the night had similar resonances. The movement of the sun, the sound of the canonical hours from church bells and, in some instances, sundials, allowed the hours of daylight to be broken up in a way that was not possible at night. Before the widespread advent of mechanical clocks, and outside monastic establishments where other means of time-keeping, such as water clocks, were employed, there was little possibility for many to divide the night in a way that was easily recognisable. Even the advance of evening was defined by reference to events such as 'the hour at which hens climb onto their perches to rest for the night'.[61] A late fourteenth-century sermon divided the night into just four 'hours', or times – evening, midnight (that is, a time when it was completely dark), cock-crow and morning – in a way that did little to distinguish anything in the hours of darkness.[62] Since many of Christ's miracles enabled the blind to see, the second mass of Christmas Day, said at dawn, commemorated the arrival of divine light in its first words, 'Light shall shine this day upon us' (*Lux fulgebit hodie super nos*). Christ's birth 'lyghtned mony þat before loket evell', that is, without sight, or in darkness.[63] The night was for sleeping, as Maurice de Wigewale believed; he and his wife, who had gone with him to Canterbury to seek the assistance of St Thomas Becket for a deformity of her face, postponed their return journey until the hours of daylight.[64]

Any nocturnal activity was the cause of suspicion. Statutes for the diocese of Worcester, of 1229, enjoined that clerks in holy orders should not wander about outside at night unless they had to visit the sick.[65] Bishop Grosseteste's injunctions of 1239 for the see of Lincoln ordered that at night the ornaments and sacred vessels of the church should be in safe and honest custody, and should not be in the houses of the laity or in their custody unless there was absolute need.[66] Among the presentments of the churchwardens of Tydd St Giles, at a session of the local eccle-siastical court in the early 1460s, two centred on night-time activities. Thomas Pantre and his wife Etheldreda were alleged to have kept a tavern to which came certain people believed to be thieves and at a suspicious time, namely at night. A second case, one of defamation, alleged that Simon Odam had come at night and taken fish from a pond.[67] It was exceptional for messengers to work at night: wardrobe and household accounts usually make special mention of the occurrence, or of the guides that might be required to escort prominent individuals at that time.[68] Other nocturnal activities encompassed the unpleasant, such as emptying latrines,[69] and the prudent. Inhabitants of Rochester were paid to keep watch at night over the possessions of Queen Joan of Navarre there in December 1419;[70] and many great households maintained either a single watchman, or a group, as did municipal authorities around the country.[71] Funerals, or vigils for the dead, sometimes took place at night.[72]

Artificial light, fires, the illumination from windows and other openings, were necessary to mitigate gloom and darkness. Access to these, however, was not distributed evenly through society: one of the marks of social distinction was access to levels or types of light or heat that were unavailable to others. In the great household, there was strict regulation of who might have light and heat and at what season. In the ordinances for the household of George, Duke of Clarence, of December 1468, wood and candles were issued only between 1 November and Good Friday, at the rate of of two shides of wood and three white (tallow) lights for every two gentlemen in the household.[73] The groom porter, who, under the direction of the usher of the chamber, was to supply wood, tallow and wax lights for the Duke's own chamber, was not to supply torches or *torteyes* (another form of candle) without direction from the head officers or ushers of the household – and each day before noon he was to bring the torches and *torteyes* to the chandlery to be weighed, as a further control on their use.[74] Ninth of the ten elements essential for a successful supper, described by Bartholomew the Englishman, was that there should be plenty of light from candles, prickets and torches, 'for it is a schame to soupe in derknes and perilous also for flies and oþir filthe'.[75]

There were few places where light was allowed to burn continuously throughout the night. One, perhaps surprisingly, was in the stables of the noble household, where the great horses were kept. In October 1312, tallow was bought for burning in the cressets in the stables at Eltham where the palfreys and dexters of Queen Isabella were housed.[76] Lights were also kept burning in the mews accommodating hawks and falcons.[77] In the monastic dormitory, it was the practice to maintain a light throughout the night. At Barnwell, the customs forbade the canons to sit by the lamp in the dormitory, to sing there or to read, neither were they to take candles in order to read in bed (a prohibition designed as much to produce silence – for reading was not usually silent – as it was to restrict the expenditure on lights).[78] This priory was comparatively well provided for when it came to light. At the evening meal, between 1 November and 2 February, there were to be wax candles, with one candle for up to three diners.[79] In the infirmary, according to the exigencies of the season and the needs of the sick, there was to be each night a fire, candles, cressets or lanterns.[80] Wax candles gave a steadier light than tallow or 'paris' candles, as well as a better smell, although there were some who argued that the fumes from wax were more harmful to the eyes.[81]

Daylight from windows was necessary both for reading and for fine work. Statutes for the diocese of Worcester, from 1229, required the windows in the choir of the parish church to be glazed; in 1287, in the diocese of Exeter, the provision extended to both the nave and the chancel.[82] Light might have a particular significance in an ecclesiastical context, but it was also a practical necessity. In 1340, a window in the chancel of the church at Glentham was adjudged so dark that it was impossible to read by its light and it was the responsibility of the chapter of Lincoln to remake it.[83] In 1330–1, the tailors of Queen Philippa had two additional windows made in their room in the wardrobe of the robes in London so that they had sufficient light for making the counterpanes for the Queen's bed.[84] At the other end of the scale, the expense and rarity of animal fats, particularly in the period up to the Black Death, restricted the artificial light available to the peasantry. After the fire had been covered, at curfew, candles were used to light the way around the house; they were rarely left burning overnight and in those cases in coroners' records where fire was caused by candles that had burned down and fallen, negligence was often implicit in leaving candles alight at all.[85] Cressets, or other forms of light that burned liquid fat or oil, were rarely mentioned and their appearance in the archaeological record is also uncommon in these contexts.[86]

If light was necessary for sight, colour was an essential component of all that was seen. In medieval scientific, philosophical and theological debate, much turned on whether colour was an integral part of light, or whether it was a material substance of its own. The former implied it had the same characteristics as light, that is, it partook of the divine, and that its presence in jewels, illuminated manuscripts, stained glass windows, vestments and so on contributed to the worship and understanding of God. If it were a material substance, a physical covering, it was not divine and, consequently, it did not have the ultimate significance of light and was mere artifice. In theological terms, this meant that it might have no legitimate place in worship and the obscurity it produced, hindering light, tinged it with immorality. This debate divided the Church during the Middle Ages, with some, such as Augustine, favouring the argument that colour was part of light, while others, like Bernard of Clairvaux, rejected the position this gave colour, and it was to have little place in Cistercian churches.[87] Another Cistercian, Walter Daniel, writing the *Apologia* for his life of Aelred of Rievaulx, noted that virtue was like light: its qualities might reach out even to the eyes of the sleeping. Iniquity (*crimen*), on the other hand, was clothed in the colour of vice, just as the image of darkness cannot be seen easily.[88] On the other hand, the description of the heavenly Jerusalem (Revelation 21: 9–27), with the foundations of its walls decorated with jewels, constituted a focus for the use and interpretation of colour in ecclesiastical contexts, as did the descriptions of the twelve jewels of the Tabernacle and the four colours of the garments worn by the High Priest of the Temple.[89]

In looking at colour in the medieval period, it is necessary to abandon the preconceptions we have, from Newtonian optics, of colour mixing and sequences. The range of colours identified in medieval theoretical texts, the terms used in literary texts or in everyday life to describe objects and art, are familiar only in superficial ways. We may have before us a fullly chromatic palate, with a scientific analysis of and names for hundreds of colours, for which there was no medieval vocabulary, but that does not mean that individuals were incapable of perceiving this range, distinguishing between them or creating them.[90] The differences are in ways of seeing. At the same time, there was a disjuncture between medieval theories of colour and the way in which colours might be produced in technical processes.

Medieval theories of colour were grounded on those of Antiquity. The purely theoretical analyses of Aristotle, identifying as basic to nature a scale of seven colours – black and white, with five intervening shades, governed by the amount of the two extremes that featured in them – had a lasting influence, but the reception of his work, through translations into Arabic and Latin, sowed considerable confusion. In the thirteenth century, Roger Bacon recognised five essential colours, with greyish-blue (*glaucitas*), red and green interposed between the extremes of white and black, red falling halfway between these poles. His teacher, Robert Grosseteste, identified no fewer than sixteen essential colours, with seven closer to white and seven closer to black.[91] Bartholomew the Englishman reviewed both the Aristotelian analysis, with white and black at the extremes, interposing yellow, citrine, red, purple and green, and the categorisation created by those who believed that light was the substance of colour. The latter produced sixteen colours, with seven ranged between white and black in a bright light, and seven more, from black to white in a dim light.[92]

Bartholomew's description of individual colours relied on the system of elements and humours: colour was created by their interaction with light.[93] Cold was responsible for creating whiteness and paleness; heat, blackness and redness.[94] Dryness and moistness were also significant. In this way, green was created by the operation of heat on matter in which moistness was predominant, as in leaves, fruit and grass. The shifting balance of the elements explained why these items changed colour as fruit ripened, or autumn approached.[95] Crucially, colour provided information about the nature of the object. To us, the argument may appear circular – that cold created white and white objects were therefore cold – but it was commonly understood as indicating the make-up of the item. Further, colour was a token of the accidents, or emotions, and passions of the soul: paleness might be generated by fear, redness by shame or anger.[96] These connotations were understood for some colours at a popular level. 'Green' was employed as an indication of fear or jealousy in the early fourteenth century,[97] although when, in 1427, a text of *Piers Plowman*, unique for its illustrations, depicted the Vices, Envy had no substantive colour. Wrath was dressed in blue and red parti-coloured clothing, and the robes of Pride were similarly coloured.[98] These examples suggest that associations between emotions and colour were not definitive: some connections, which are now commonly accepted, such as 'livid', or 'lividly', for furiously angry, pale with rage, are late, not occurring before the nineteenth century.[99]

A further colour system evolved in the medical school of Salerno in the eleventh and twelfth centuries. This had white and black at the extremes, but rather than placing red at the mid-point between them, this position was ceded to green; red was placed between green and white, and blue between green and black. The reason for the position of green was that it was the most pleasing of colours.[100] Texts containing the principles of Salernitan colour theory were available to Englishmen by 1200. Their impact, however, was varied and there is less evidence of reference to them than there is on the Continent: there is no parallel for the remarkable horse given to Enide in the first great romance of Chrétien de Troyes, which was black on one side, white on another, with a green stripe in between – a match for the well-balanced love she held for Erec.[101]

What were the colours represented by these terms? There are two separate issues here, one to do with the perception of colour, the other to do with vocabulary. Medieval colour perception had very different biases to our own. Description of colour is now based on three elements – hue (the name of the colour, 'green'), brightness or value (reflecting the amount of black or white added to it) and saturation (its purity) – and further factors such as lustre, texture, transparency or opacity. We place much emphasis on hue, but medieval analysis focused on the relative brightness of the colour, or the strength of its saturation, and was more vague about hue. There was a closer association between a bright red and a bright blue than between a pale red and a bright red.[102] As noted earlier, there was a particular interest in lustre, or shine.

Further dimensions to this question can be seen from medieval vocabulary. Individual terms for hues covered a broad range – and one should not assume that the boundaries between terms equate to those of modern colour hues. The vocabulary of Anglo-Saxon England shows that it turned essentially on brightness rather than hue, although not failing to distinguish hue.[103] In Anglo-Saxon, *read* encompassed some yellow hues as well as some reds. At other points, the Latin *rubeus* might cover hues that we would now identify as falling between yellow and purple; and the popular, late medieval *perse*, anything from dark blue-black to turquoise or even crimson.[104] The late thirteenth-century *Tretiz* of Walter de Bibbesworth, designed to provide the rudiments of French vocabulary for estate owners who only knew English, offered a sole English word – *reed* – to translate no fewer than five words in Anglo-Norman, *rous, sor, goules, rouge* and *vermaille*. Bibbesworth's contention was that Anglo-Norman was the richer language in colour terms, that there were important distinctions to be made. Indeed, by the end of the Middle Ages, English did have lexical equivalents for the Anglo-Norman reds.[105] There were, however, some important deficits. In literary texts, there was a comparative paucity of colour terms describing hue, particularly in the twelfth and thirteenth centuries.[106] In French literature of this period texts are commonly confined in their vocabulary to no more than six or seven basic colours: white, black, yellow, green, red and blue, with some use of brown as an analogue for black.[107] This pattern corresponds in some ways to the more general development of basic colour terms observed across a wide range of languages, with an orderly expansion of terms. It has been argued that if a language contains terms for seven basic colour hues, they will be black, white, red, green, yellow, blue and brown. In contemporary languages, the basic terms then commonly expand to eleven to include words for purple, pink, orange and grey.[108]

In historical terms the picture is more blurred, probably because it is closely bound to a transition from a way of seeing that emphasised tonality and colour saturation, to one in which description based on hue was to predominate. The pattern of acquisition of colour terms was affected by context and it almost certainly expanded in this way. It was richer in its breadth, supported by specialist vocabularies that had grown up for trees, plants, animals, furs and textiles, and by words of attribution – 'coloured like' – such as 'sanguine', for a blood red or purple. The specialist vocabularies supplied words that then passed into more common usage. Although the distinction of purple was known from the Anglo-Saxon period as a term for the garments of the Roman emperors, it was not in current usage to mean the colour that we now associate with that hue until probably the fifteenth century, when it is used to describe textiles of this colour, particularly those to be worn by the royal family. The colour of the fruit of the mulberry – murrey – was close to the modern hue of purple and in use from the late fourteenth century as a colour term, as was sanguine. In the late fifteenth and early sixteenth centuries, there were notable additions to words in the red area of the spectrum: 'orange'; 'rose', for a pale crimson colour; and 'carnation' or 'incarnation' for a flesh colour of a slightly darker hue (the separate designation of pink, derived from the colour of the flower, was a development of the late seventeenth century). 'Tawny' had been used from the late fourteenth century to describe a mottled cloth of a yellow, orange and brown colour. The orange – the citrus fruit – was extremely rare in England before the end of the fifteenth century. By the 1510s, however, the word was in use to designate the colour. In June 1518, the Duke of Buckingham's wardrobe bought from a Bristol merchant two ounces of 'orenge colour silk', with a further ounce from a cap-maker; and the new colour is found alongside tawny in the wardrobe of Henry VIII. 'Grey' was present in Anglo-Saxon in the sense of an indeterminate colour between white and black; by the thirteenth century, it was employed for the colour of undyed, woollen fabrics, but it acquired richer and more positive connotations from the specialist vocabulary that was employed to describe the fur from Baltic squirrels (*gris*).[109]

At the same time there evolved a series of much more specialised vocabularies for colour. These are particularly in evidence where there was a need to distinguish between closely related materials, for example textiles, or animals. The four essential things to know about a horse were its shape, beauty, temperament and colour, the last a crucial determinant of the horse's character. According to Isidore of Seville, the white horse was suffused with pure light, and a bay was stronger than others. He then described the equine merits of eleven further colours.[110] There are long lists of horses in thirteenth- and fourteenth-century royal wardrobe accounts, especially those that encompass the replacement of animals lost on service. Each is described carefully, distinguished by colour, and occasionally by name. The eighteen mounts that replaced the losses of Henry Burghersh, Bishop of Lincoln, and his retinue in the service of Edward III overseas in 1338–40 included a black horse with a star on its forehead and two white forefeet; a bay with one white rear foot; a dun with a white nose and one each of its front and back feet white; a white horse; a clear bay with a white forehead; a morel (dark brown); a dapple grey; and a further morel with a star.[111] Similar principles applied to the description of livestock. Nicholas, the son of John the Fisherman, giving evidence in 1307 about his own, miraculous resuscitation after falling into the River Wye, recounted how he was deputed to look after one of his father's cows. It was was red, with white on its head, and a creature that was less than docile.[112]

Medieval taste favoured vivid, bright colours. How were they produced? The division in observation of colour, between the theoretical analysis of light and the day-to-day perception of individuals, was repeated when it came to creating and mixing colour. In principle, according to Aristotle, all colour was a mixture of light and

51 An artist, with dishes of colours. From the *Omne bonum*, late 1330s and 1340s.

dark. From this and the primary colours he had identified, mixtures might be generated in three ways: first, like an Impressionist painting or photogravure printing, with small spots of colour positioned side by side, mixed by the eye; secondly, by superimposing one colour on another; and thirdly, by blending.[113] In practice, creating or mixing colour was akin to alchemy: it changed the nature of things, of which colour was an important indication. The way in which colour was created, therefore, was different to the way in which we might achieve chromaticism: the mixing of blue and yellow to create green, for example, was a procedure that was not expected – and not practised – in the medieval analysis of colour. On most occasions, pigments were employed that did not require colours to be mixed: they already possessed the desired colour, although a chemical or other process might be required to create them.[114]

The vegetables dyes and minerals used for the principal colours were comparatively well known, although the processes by which they might be employed required in some cases a good deal of attention and, therefore, expense. Investment in materials was also substantial. About 1412, John Swift, a dyer of Salisbury, was working with woad and wood ash to dye both woollen and linen cloth in the same vat, when suddenly the process went awry, changing colour in an unexpected way (*extra naturam suam*). Believing he faced ruin, John prayed to God and St Osmund: within two hours, all had changed back to the correct colour, with a better result than was normally achieved, and he went on to offer a vat, made of wax, at Osmund's tomb in commemoration of the event.[115] *The doctrine of the hert* noted that cloth that was dyed in the wool never lost its colour, 'but þe cloth þat is died in cloth, it wille oft tymes chaunge colour'.[116] This distinction reflected the properties of the different dyestuffs. Madder, brazilwood and kermes, or 'grain', an insect-based dye, produced a range of reds and some browns suitable for dyeing. The use of these reds required the addition of a mordant, such as alum, and careful monitoring of the dyeing processes – the colour depended on the pH of the solution – and they were generally suitable only for dyeing pieces of cloth, not in the wool. Woad and indigo were employed as blue dyestuffs, requiring no mordant, but the addition of wood ash or potash. The blue dyes again needed careful supervision, but could be employed directly on wool. It was then possible, after the cloth had been woven, to re-dye the blues with the red dyes or weld, which produced a yellow dye, to create a range of greens and browns.[117]

Different pigments were used for painting and manuscript illumination.[118] Bright reds could be produced from vermilion (sulphide of mercury);[119] minerals, such as ultramarine (from lapis lazuli) or azurite (a copper carbonate), were employed for blues. But, as with dyeing, most work was done by employing a single colour pigment, rather than mixing colours on a palate. For green, most colourists, whatever the medium of their work, would have started with a green pigment, although changes to colour were created, particularly in wall paintings, by under-

52 Dyers at work: from what was probably Edward IV's copy of Bartholomew the Englishman's encyclopaedia. The presence of both red and blue cloth in the same workshop was exceptional in the Low Countries, where this volume was produced, as dyers were usually licensed to work with only one of the two colours. The dyeing vat has its own furnace at the base.

painting in one colour, then overpainting or adding detail in a second. This technique might be used to create shadow – an underpainting in green earth, for example, was used for shadow on flesh tones in some instances in twelfth-century wall paintings.[120] A late twelfth-century treatise on colours, from the Cistercian monastery of Rufford, confirms the way of working with a principal pigment, but conceded that some made admixtures. The greens here might be produced from a green powder, possibly verdigris; it was recommended not to use saffron to brighten it, although this happened elsewhere. Aside from mineral pigments, green could be produced from vegetable dyes, from the berries of buckthorn.[121]

Colours, both synthetic, created at vast expense or comparatively cheaply, and those of the natural world, had, aside from links to specific emotions already noted, symbolic connotations that are not now easy to recover. Where that is possible, we can see that there was a good deal of variation in their meaning.[122] The association of white with goodness and purity, and black with misfortune, was of long standing, found in Antiquity.[123] To these, patristic theology and the Bible added further dimensions. The Hebrew, Aramaic and Greek texts of the Bible had used colour terms with comparative rarity and many of the colour words that we now associate with these works were first introduced in Jerome's Latin translation.[124] It was here that many words to do with qualities of light or dark were transformed into colours, or into words that came to express colour. The Hebrew for 'shining' was often translated into Latin as *candidus*, a shining white, the opposite in Latin of *niger*, a shining black, both to be distinguished in Classical Latin from *albus* and *ater*, white and black in a form without lustre, although the Vulgate never uses the last, only *niger*. Once words had an ethical colour, moral opprobrium or sanctity was a small step away from chromaticism. By the end of the sixth century, black skin was an indication of evil, even if, for St Jerome, some two hundred years earlier, it had only indicated a state before becoming a Christian, before baptism.[125] The black man or 'Ethiopian' had become synonymous with the Devil and was to remain so throughout the Middle Ages, although sometimes any dark colour, however improbable to a modern mind, sufficed to delineate evil. In the *Vitae patrum* there is a tale in which Abbot Macarius watched devils running through a church, 'small, black, Ethiopian boys'; by the fifteenth century the story had mutated, the author of *Jacob's Well* referring to 'dyverse feendys smale as chylderyn, blewe [blue] as men of Inde'.[126]

In contrast, the bodies and clothes of saints and celestial beings were white, often a shining or radiant white. Hagiographic tradition ensured continuity of this trope. According to Walter Daniel, when the body of Aelred of Rievaulx was laid out, it shone more than the white of snow; his flesh was radiant and purer than glass. These features had as their model the life of St Martin of Tours (d. 397) by Martin's friend, Sulpicius Severus.[127] Some thirty years later, another hagiographer, Adam of Eynsham, also made the comparison with St Martin: when the body of Hugh of Avalon had been washed by Adam, its brilliance was scarcely credible; it shone with a brighter white than milk, his internal organs more pure than glass.[128] A blind girl, cured at the tomb of St William of York, had a vision of a man dressed in clothing so white that it made snow seem black[129] – and manifold descriptions of saints in visions or hagiographic literature have them similarly clothed, shining and radiant.[130] In the same way, angels were clothed in white.[131] A late fourteenth-century Easter sermon described the clothing of the angel, seated at the empty tomb, as it appeared to the two Maries – according to St Mark, it was a youth covered in a shining white stole; according to St Matthew, its face was like lightning and its clothing like snow – showing the way to Heaven and symbolising the clean victory men could have over their spiritual enemies.[132]

In its righteousness, the resurrected body had four attributes or 'dowers', one of which was clarity or a shining soul.[133] Facing judgment, souls appeared in white, black and shades in between, according to their virtue. In the vision of the monk of Eynsham, of 1196, on the edge of the field on which the righteous souls were gathered were others whose clothes were not of such brilliant white – there did not seem to be any blackness or stain about them, but they nonetheless had less of the shine of grace.[134] In Thurkill's vision of 1206, the souls assembled in the basilica of the Virgin were grouped according to their state of grace, the most righteous gleaming, those that faced the rigours of Purgatory – from which they might emerge immaculate – speckled with white and black, some appearing more white or black than others.[135] A similar metamorphosis, from the degrees of black through to white, marked the transition towards salvation of the souls of those who appeared as ghosts.[136]

White and black were rarely absolutely pure colours in the Middle Ages. The undyed woollen garments of the first Cistercians at Rievaulx would not have appeared pure white to us, but there was no difficulty for Aelred, who considered that these white monks were clothed just as one might expect angels.[137] Other undyed, ecclesiastical garments might also be described as shining or bright, whatever their colour. Byss – believed to be a cotton or a linen, listed among the textiles that God indicated to Moses would be acceptable as offerings and probably therefore always a fine textile[138] – may have had a greyish colour. Although the pallium of St Edmund of Canterbury, bequeathed by him to the nuns of Catesby, was made of the cloth popularly known as camelot or camlet, it was described in one list of bequests as having the colour of ash and separately, in the Anglo-Norman metrical life of the saint by Matthew Paris, as having the colour of byss.[139] An elderly man, with the face of an angel, clothed in byss that shone with more brilliance than snow, led the monk of Eynsham on part of his vision.[140] Stamin, a coarse woollen fabric, often worn by ascetics, was described as white, with a similar connotation.[141]

White was not the only colour that came to be considered good. Saints were sometimes described as 'well coloured' (*bele culur*)[142] and ruddiness in complexion also formed part of standard description, echoing the words of the Song of Songs, 'My beloved is white (*candidus*) and ruddy (*rubicundus*).'[143] Benedict of Peterborough described Thomas Becket, whom he had seen in a vision, in almost exactly the same words: 'Behold, our beloved, white and ruddy.'[144] The body of St Hugh of Lincoln had red cheeks, like roses, as if he were sleeping, having come happily from a bath.[145] The late thirteenth-century life of St Richard of Chichester records that his body was, in death, as it was in life, the colour of roses and of lilies.[146]

Not far removed from the merits of this basic chromaticism was a general association of plain, clear and sober colours with truth and righteousness. The *Book of vices and virtues* advised full, meek and frequent confession, which it likened to washing cloth. The water of confession would run clear, acquiring more grace and cleanness, like linen cloth growing whiter the more often it was washed.[147] For some, the converse was true: nothing was a surer sign of vanity and baseness than bright, multi-coloured attire. Perhaps with the intention of pleasing her husband – for she had exquisite feet – an English noblewoman reworked her footwear with extravagance, adding not only gold but also fur. She only recovered from the affliction that struck her after vowing to St Thomas Becket that she would renounce the practice.[148] Margery Kempe was in time to forsake her vain attire, the gold pipes in her head-dress and the dagged hoods (that is, with stylish cuts into the edge of the fabric) with tippets. 'Hir clokys also wer daggyd & leyd wyth dyvers colowrs be-twen þe daggys þat it schuld be þe mor staryng to mennys sygth and hir-self þe more ben worshepd.' Her decision to follow the guidance of Christ and to dress in white – a decision that may call on symbolism in various ways, perhaps in recognition of the purity of her soul, the remission of her sins or her desire to live chaste from her husband – was to cause her considerable trouble.[149]

The Lollards commonly dressed in russet or grey, forsaking bright colours and the earthly riches that could provide them – although not all did so and others who were not heretics dressed in this simple style.[150] According to a late fourteenth-century sermon writer of Wycliffite persuasion, the single, white colour of Christ's clothes at the Transfiguration signified the stability of his virtue, whereas the medley or mixed cloth worn by the friars was indicative of the instability of their orders.[151] Archbishop Arundel, examining William Thorpe in 1407, poured scorn on the assumption that all priests should dress like the Lollards. Thorpe might not judge a man's pride: one who dressed every day in a scarlet gown might be more meek than Thorpe in his threadbare blue gown.[152] The contrast might also make a political point, or one of virtue, between the honest array of a labouring man and the dissolute dress of others.[153]

The dress of the clergy was a particular focus for the discussion of colour. The clothing of monastic orders was intended to be spartan, to reflect the humility of their life, but it was not always as austere as it might seem. To the mischievous eye of Walter Map, although the Benedictine (black) monks only used the poorest cloth of the district, they might, with special dispensation, wear lambskins. The Cistercians, who might taunt the Benedictines over the lambskins and who themselves were supposed to wear undyed cloth, had themselves ample garments made of cloth which, had it been dyed, could have served as scarlet for kings or nobles.[154] The Council of Westminster of 1200 prohibited Benedictine monks and nuns from wearing hoods of a colour other than black, and prohibited the use of skins other than lambskin, cat and fox.[155] Conciliar and synodal legislation similarly addressed the clothing of priests. It was to be of a single colour, not mixed, with appropriate footwear, as all ornament was unsuitable for the clergy.[156] Following the injunctions

53 Clerical clothing: what is proper to be worn by clerics – the short tunics and swords of the group on the right are contrasted with the correct attire of the group on the left. From the *Omne bonum*, late 1330s and 1340s.

of the Fourth Lateran Council of 1215, the wearing of either red or green cloth, or any silk, was forbidden in a succession of English dioceses.[157]

The persistent complaints of anti-clericalists against the fine attire of the priest-hood show that these restraints on colour and ornament were not always effective and needed reiteration.[158] In 1351, the prohibition for the sisters of St Katherine's Hospital, next to the Tower of London, included striped material as well as red and green; the Council of Vienne, 1311–12, had added this to the general prohibitions on clerical clothing, although it had been banned intermittently before.[159] A similar injunction against coloured and striped cloth was included in 1382 by William Wykeham in the statutes for his new school at Winchester.[160] Striped material attracted particular opprobrium.[161] Striped hoods were required by London prosti-tutes in 1382 and striped fabric was in fact used throughout Europe to mark those who practised this trade.[162] But these injunctions were unusual in England: little in English sumptuary legislation was founded solely on colour, save the Yorkist and Tudor statutes against the wearing of purple by anyone other than the royal family,[163] although distinctions were made on the basis of the cost of the fabric.

54 Princess Cecily, the mother of Edward IV and Richard III, wearing purple. From the 'Royal' window in Canterbury Cathedral, *c*. 1482–7.

It was not uncommon, therefore, for peasant and urban households to include a range of brightly coloured cloths and other materials.[164] In 1272, Robert Atwater was accused of an assault at Little Cainhoe, followed by the theft of a hood of perse, a red belt with brass accoutrements, a white sheepskin purse, as well as shoes, gloves, knives and an axe.[165] The commissioners enquiring into the sanctity of Thomas Cantilupe heard that Joan le Schirreve's mother, Cecily, had been able to identify her young daughter's body solely from her new shoes, with their red straps.[166] The will of Thomas Hamunde of Tydd St Giles, of 1467, included brightly coloured bed-coverings, a yellow one bequeathed to his son Richard; two for his daughter Isabel, one green, the other green and red; and, for his son Robert, one that was blue and red.[167] On the other hand, the Stafford blue, with which a fifteenth-century Noah's wife threatened to clothe him, was no textile, but the result of a beating.[168]

A variety of symbolism would have been understood by the use of other colours, particularly in decorative art. In the highly stylised vocabulary of early medieval art, certain colours had a significance far beyond their hue, for example in the use of red to symbolise light.[169] Theophilus' treatise *De diversis artibus*, of the twelfth century, gave instructions on the employment of certain colours, for example for tree-trunks or mountains.[170] High medieval art, with its defined iconography, focused on a way of presenting images that is alien to us: each figure, each colour, was deployed as a means to access a higher truth. The naturalistic style of the later medieval period, pioneered in Italy and the Low Countries, marked a way of seeing and representing the world that was different from the stylised decoration still current in England.[171]

The significance of colour might be local and occasional, or it might have a more widely recognised significance that varied over time as well as context. The Northamptonshire meadow where St Rumwold was baptised retained its green colour, without withering or fading, according to the late eleventh-century life of the saint.[172] The burning bush witnessed by Moses in the N-town play cycle kept its fair and bright colour, 'fresch and grene withowtyn blame'.[173] Black, for all its associations with evil or ill-repute, had a wider repertoire of use. Ebony was to be employed in child's rattles, according to Isidore, or in cradles, according to Bartholomew, almost as an inoculation, so that infants might not be frightened at the sight of black.[174] The late fourteenth and fifteenth centuries saw a major change of preference in colours for clothing, with a fashion for black clothing among the upper classes. The sons of Margaret, Duchess of Clarence, by her marriage to the Marquess of Somerset, travelled to Normandy in her entourage in November 1419 with a substantial proportion of black garments among those specially tailored for the occasion, where they would not have looked out of place in contemporary fashion.[175]

The use of the colour purple for royal clothing was of long standing, replicating its use in ancient Rome for the garments of the most senior officials and, ultimately, of emperors, although the exact shade that 'purple' denominated varied considerably and the hues concerned were in some instances red.[176] In the Vulgate, *purpureus* had been used in relation to biblical vestments to translate the Hebrew for 'rich', rather than a colour; and this, rather than an exact hue, is to be understood as its meaning in references prior to the fifteenth century. When King William of Scotland sent the Bishop of Glasgow and an archdeacon to investigate a Becket miracle, the case of John of Roxburgh, who had been thrown from his horse into the Tweed, it was because those wearing the purple, that is, the rich clothing of the King, might not enter humble taverns.[177] According to tradition, the Virgin Mary had been one of the eight virgins selected to make the veil of the Temple. The colours were divided by lot and purple was allocated to Mary.[178] When a lady, dressed in a cloak of purple, appeared to a nun of Sempringham who was suffering from leprosy, and laid the garment around the tomb of Gilbert of Sempringham, there can have been little doubt that it was the Queen of Heaven.[179] Christ, dressed in a mantle of purple silk, appeared at Margery Kempe's bedside.[180] Purple cloth was used to wrap relics;[181] and it was an integral part of the wardrobe and furnishings of the Kings of England.[182] Henry VII expected to wear either purple or red velvet on the four principal feasts of the year.[183]

In the twelfth and thirteenth centuries, clothing made of polychromatic material had a strong association with sanctity and immortality. In the miracle of the young daughter of Salomon, a weaver of Northwood, the news that she had fallen into a pond was brought by a boy wearing a tunic of a fabric described as *polymita* – of many threads – which we are to understand as being coloured, or shining like damask. It was this term that was applied to the different coloured fabrics used for making the veil of the Tabernacle in Exodus[184] and it was in a tunic of a similar kind that Adam of Eynsham recorded the nineteen-year-old St Hugh to have served as a deacon.[185] The vision of Thurkill, of 1206, drew almost verbatim on the text of another vision, of a Cistercian novice, Gunthelm, of *c.* 1153, to describe Adam's multi-coloured cloak, the gown of immortality and glory. In both cases an archangel explained that it grew more complete as the virtuous added to it, their variety being the cause of the polychromaticism; when it was finished the number of the elect would have been completed and the world would end.[186]

Third among the raiments given by Christ to the devout in *The doctrine of the hert* were clothes of diverse colours, that is, a diversity of virtues that dressed the soul marvellously. They were probably to be understood as a suit from which the one with the most appropriate colour for the occasion might be employed, such as clothes of penance, of mourning, in Passiontide.[187] But what were these colours? The colours of the priestly vestments (Exodus 27), along with that of the veil of the Tabernacle (Exodus 37), were among the principal sources for the discussions of liturgical colours.[188] Before the twelfth century, there was a good deal of variation in colour, some of which continued in the later period. Although the work of Innocent III, attempting to establish a framework for the liturgical use of colours, did not become a part of the legislation of the Fourth Lateran Council in 1215, it had an enduring impact.[189]

Innocent defined four principal colours – white, red, black and green – for vest-ments. In general, white vestments and hangings were to be employed for Easter and the feasts of Christ, the Virgin, the confessors and virgins, as well as those of the invention and exaltation of the cross; red for those of the Holy Spirit, apostles and the martyrs (who had formerly been venerated in white), Whitsun, and feasts of the cross associated with suffering; black for days of penitenence and abstinence, throughout the long Lent (from Septuagesima through to Easter Saturday) and in Advent up to Christmas Eve, and the offices of the dead, although there was also some limited use of violet, for example on the fourth Sunday in Lent; and green was to be used for other days, 'because it is the middle colour between white, black and red', possibly making a link to the Salernitan system of colours. Innocent added that some used scarlet instead of red, violet for black, yellow for green, while others used rose for the feasts of martyrs, yellow for confessors and a lily-white for virgins.[190] The expressions of colour here were schematic – as with heraldry – and did not delineate hue. Beyond that, significant local variation persisted. At Westminster Abbey, for example, from the abbacy of Richard de Ware (1258–83) through to at least the late fourteenth century, red vestments were worn on more than thirty-three weeks of the year, including the period from Septuagesima to the first Sunday in Lent and almost continuously from the last fortnight of Lent through to Advent, as well as on the feasts of martyrs occurring at other times.[191] In the chapel royal of Henry VIII, the vestments for Lent were of white damask, with red crosses on them.[192] These coloured vestments were worn over undyed linen, just as, the author of *Jacob's Well* pointed out, Aaron and his children (the priestly lineage of Exodus 27) were to be clothed, 'for as lynen cloth muste be beten in þe wasschyng to make it whyȝte & softe so schulde þe bete here flesch with doctrine & lyvynge to wassche hem whyȝte fro fowle thouȝtys and wyked desyres trugh verry schryfte & repentaunce.'[193]

Within the great household, colour of clothing, as much as quality and quantity, was used to mark out the different ranks.[194] At Whitsun in 1286, Bogo de Clare, the clergyman brother of Gilbert de Clare, the Earl of Gloucester, had clothes made of perse and bluet, each cloth costing £6 or more. Members of his household were attired in a variety of cloths: Master Gilbert of St Leophard had a cloth of perse at slightly less expense; the two knights of the household, Hugh de Turberville and William de Monterevell, had a cloth of bluet that cost £6 between them; and the esquires of the household had a scarlet ray (probably striped in the weave), along with yellow cloth.[195] The vast array of gentle ladies and noblewomen who came to the court of Lady Loyalty, all clad in blue, with their mottoes on their sleeves, in the late fifteenth-century poem *The assembly of ladies*, marked a transition to the use of livery almost as a uniform.[196]

The clothing and make-up in stage directions for medieval drama underscores many of the popular associations between colour and morality. *Wisdom*, the theme of which is the struggle between good and evil for possession of man's soul, probably written in 1465–70, is unusual in giving directions for the clothing of each character. Wisdom was dressed as a king, in rich purple cloth of gold with a mantle of the same material, lined with ermine, a royal hood furred with ermine about his neck, a wig with eyebrows, a crown, with rich imperial (a cloth) and jewels on his head, a gold beard, an orb and sceptre in his hands. The soul, Anima, was dressed as a maid, with white cloth of gold edged with miniver, a black mantle, a wig like Wisdom's, with a rich chaplet laced behind and at the side with knots of gold thread and tassels. The Five Wits and the Three Mights were dressed in white, and Lucifer was dressed as a devil, with six of his eighteen retainers having red beards, an example of the common connection between those with red hair and evil. The connotations of colour are made explicit, as Anima is addressed by Wisdom:

> Thes tweyn do sygnyfye
> Yowr dysgysynge and yowr aray,
> Blake and wyght, fowll and fayer vereyly . . .[197]

In the *Castle of Perseverance*, the Devil was dressed in black and blue, and the four daughters of God dressed in cloaks: Mercy in white; Righteousness in red; Truth in 'sad greene'; and Peace, all in black.[198] In *Mary Magdalen*, probably written before *c.* 1515–25, Curiosity (a gallant, that is, Pride) comes to charm Mary:

> A, dere dewchesse, my daysyys iee!
> Splendaunt of colour, most of femynyte,
> Your sofreyn colourrys set wyth synseryte![199]

That colours could be sincere might strike us as a curious concept, but it was wholly consonant with a way of perceiving the world that considered the virtue of everything – and colour was an important indicator. In every description where colour is mentioned, therefore, we have to be alert to the potential for further significance; but, at the same time, there was often a plurality of meanings and explanations for these links.

Lapidaries, when considering the colour of stones, frequently made reference to the heavenly Jerusalem and the twelve kinds of precious stone that adorned its foundations, but also to a range of other sources.[200] Green jasper, the first of the stones, was also used for the walls: civic pageants, particularly those prepared for the occasion of a formal, royal entry into a town or city, commonly made use of castles with green walls in direct allusion to this text.[201] St John's vision of Heaven included a throne of jasper and sardonyx, that is, green and red. A late fourteenth-century sermon explained the meaning of this text as telling men that Christ was full of comfort for them, as green makes men glad and brings comfort to their eyes, and the red, in recognition of Christ's charity, inspires men to martyrdom and to shed their blood for his love. In another interpretation, Wycliffe saw the jasper as God's mercy, its green comforting the inner eye of man, and the sardonyx as divine wrath.[202] The fifteenth-century *Jacob's Well* described sardonyx as a stone of three colours, black, white and red, the last signifying the red fire burning in charity towards God.[203]

In miracle collections, change in colour was indicative of change in virtue. The face of an infant boy from Eye, who had swallowed a ring, had turned a foul, black colour, to be cured at the intervention of Thomas Becket. He was one of many children who had turned black, dark blue, grey or the colour of lead, *in extremis*, for

55 The Last Supper, from the Hours of Elizabeth the Queen (the wife of Henry VII, who bequeathed this book to Edward, third Duke of Buckingham). Probably written and decorated *c.* 1420–30, the red hair of Judas Iscariot marks him as evil. Seated on the corner of the right-hand bench, in front of Christ, he is also hiding under the cloth a fish that has been taken from the dish on the table before Christ, a feature that has already indicated his traitorous tendencies.

their evil coloration to be swiftly dispelled at the intervention of a saint.[204] In a further case, Atheldrida, a grown woman living at Canterbury, died from a quartan fever and her body had turned a dark blue. A monk brought her some of Becket's blood to drink, which he mixed lest the taste or colour of blood cause revulsion. The woman revived and returned to her natural colour and strength.[205] In other miracles, parts of the body that had been afflicted changed colour. Mabel, a nun who had dislocated her foot by tripping over some wood concealed by the straw on the kitchen floor, and whose foot had turned as black as her veil, was spared by St Gilbert from amputation.[206] St Richard of Chichester came to the aid of a pregnant woman, one of whose breasts had turned black and who despaired of a cure.[207] Eustace, from Powick, near Worcester, had a white (*albus*) horse which he left tethered in a field to graze. Returning in the evening he found that the horse's head had turned black and was badly swollen. He thought the horse might have been bitten by a snake or that it had eaten a harmful worm along with the grass. He vowed a penny to St Wulfstan and made the sign of the cross over the horse: the horse's swelling abated and the colour of its head changed back to white, this time, a shining white (*candidus*), like wiping off mud with a linen cloth.[208] Discharges – yellow, green, black or of a variety of shades – from wounds and disease were often a prelude to relief, which came with the dissipation of these alien colours and, frequently, evil smell.[209]

Colour change in natural phenomena provoked a variety of responses. The colours of the rainbow were a subject of great curiosity. Newton's perception of seven colours – an analogy between the spectrum of light and the notes of the musical scale – is completely alien to the medieval period, which set out to analyse the chromatic range from a different viewpoint. Rainbows commonly appear in manuscript illuminations with between one and five colours, not necessarily in an order that we might recognise; they were described in texts sometimes simply as having many colours.[210] Aristotle's work on meteorology was the most common source for the explanation of the rainbow and it was on this, along with Bede, that Bartholomew the Englishman drew for his explanation, noting that the rainbow took on the colours of the four elements, red in the outermost part (from fire), green from earth on the innermost, a hue of brown from the air and blue from water in the middle.[211] This position was also adopted in a set of prose Salernitan questions that circulated to England, the balance of the colours here depending on the equilibrium of the elements concerned.[212] To Wycliffe, following Genesis 9: 8–17, the rainbow symbolised the bond between God and earth. The purple colour (*puniceus*) at the top and the green, closest to the earth, signified the benign and palliating government of the Church by God.[213] But the link that the rainbow made with Heaven was not always a virtue. In the York plays, one soldier gave Pilate his view of Christ:

56 On the ninth day after the birth of St Fremund, over the palace a rainbow appeared, the three colours of which had particular significance: watery green, for the saint's chastity; sapphire blue, for his assured position in Heaven; and hardy red, for martyrdom. Although the illumination of the rainbow has four colours, the text of Lydgate's life of the saint describes the significance of only three. This work is typical of the allegorical view of iconography that was superseded by a more naturalistic form of image. From the Lives of St Edmund and St Fremund, written by John Lydgate, between 1434 and 1439.

On þe raynebowe þe rebalde it redis [the ribald it reads],
He sais he schall have us to Hevene or to Hell
To deme us aday aftir oure dedis.[214]

Changing colour was a sign of duplicity. The fourteenth-century *Book of vices and virtues* likened the liar to the butterfly, for 'at every colour þat sche seþ sche chaungeþ hire owne'.[215] A liar was like a chameleon, according to *Jacob's Well*, a bird that lived by air and had nothing in it but wind: 'he wyl chaungyn hym to alle colourys þat he seeth.'[216] Living by one of the elements alone, this fearful creature both beguiled and confused authors who had not seen it. Some held that it could change colour because it had little blood.[217] Like many other explanations of the medieval world, those for the colour of the peacock's tail were based on humoral theory. The colours were generated by superfluities of the humours, each producing a different colour or combining to produce others.[218]

If colour was a manifestation of the humours, it might also then have a role in medical diagnosis and in the assessment of character or morality.[219] Guy de Chauliac's treatise on surgery noted that a good colour resulted from a favourable balance of the humours in the blood coming to the skin; an evil colour, from evil humours, such as the black of melancholy, white of phlegm and yellow of choler. He went on to note how these colours might result: not bathing, using vinegar and poor water might make one black; cold, lechery, heaviness of heart and loneliness led to whiteness; and eating yellow things, such as cumin, or salted items, led to a preponderance of the choleric.[220] In terms of diagnosis, analysing the content and colour of bodily fluids, especially urine, would allow curative steps to be taken, steps that might encompass moral changes as much as physical ones.[221]

Observation of the features of the individual – colour of skin generally, or of cheeks, lips, hair and nails – revealed much about character.[222] This gave a particular cast to the use of cosmetics, substances which might change the colour of the body and hence its character, an intervention that changed or concealed the natural order. From the twelfth to the fifteenth centuries, the young, female body was considered at its most beautiful when it was white of complexion, with a ruddiness to the cheeks.[223] In the late fifteenth-century romance, 'The squire of low degree', the king was dismayed at the state of his daughter: pining for her lost love had transformed her from this epitome of beauty.

> Ye were whyte as whales bone;
> Nowe are ye pale as any stone.
> Your ruddy [complexion] read as any chery,
> With brows bent and eyes full mery . . .[224]

The N-town play expounded the ninth commandment with similar regard for the ideal:

> Desyre not þi neyborys wyff,
> þow she be fayr and whyte as swan
> And þi wyff brown . . .[225]

And in its play of the marriage of Mary and Joseph, Joseph was told that he was not to strive against God, but to marry Mary, who was a fair maid, obedient and 'white as a loaf'.[226]

How did cosmetics effect these changes and how widespread was their use? Recipes for cosmetics, aimed at the upper classes, circulated in England in the thirteenth century. These texts, in both Anglo-Norman and Latin, show close links to the Mediterranean, not only in terms of ingredients, but in their connection to Salernitan medicine. If it may seem curious to find clerks copying Latin treatises on cosmetics in England at this period, it is less so if one recalls that the use of cosmetics was not separate from medicine, but part of a range of practices that, by physical intervention, looked after the well-being of the body and the soul. To this group one should also add cookery and the changes made there by spices and colorants. The application of humoral theory was relevant to all.

Some ingredients in cosmetic recipes are esoteric or unusual for England, such as peach stones, and may not have been widely available, although the aristocracy had good links to southern Europe in general, and to Sicily and southern Italy in particular; and the connections can be seen in other upper-class practices, from hunting to cookery. Some of the royal spicers and apothecaries in the thirteenth and early fourteenth centuries had family connections to Montpelier or Paris.[227] Thirteenth-century texts principally contain recipes for creating cosmetics: they were rarely about the application of preparations that had been purchased ready-made. Typically, these included recipes for colouring the face or for making it white, usually with mixtures of herbs, such as Aaron's Beard, or their roots, with ceruse – white lead, commonly a mixture based on carbonate or hydrate of lead – boiled together and dried, presumably to create a powder, which, mixed with rose-water, could then be applied to the face with water, soap or barley water.[228] Other whiteners included asses' milk,[229] lye[230] and fats, for example that of the stork, presumably in the hope of transferring some of the whiteness of the bird. These grease-based make-ups might be coloured in other ways: when it was desirable to achieve a red colour the recipe might be modified, using brazil as a colorant, one of several ways of creating rouge.[231] There were also depilatories and remedies for hair loss, instructions for the removal of freckles and precautions against tanning in the sun;[232] recipes for dyes, to create blond, red, black and auburn hair; and remedies for split ends.[233]

As noted when discussing smell, there was a significant change to the trade in cosmetics in the fourteenth and fifteenth centuries, with increasing numbers of finished products reaching the market in England. Among these were sweet powders, such as citrinade, a confection made of sugar and citron, the precursor of the lemon, that was applied to the face as a whitener.[234] They were expensive: witness a group of purchases of sweet powders – two pots each of succade, coinade (made with quince), citrinade and pomade for 37s. 8d. – shipped out of London to Rouen for the use of either Margaret, Duchess of Clarence, or her husband, in 1420–1.[235] By the end of the fifteenth century, England was closer to continental practice in the volume of cosmetics that might be employed[236] and they might be found further down the social scale. Peter Idley lamented the progression of the sin of pride:

. . . and she bee as oolde as Abrahaums modire
She woll be lothe in arraie onythyng to loose;
She woll depeynte hirsilf as fresshe as ony rose,

With wymples and tires wrapped in pride:
Yelow undre yelowe they covere and hide.
For if the beawte begynne to amminysshe and fade,
Then can they werke with a certen floure
To pullisshe hem othir wyse than ever God made,
And set uppon hem a more fresshe coloure,
And cover yt with goolde the synne to socoure,
That man wote not whethir is whethir–
The yelow goolde or the tawny lethyr.

The problem was no longer restricted to the upper classes and urban patriciate:

It is now harde to discerne and knowe
A tapester, a cookesse, or an hostellers wyffe
Fro a gentilwoman, if they stonde in a rowe;
For whoo shal be fresshest they ymagyne and stryve . . .[237]

Others condemned the use of cosmetics as an offence against the natural order. The wife who hid her wrinkles under cosmetics and her grey hair under a wig deceived her husband.[238] The *Book of vices and virtues* condemned men as well as women for vanity with their hair, for dressing it 'oþer þan þe riȝt kynde askeþ'.[239] God punished a lady of great estate who 'popped [patched with cosmetic], painted, plucked, and fared her hede', and the Knight of La Tour-Landry issued the monition to his daughters that they should not attempt to alter the face that God had given them, nor wash their hair in anything other than lye and water.[240]

For the natural order to be clear by sight, the assumptions about how it should look – and how it should be seen – had to be plain. In hearing the litany of sin, the confessor paid particular attention to sins that might come through the eyes, concentrating on things that should not be seen or actions which, if observed by others, would also constitute sin. A confessional formula of the thirteenth century, used by the Dominicans, enjoined the confessor to ask if a man had seen and desired a woman, if he had observed fornication or watched women performers.[241] Females dancing were long considered to be evil, a pejorative that can be traced back to St Jerome.[242] In another work from the first half of the thirteenth century, probably addressed to Cluniac monks, Robert Grosseteste produced a more extensive list of the ways in which a monk might come to sin through sight: by enjoying things of earthly beauty – men, horses, clothes, jewels – and wanting to make himself beautiful; through delight in the beauty of women, watching animals mate, especially anything that aroused the desires of the flesh; and, thirdly, through watching games, horse-racing, minstrels, animal-baiting, wrestling, women dancing and similar events; and observing with unjustified suspicion the actions of others.[243]

57 The dangers of women dancing, described by St Jerome as 'the swords of the Devil'. From the *Omne bonum*, late 1330s and 1340s.

A late fourteenth-century sermon writer focused on the sins of lechery and covetousness of worldly goods. Sight, the highest among the external senses, was given to us to serve God and our souls.[244] In 1354, Henry of Lancaster lamented his ability to look with too great delight upon the follies of man, at books, paintings or beautiful jewels; and the misplaced pride he had shown in the way his feet looked in stirrups, shoes or armour, his light-footed display of dancing and the garters he had worn, that were beyond compare.[245] The secular audience, listening to the sermons of *Jacob's Well* on the virtues of temperateness, was advised to avoid unseemly behaviour such as 'twinkling', looking and staring about, injunctions that would not have been out of place in a courtesy book of the same period.[246]

Monastic regulation indicated how monks should behave with their eyes. St Benedict reminded his monks that humility required appropriate bearing: the monk should always have his head bowed and his eyes cast down.[247] At the Augustinian priory of Barnwell, the customary advised the canons that, in church, they should not let their eyes wander around aimlessly, 'because the impure eye is the messenger of an impure heart'. Nor, in the dormitory, was one canon to stare at another.[248] The Cistercian novice, in choir, was advised to fix his eyes on a single place: wandering eyes were a great harm to the stability of the heart. He was advised to turn his eyes to internal contemplation, to Christ lying in the stable, the cross and the elements of suffering of the passion, to the heart of God, to Christ's wounds and to similar subjects.[249] This tradition, particularly strong in the monasteries, distinguished between corporeal things that might be seen, and eternal things that might not be seen with the physical eye, but might be the subject of contemplation of the internal eye.[250] The Cistercian tradition insisted on avoidance of stimulation of desire in the eye, at the expense of contemplation. Aelred of Rievaulx included in the category of 'unnecessary beauty' paintings, sculpture, decorated wall hangings and over-large buildings.[251] This sight of eternity and the knowledge it brought of realms beyond the corporeal was not restricted to the cloister. To a Lollard sermon writer of the early fifteenth century, 'ghostly sight' was the most important thing to ask of God, the eye of discretion in the soul, and the ability to choose between good and evil.[252]

The act of looking in itself was one of great power. Prolonged contemplation of the divine enabled one to perceive divine truth.[253] Even between mortals, the contact between the viewer and the observed conveyed a moral force far beyond our concept of vision, allowing the one to read the innermost soul of the other or to convey to the observed elements of the make-up of the viewer. Prolonged contemplation, or gaze, reinforced these elements. The effect might be multiplied by great numbers of people looking at the same things – the look of crowds, for example – in both positive and negative directions.[254] Numerous set pieces of medieval display allowed monarchs to be seen in formal crown-wearings, or aristocrats and clerics to appear before their households – and these events must be seen in this context. The ceremonial of the court of Richard II gave that king important occasions for display, for example at the formal crown-wearings that took place three times a year. In 1398, after the installation of Roger Walden as Archbishop of Canterbury, Richard observed these occasions in a more exaggerated fashion than his predecessors, sitting on his throne, in his chamber, without conversing, in his regalia from lunch (*prandium*) till vespers. His courtiers were expected to kneel when his gaze fell upon them.[255] He attended Westminster Abbey with his chapel royal for all the principal services of the day on the feast of the Translation of St Edward 1390, sitting in the choir at high mass, wearing his crown, accompanied by his queen, also wearing her crown.[256]

If gazing gave one the power of seeing beyond the corporeal, there might be ways in which that sight could be enhanced, for example with mirrors, to see into the future. Lapidaries alleged that Nero had a mirror of emerald, the powers of which were redoubled by the virtue of the stone, which allowed him to see all he wished to know.[257] The *Gesta Romanorum* told how a clerk saved the Emperor Felicianus, whose wife had employed a necromancer to kill him while he was on a pilgrimage to Rome. The Emperor was instructed to bath and, while he was in the bath, the clerk brought him a mirror in which he was able to see what was planned.[258]

The notion that a look might harm an individual, the 'evil eye' – *fascinatio*, an enchantment or bewitching by the eyes or by words – was discussed by authors including Alexander Neckham and Roger Bacon, the evidence for its action reinforced by the theories of sight; but it is difficult to tell how far this belief was a motivation in late medieval England.[259] The Wisdom of Solomon held that the engagement of the eyes (*fascinatio*) with trifling things hid the good from the eyes,[260] a view that fits closely with the visual concentration enjoined on monks during services. Early Christian authors and ascetics had been concerned about the deleterious effect of the gaze that might arise from someone with impure thought or intention. Amulets were traditionally held to deflect this unwanted gaze and the walls of the cloister or the anchoritic cell protected the inhabitants against lustful eyes.[261] These practices are difficult to trace in the later periods, when most amuletic material had a link to Christian traditions, although there may be a case for them in the Saxon and early Norman periods.[262]

There was, nonetheless, a concern for the secret things of a monastery, and that they should be shielded from inappropriate scrutiny by outsiders. In selecting a tailor, the Augustinian canons of Barnwell realised that a secular must penetrate the interior of the priory frequently on account of his business 'where he might not only hear but also see the secrets of the brothers'. The customary therefore enjoined that he be a compliant and confidential person, whose employment should not be taken on, or terminated, lightly.[263] The servant of the master of the infirmary was similarly placed in terms of confidentiality, although the qualities required of him were primarily those of enduring with equanimity the rigours of the sick and dying.[264]

While it was, as we have seen, a sin to observe some things, there were also things that should not be seen, which created shame. The corollary, that looking in these circumstances generated unfortunate effects, may also have been true. When a Benedictine monk was beaten in the daily chapter, all were to bow their head in compassion; only the senior monks who might intercede on his behalf were allowed to look at him.[265] Some of the medical texts known as the Trotula advised those who assisted a woman in childbirth not to look at her face, as that might cause her to feel shame or bashfulness during and after birth. Versions circulating in England made it clear that this applied to women present at the birth, although other recensions were not specific about the sex of those present.[266] Continental manuals discouraged confessors from looking at the penitent during confession, which normally took place in locations where both parties might be seen.[267] These examples, however, are not large in number.

If the gaze saw into a person, perception might also gather a good deal of information from external signs and gesture. The use of visual codes and special signs played an important role in the later Middle Ages. If they are now apparent to us mainly in terms of clothing, paintings, pictures, sculpture, heraldry, in the badges and marks enjoined by ecclesiastical legislation, and in set-piece spectacle and pageantry, other signs, much less in evidence, would have wrought the fine distinctions that mattered so much to medieval people. In the absence of widespread literacy, these were marks that immediately conveyed to the viewer important information. Sight ensured that the viewer take cognisance of the array of signals such events might proffer.

Each event reinforced the meaning of these signals: on every occasion, the act of seeing them displayed conveyed their meaning and their ultimate significance or benefit. In ecclesiastical processions, for example, the presence of burning candles, crosses and vestments of specific colours were designed to invoke in the watcher the mysteries of the Church. Liturgy was performance and sight ensured participation. In the cathedral at Old Sarum (see Illustration **2**) – as in many other places – the purpose of Sunday processions was to visit each altar and to reinforce its sacral quality, renewing it with holy water, an action clearly visible to all. The ejection of penitents from the west door of the cathedral on Ash Wednesday was mirrored all over Western Christendom.[268] The signs and gestures of the Church liturgy created participation: much more than re-enactment, they were a creation of the original event itself. On the one hand, the barefoot monastic processions of Lent were an obvious mark to watchers that the purpose was penitence; on the other, the Palm Sunday procession at Christ Church Priory, Canterbury, was a dramatic staging of Christ's entry into Jerusalem, an event that was designed through sight and participation to create in the monks, both physically and symbolically, the reality of the biblical event.[269]

The monastic habit was likewise both a sign of its wearer's status and his guarantee of salvation. Peter of Cornwall's *Liber revelationum* of 1200–6 recounts a tale of a shepherd who had served the Cistercian monastery of Stratford Langthorne as a lay brother. The shepherd decamped to the secular life, but subsequently changed his mind and returned to the monastery. He was allowed back, but because of his apostasy he was only allowed to wear the tunic of the lay brother and not the hood (Cistercian lay brothers wore black, the monks white, undyed cloth). The lay brother died and was buried without the hood: as the Abbot was overseas no one, not even the Prior, was allowed to mitigate his sentence. In Heaven, the lay brother was judged and told that he might tell the Abbot that he could have his full habit and thus would be saved. The Abbot saw him, with the hair of his head burnt and similar burns on the shoulders of his tunic, which the erstwhile shepherd told him came from drips of burning pitch, sulphur and lead from a cauldron under which he had been required to pass as part of his punishment and which had fallen on his head – as he was unprotected by the hood of the lay brother's habit. The Abbot conceded that he might have it again; the lay brother returned to England where he appeared to the Prior and repeated his tale. The following day, the monks exhumed his body and dressed it in the full habit of the lay brother.[270]

Dress might signal a number of distinctive arrangements. Before 1300, those who had abjured the realm were to be dressed in a single garment of sackcloth, although subsequently this was relaxed to allow such people their shirt, breeches and coat; they were to go bareheaded and barefoot, carrying a wooden cross.[271] Penitents might be recognised in a similar way: Agnes Doune, confessing to adultery, at Searby in Lincolnshire in July 1339, was sentenced to be beaten by the priest each Sunday until the following Michaelmas, wearing hanging crosses on her shift in the manner of a penitent.[272] Elements of the dress of pilgrims were distinctive – and Henry VI was able to combine these with his equally distinctive blue velvet gown in posthumous appearances to his devotees.[273]

Heraldic devices, badges and mottoes showed identity and allegiance, and we can trace a little of their development into a 'visual vocabulary'. Heraldry appears first in the very late eleventh century and increasingly became employed as a way of distinguishing armed fighting forces in the twelfth century. Its development throughout the Middle Ages was closely linked to the processes of warfare and chivalry, in creating a means to identify those whom one could not easily recognise in the mêlée of combat. The difficulties of this were increased from the mid-twelfth century as patterns of armour – particularly helmets – changed, giving more protection to the face while obscuring it.[274]

From the second quarter of the twelfth century, devices on seals are found which were used subsequently by descendants of those who originally employed them: the implication is that these devices were inherited or became indicators of family in a customary fashion. Many of these designs are distinguished by their simplicity and their clarity. There was a broad change in types of heraldic device: there were many more symbolic or religious devices employed in the twelfth century than sub-sequently. We know little, until later, of another aspect of these designs – colour – although literary works from this period imply that colours were often associated with them. These texts first describe devices painted on helmets; by the mid-twelfth century, the descriptions extend to the elements or 'charges' on the shield. It is almost certain that such changes did not happen at random, but were brought about by a strong, organising force. That force is most apparent in the compara-tively sudden emergence of blazon, the language used for the description of coats of arms, around the middle of the thirteenth century, although that it was present earlier is implicit in the regularity of twelfth-century heraldic devices. Blazon shows us that the way in which heraldry was used had become highly stylised and regular, so much so that it is still possible to recreate from it the coats of arms described.[275]

Blazon required there to be rules for the use of colour ('tincture') and for its employment in combination with the metals and furs commonly used in heraldry. The description it gave of colour and these other elements was indicative of their general nature, rather than a precise indication of hue. A defined vocabulary had the advantage of conveying a pattern verbally, without having to address the practical difficulties of ensuring that colours corresponded exactly to a precise shade. It may be anachronistic for us to seek a precise identification of colours: the limited range of strong colours employed in coats of arms always ensured that there was signifi-cant contrast. The earliest English roll of arms, Glover's Roll, originally compiled

58 A knight with a heraldic shield, with the chevrons of the FitzWalters, and a further shield with the lozenges of the Quincy family, their cousins. From the seal matrix of Robert FitzWalter, *c.* 1207–15.

around 1254 (now surviving only in a sixteenth-century copy), shows this ordered pattern of description, progressing through the elements of the arms in a methodical fashion, with three colours and a range of technical terms.[276] What is striking about the 218 coats of arms described in Glover's Roll is that they scarcely overlap; and, in 1300, in the poem the *Siege of Caerlaverock* it was a point of major concern that they should not. Heraldry created a distinct identity embodied in a device and the intention was that it should be unique to its bearer. Much as distinctions were made in language itself where there was a need to differentiate closely similar items, such as the colours of horses, the vocabulary of heraldry was able to create distinctive patterns for the knightly class. What is different in the case of heraldry is that it was organised systematically. This is implicit in the orderliness of blazon and explicit in references to the marshal and the constable, the principal military officers of the English Crown, which show that they met in the fourteenth century to adjudicate cases relating to arms in what was to become known as the Court of Chivalry. Heralds had probably played a key role in the development and promulgation of the rules relating to devices on arms and, in 1395, expert evidence was given to the constable and marshal in 1395 by two heralds, kings of arms.[277]

At the same time, the images of heraldry had a wider impact on society. Much as patterns of meditation – as we shall see – were designed to use objects and settings to call to mind elements of devotional practice, heraldry recalled a sense of identity, of family ties, remembrance of deeds of valour and much else besides. Heraldic devices were commonly understood as a shorthand for ownership or association with individuals of the armigerous classes. At a time when the written text would have made little impression as a mark of distinction, the use of heraldry in ornament and decoration was especially effective. The royal arms seem first to have been used in this way on the window shutters of Henry III's great chamber in the Tower of London in 1240.[278] Heraldic devices became enormously popular as decoration and we cannot look at descriptions of the tapestries or the sculpture that decorated the halls of late medieval England without seeing these allegiances at many points. Among the goods of Henry VIII on his death in 1547 was a tapestry showing the pedigree of the Duke of Buckingham, with the Duke's arms in its border, an item forfeited to the Crown on Buckingham's execution in 1521.[279]

59 The Lion Tower at Warkworth, so called after the heraldic emblem of the Percies over the doorway into the hall in the outer bailey. Late fourteenth or early fifteenth century.

Sight, however, was a vulnerable sense and it had to be nurtured. Several authorities, deriving their views from Arabic medicine, agreed that after birth an infant ought to be kept in the dark, or not exposed to strong light.[280] Salernitan medicine explained the differing abilities of sight by reference to the amount and clarity of 'visible spirit' in the eye: those with good quantities of clear spirit could see well objects both close at hand and in the distance; those with a small quantity, but clear, could see well close at hand, but not things in the distance; and those that had both a small quantity of the spirit and one which was murky, could not see well either close at hand or far away.[281] Some relief for short and long sight was available in the form of spectacles, invented in Italy at the close of the thirteenth century.[282] In 1436, the priest who was Margery Kempe's amanuensis set a pair on his nose, which made his sight worse. Undeterred by his complaint, Margery told him that he should do God's work and not stop; when he returned to his task, it transpired he could see as well as he ever did before, both by daylight and candlelight.[283] The inventory of Henry VIII's goods is testament to their ubiquity, with lists of spectacles and their cases, some for reading placed with books and two pairs that might have been designed to go with a backgammon set.[284]

Without modern treatments for bacterial infection, eye disease was widespread. John, the son of Hugh le Chaundler had beautiful, healthy eyes when he was born; but six days later he could not open his left eye; his right eye then became so infected by humours weeping from it that he could not open that either. The child would not stop crying, nor could it sleep. Hugh, his wife and his mother, had no wish to raise a blind child and prayed to Thomas Cantilupe. In the middle of the night John stopped crying and, on waking the next morning, had overcome his infirmity and could see with both eyes.[285] The extremes of weeping accompanying religious fervour were another potential impediment to sight.[286] Miracles that restored sight were prominent among the works of saints. The texts of these accounts had for their models not only the works of Christ, but that of Tobias, who cured his father's blindness with the advice of the archangel, Raphael.[287] Bringing divine light to the blind was particularly meritorious,[288] for 'light' was what they had lost. When Robert Neton, a vicar in the cathedral church of Salisbury, recounted the miracle of his sight restored by St Osmund, he noted that he had been deprived of 'the light of the left eye'.[289] Most accounts dealt with those who had been blind since birth or had become so in the course of life, through illness or injury. An account of a miracle, recorded in the late eleventh century, told however of a boy from the Isle of Wight. He had had an eye put out as punishment, came to Winchester and, after the intervention of St Swithun, found that another eye grew in its place over a period of a fortnight.[290] These reports of cures are particularly interesting for the ways in which they assess restored sight.

What did those that had their sight restored see? One test was whether they might see colours. Godiva of Chelmsford recovered her sight and saw a shaft of sunlight at

60 Christ healing the blind. From the *Omne bonum*, late 1330s and 1340s.

Becket's tomb. Although she could see people and colours, her cure was deemed imperfect.[291] In two other Becket miracles, there were difficulties in using colours as a test. While they could be distinguished, those blind since birth did not know how to describe them. Nonetheless, one of the newly-sighted could correctly identify white.[292] Those cured at the tomb of Thomas Cantilupe were also asked if they could see colours: Agnes de la Brok, who had previously been sighted, identified a range of colours, as well as describing objects put before her.[293]

The acuity of the restored sight was another important consideration. A fifteen-year-old boy, blind since birth, brought to the tomb of St Milburga of Much Wenlock, had his sight restored. At first, however, he could not see clearly: although he could see hands and fingers held out to him, for other things, such as stones and leaves, he made a preliminary examination of them with his fingers. Still he was able, without a guide, to follow the Prior to the altar on which the bones of the saint rested.[294] In 1139, a man cured by the intervention of St Erkenwald demonstrated how he could now pull out the central pin that held together a cross and replace it.[295] Reports on the miracles of Thomas Becket noted those who could again thread a needle.[296] Hugh le Barber, giving evidence in 1307 to the commissioners enquiring into the sanctity of Thomas Cantilupe, described how, while visiting London some fourteen years previously, he had become blind in both eyes, as if they had filled with ash. He had consulted a surgeon in London, who had applied a

61 Much Wenlock
Priory, looking west
from the Lady Chapel
to the area where the
shrine of St Milburga
was located.

plaster, but this made things worse, and his other medicines were also to no avail –
all of which had cost Hugh 4s., without a cure. He remained blind for three years,
seeing no light, nor clearness, nor any physical object. He subsequently prayed to
Thomas for a cure and was measured for him, sending the candle and some wax
eyes to Hambleden, where Cantilupe was born. Within two days of this he recovered
his sight sufficiently to see chickens coming into his house, and within a week he
could go around without a guide. He could again see to play chess and dice. The
commissioners tested him: he could see a cap held in the hand at five ample paces
and more; closer to, he could see small and fine things, such as a dried rush or a pin,
and confirmed that he could see the points of a dice.[297]

Death drew a veil over sight. Among the signs of death enumerated by John the
Fisherman, considering the body of his son, Nicholas, recovered from the River
Wye, were his eyes, open and shrouded as if by a web, but as John closed his son's
eyes – for he considered it improper to bury the child with his eyes open – his son
called out for Thomas Cantilupe's assistance and he recovered.[298] Although we do
not know if their eyes were closed, it was traditional for the faces of monarchs and
the senior clergy to be left uncovered, right until the point of burial, when,
according to the household ordinances of Henry VII, the king's face was to be
covered with a silk cloth.[299]

Like all the external senses, sight was linked to the internal ones by way of the
common sense. Special connections between sight, the imagination and memory

were developed, particularly in the monasteries and the universities, and became an integral part of the process of religious contemplation and thought. A fertile imagination allowed medieval man to conjure before him and create the reality of the events of past, especially the scriptural past. Reference to the detail of texts was brought out in quotation, or allusion, on an extensive scale: biblical texts and ideas were everyday reality. A way of thinking that placed considerable emphasis on memory built a common imagination and way of seeing things, as the common features in descriptions of miraculous events or the lives of saints have shown. One of the merits of artistic works was that a constant iconography could be used to inspire particular recollections.[300]

62 Funeral of a king, *c.* 1380–1400.

Prodigious memories are well documented and appear as an element in much hagiography. A good memory was a question of ethics:[301] individuals who had absorbed – through sight and sound – large quantities of sacred texts had also absorbed the virtue that passed with them. That memory was cultivated by a particular use of sight, based on mnemonic techniques known from Antiquity. The creation of sites within the imagination to focus memory and meditation was an essential part of this practice. These sites had specific characteristics. Architectural frameworks were a common form, sometimes of fantastic proportions, but some might be expanded from the pattern of the cloister or the church.[302] Around 1325–35, Thomas Bradwardine (later Archbishop of Canterbury), in his work on training the memory, described how to construct ordered sets of places in one's memory in which one might locate the things one wished to remember. Favoured here might be a location without distracting detail, such as a field, that one might visualise at different seasons; or buildings with different roofing materials. It was important to visit such locations, either physically or mentally, to reinforce the pattern of memory.[303] The texts to be memorised were themselves frequently constructed with features to assist the mnemonic process, such as illumination and grotesque characters.[304]

Any place, physical or imagined, might form a locus for establishing memory and, following on from that, might act as a key to meditation. In a similar way, allegory might be used to conjure up deeper meaning, an inner sight. Domestic settings were one popular genre in devotional works in the later Middle Ages, sometimes in those addressed to women with a religious persuasion,[305] sometimes to both sexes without differentiation. A late fourteenth- or early fifteenth-century text enjoined the reader to learn how to make a good bed, serving as an aid to memorising elements of the catechism and common formulas of late medieval belief. It progressed, associating each step with these thoughts: the litter for the bedstraw (shaking out the dust of sin with a two-pronged shakefork, one prong to keep one from sin, the other for devout prayer), the canvas next to the straw (bitter sorrow for one's sins), the mattress (holy meditations to put out sin), the two blankets (abstinence against gluttony, chastity against lechery), and so on. The chamberlain was one's conscience, a fearful usher, letting in nothing by the door of the chamber, the gate of the house or the windows of the five senes that might disturb the resting Christ, his chamber or his bed.[306]

In these ways, therefore, the sensory environment could be much richer than its superficial appearance. This applied not only where there was a clear iconographic link, an obvious visual clue, such as meditation in the presence of images or recollections inspired by heraldic devices,[307] but in many other day-to-day locations and practices, which memorial practice, the imagination and inner perception could associate with meanings of divine truth. Vision extended reality as we might perceive it. It was a powerful sense that brought ultimate realities to the visual present and created communities by perception.

* * * * *

The first part of this book has focused on ideas and beliefs about the senses in a wide range of contexts. The concern of the second part is to relate this evidence about perception to the daily realities of life. Making this link is not easy, and at times the association can seem obscure: although we have vast amounts of information about daily life in the Middle Ages, the occasions where we can make direct connections with sensory perception are limited. How, for example, can we show that ideas about the senses influenced diet, hygiene, dress, architecture or music? They surely did so, yet the means to demonstrate this can at times seem speculative. To provide the best evidence, I have selected for closer examination three similar but distinctive environments, all well documented: episcopal households of the thirteenth century and the first part of the fourteenth century, with a longer-term view of their accommodation; the households of Queens of England through the late medieval period; and the aristocracy at the end of the Middle Ages.

The domestic metaphor that often accompanied moralising works on the senses makes this study of the household context apposite, and we have just seen how contemplative works might encourage a richness in perception in the daily performance of the domestic routine. In the earliest domestic moralising treatise in this genre in English, *Sawles ward*, written probably between 1200 and 1220, the house is man; his Reason is master of the house, but Will is his wife, whose waywardness creates chaos when the servants follow her lead. There are two groups of servants. Those that serve out of doors, the external senses, quickly become unruly without Reason's supervision. Internal servants, the four cardinal virtues, are positioned to counteract their incursions: Prudence, the door-keeper, and her three sisters, Fortitude, Temperance and Justice.[308] Henry of Lancaster's devotional treatise of 1354 employed objects in an aristocratic, domestic setting to recall sin and to conquer it. The fox skin that hung in the hall, for example, so that the lord and all others could see it, was to bring sin to mind: it was there for all to see, with the eyes of the heart.[309] These texts suggest that we should look to the household for a valuable set of examples of the sensorium in practice.

Chapter 9

SENSORY ENVIRONMENTS AND EPISCOPAL HOUSEHOLDS OF THE THIRTEENTH AND EARLY FOURTEENTH CENTURIES

At the end of the fourteenth and start of the fifteenth centuries, the households of prelates were the target of moralists, who condemned their false, leisured life. Lollard sermons drew attention to the sin of pride. These clergy surpassed temporal lords in strong castles, exceptionally fine manors, proudly apparelled within, in halls, chambers and all other facilities; in their clothing, costly and furred; in their fat horses, gilt saddles and shining bridles; and in the richness of their households, 'sittynge atte mete eche day schynygeli', with vessels of gold and silver and a royal cupboard.[1] These were features of a sensualism that we have seen commonly condemned, but the effect of the Lollards and their precursors was to polarise opinion and their assessment of this way of living was deliberately extreme.

A range of sensory practices was current in households of bishops, some of which would not have been unfamiliar to those who made these plaints. Others employed the senses not in pleasure, but in devotion, notwithstanding the outward appearance of their establishments. A sermon on the feast of St Thomas Becket, for example, drew a contrast between his household as chancellor of England – with the future saint himself clothed in furs and the richest cloth that might be found, the best horses in the realm, and a hall which was strewed anew each day, in winter with fresh hay and in summer with green rushes – and the transformation that occurred once he had become Archbishop of Canterbury: wearing black cloth of middle price, he rejected silk and wore next to his skin a hair garment (which he changed only every forty days) and hair breeches infested with vermin.[2] The sensory contrast was intended to be stark: these were signs required for the progression to sanctity. Their retelling at a distance of two hundred years, however, should not obscure the evidence for arrangements in earlier episcopal households and the modifications that were made to meet the sensory requirements of the individual.

That the management of the senses could be central to the life of the episcopal household, or at least to the bishop, is apparent from the submissions made in support of the canonisation of Edmund of Abingdon (Archbishop of Canterbury 1233–40). One of our primary texts is a series of reports by four of his servants, showing how he managed sensuality in support of his devotion.[3] Richard of Dunstable, a Dominican friar, reported that for about ten years he had known Edmund well. Stephen the Subdeacon had been Edmund's chamberlain and private servant (*secretarius*) for six years. The Cistercian priest, Robert, formerly an Augustinian canon, had also been his chamberlain and private servant for many years. Eustace, a monk, had been Edmund's chaplain and chamberlain for a lengthy time.[4] A further testimony, of the same nature, was recorded by Brother Robert Bacun and appears in Matthew Paris's *Vita Sancti Edmundi*.[5]

Richard of Dunstable often ate the main meal of the day, *prandium*, with Edmund, but he had never known Edmund eat to satiety. He only deviated from this stance, eating the heavier meats (*grossis cibariis*) if he was markedly unwell, or if there was a great feast or an important guest. He rarely touched delicate foods; the more delicate or precious the food, the less he ate of it. During Lent – the long Lent, starting at Septuagesima – he abstained from fish (the food that was commonly eaten at this season), eating only bread and the accompanying dish, probably of cereals, vegetables or pulses (*pulmentum*).[6] He also abstained from meat at Advent.[7] At other times he avoided meat – and hence the carnality it induced – for up to a fortnight at a time, especially after he became a priest.[8] On days when he did partake of meat, he usually did no more than taste it, rather than consume it, making do with bread and *pulmentum*.[9] He very rarely ate the second meal of the day, supper (*cena*).[10] On Fridays throughout the year, he fasted on bread and water; and he never ate meat on Mondays, or on days when he was to say mass.[11] Sometimes, when fasting, he was content with bread alone, not tasting water the whole day.[12] He rarely drank before *prandium*, although he did sometimes in summer take boiled water; and it was suggested that this lack of liquid humour caused him to lose the hairs of his head and beard.[13] He also avoided spices and electuaries, shunning things that stimulated taste.[14]

His dress was designed to minimise sensuality and to mortify the flesh. His clothing was mostly of a grey colour, fitting for a priest: not of a luxurious cloth or one that was too poor, but of a middling quality. Against the flesh, he wore a hair garment which he carefully kept secret. The hair cloth was not woven, but knotted in the manner of a closely meshed net.[15] He would not allow either his body or his head to be washed.[16] Brother Robert Bacun recorded that when Edmund was a student at Paris, his mother used to send him a hair garment along with linen cloth. Piety appears to have been a hallmark of the women in his immediate family: his mother had used a garment of mail for mortification;[17] and, on her death, Edmund arranged for his two sisters to become nuns at Catesby.[18]

Edmund slept clothed and belted, covering his hands and bare neck with the hair cloth. He never slept in bed, but on a bench in front of the bed or leaning on the bed. Nor did he have a bedcover, sheets, mattress or a featherbed; instead he covered himself with his cope, pallium or rug (*tapetum*). He would get up after the first period of sleep and spend the time in prayer, meditation or reading, sometimes after matins and sometimes for the remainder of the night: it was rare, both in summer and winter, for him to sleep after matins. When he was treasurer at Salisbury, he used to remain in the church from matins through to the first mass of the day. If tiredness overcame him, he did not go to bed, but slept briefly in the place he was praying or seated, bending his head.[19] Rarely in church did he sit to pray; he stood up or knelt. His knees were calloused with the frequency with which he knelt.[20]

His first concern was the study and reading of the sacred text; he spent all morning until *prandium* in his cell. He regarded eatintime, as he did any time that was not spent in the service of God.[21] He was assidg, sleeping and riding as lost uous in his devotion and no mortal sin stained the innocence he had received at baptism.[22] Before opening his bible he used to kiss it.[23] There was no doubt about his chastity, his firm faith and hope.[24] His prayer and study were often accompanied by a flood of tears, sighs and groans as he poured out the most sincere devotion before God.[25] He did not wish to hold more than a single benefice, where he might reside.[26] The strength of his body became the strength of his spirit; his face was pale from fasting.[27] In death, his body was illuminated from Heaven and his face, which before had been ashen, became ruddy.[28]

He had no time for wordly affairs, or for money; he always spent more than he had.[29] Robert Bacun reported that when Edmund was given money by scholars he used to throw it out of the window, with the words 'Dust to dust, ashes to ashes.'[30] He was generous with his hospitality;[31] humble, patient and mild.[32]

These accounts show that Edmund intended to make the impact of his external senses as low as possible, not allowing them to admit through their gateways bad influence that would alter his soul. In contrast, his inner senses were turned to God and the messages that were conveyed out through the gateways of the senses reflected the goodness of the inner man. His external appearance – his modest clothes, the ruddiness of his complexion – the mortification of the flesh and his dietary pattern all expressed his virtue, suppressing any stimulus to the senses that might have a negative value. The reports, however, contain comparatively little information about the context in which Edmund lived, his household. We can see more of this in the case of his contemporary, Robert Grosseteste (Bishop of Lincoln 1235–53). He prepared an ordinance for his own household and one of his confessional treatises may have been addressed to the same group.[33] These works pay careful attention to the good order of the establishment and its moral discipline. The domestic statutes required all members of the household to be just, clean, honest and useful, as well as prudent in carrying out their offices. There was to be no dissension in the establishment; alms were to be given faithfully, liberally and prudently; guests were to be received courteously and affably; service at table was to be courteous, with moderate speech and gesture. The confessor was to look for sins of pride, in dress and in seeking good repute, and sins of the tongue such as murmuring against bad food, hard labour, sharp correction, permissions denied or the prolixity of fasting or vigils. He was also to watch for gluttony and drunkenness, as well as the taking or giving of bribes; evil thoughts or false suspicion; unclean delights; and a whole litany of sins ranging through the Ten Commandments to secretly taking the wax from candles or leaves from ecclesiastical books.[34] Grosseteste wished to create a wider environment in which virtue might be recognised and in which his senses, like Edmund's – and as many of his own household as possible – might be devoted to higher concerns.

The evidence taken in depositions by papal commissioners in 1307, in support of the case for the sanctity of Thomas Cantilupe (Bishop of Hereford 1275–82), is much more extensive. While it reflects a similar pattern as that of Edmund of Abingdon, of a bishop managing his senses for inner virtue, it does at the same time provide more of the domestic context in which the prelate lived. Cantilupe was born at Hambleden, near High Wycombe in Buckinghamshire, probably between 1218 and 1222. His father, William, was a prominent baron, steward of the household of King Henry III, with lands estimated to have produced between £1,000 and £2,000 in rent per annum, a level of income which would have placed him among a small number of the higher aristocracy; his mother, Milicent, was the relict of Amaury de Montfort, Count of Évreux; his uncle, Walter Cantilupe, was Bishop of Worcester (1237–66) and had a reputation for great sanctity. Thomas had four brothers and three sisters. He had an academic career: educated at Oxford, Paris and Orléans, he taught canon law at Oxford from c. 1255, where he was chancellor in 1261. He was a supporter of the Montfort cause and chancellor of England, under the Provisions of Oxford in 1265, returning to Paris and academic life after the battle of Evesham. He came back to Oxford in 1272 and again became chancellor of the university. He held under licence in plurality a series of benefices including Hampton Lucy and Snitterfield near Stratford on Avon, Dodderhill in Hereford and Worcester, Deighton in North Yorkshire, Bradwell in Essex, Sherborne St John in Hampshire and the archdeaconry of Stafford. He was appointed to a prebend of Hereford Cathedral in 1274, succeeding as bishop the following year.[35] Following a dispute with Archbishop Pecham, in which the Archbishop had excommunicated Cantilupe, the latter went to Italy to present his case to the Pope. Cantilupe died there in 1282. His bones were brought back to Hereford where they were to form the focus of a major cult, with further centres under the tutelage of Edmund, Earl of Cornwall, at Ashridge in Hertfordshire, where Cantilupe's heart was buried, and at Hambleden. At the last, the Earl of Cornwall built a chapel, decorated with a painting of the saint, and it was here that the knife Thomas used at table was venerated as a relic.[36] The final dispute with Archbishop Pecham, over Cantilupe's excommunication and an arrangement that might allow his absolution, delayed the return of his bones to Hereford for burial until 1283.[37]

The depositions lay considerable stress on Cantilupe's virtue in performing the

63 The shrine of Thomas Cantilupe, Bishop of Hereford, in the north transept of Hereford Cathedral. The knights around the base are probably soldiers of Christ, fighting sin. Cantilupe's bones may always have rested in a separate reliquary, perhaps on top of the shrine.

word of God, sustaining his clerical offices, and the impact this work had on others. Implicit in all the reports is the way in which he managed his senses. To this end, they note his baptism at Hambleden[38] – his entry into the Church and its sensory world – and the manner and frequency with which he conducted his devotions, maintaining the presence of holiness and its virtue. Witnesses recalled that, as a priest, he took part in a continuous round of services. Hugh le Barber deposed that there was a chaplain with him all the time in his household, even while studying in Paris, and that – quite unusually – he celebrated mass each morning before going to the schools, as well as all the other canonical hours; all the services were sung on feast days.[39] Several witnesses noted that he was greatly affected by the saying of mass, his tears were frequent and had to be wiped from his eyes, but in his devotion he did not behave as many did, like a hypocrite, waving his arms to Heaven.[40] His nephew and chamberlain, William Cantilupe, recalled seeing this especially at Bosbury, Snitterfield and Dodderhill. Richard de Kymberle noted that he rose to say matins after the first sleep – that is, during the night – both in winter and summer.[41] He was held to be pure and chaste, a virgin.[42] Others recounted details of his ministry, preaching the word of God. Ralph de Hengham had often heard him preach on feast days in the presence of the King and many magnates.[43] On visitation, he performed the duties of a bishop sedulously, preaching every day, hearing confessions, baptising children, celebrating mass and correcting delinquents.[44] In three villages which were disputed between the dioceses of Hereford and St Asaph

64 Bosbury: the entrance range, probably of the early fourteenth century, of the manor house of the Bishop of Hereford. To the left of the full-height arch is a small, pedestrian one, now blocked.

– under the control of Llywelyn, who believed they were Welsh – Cantilupe also preached, particularly against sorcery; on this occasion he spoke through a translator as he did not understand Welsh.[45] Brother Walter de Knulle considered that he had all the abilities and virtues requisite for a bishop outlined by Paul in his first epistle to Timothy.[46]

The Bishop was often to be found deep in devout meditation. One witness recalled frequently seeing him at Earley, the episcopal manor just outside Reading and a convenient staging post between London and his diocese, sitting at table or standing in his chamber, with his eyes looking up, so enrapt that it was believed he would not have seen anyone unless they came really close to him.[47] He also spent a good deal of time in the study of the word of God. William Cantilupe reported that Thomas slept little, often burning two candles a night while studying.[48]

As we have seen, this constant contact with Scripture and liturgy was recognised as creating a positive sensory environment, a moral force for good. It is not surprising, therefore, that Thomas attempted to avoid all disturbance to the recitation of services. Others were forbidden to speak at these times or to interrupt him: no matter how important they were, he would not turn his head.[49] At Paris, he frequently celebrated mass secretly – that is, privately – in his chamber. All strangers were excluded from the room while he performed the liturgy.[50] When anyone failed to say or sing the text correctly, he would make a sound or noise with his hand and make them go back to say what had been omitted or incorrectly spoken.[51]

In the account of his death at Ferento near Montefiascone, given by Robert Deynte, who had served in his pantry and buttery but was his chamberlain at this time, the fact that he had confessed to his chaplain was carefully noted. There were only three people present with him: his chaplain, the keeper of his wardrobe and Robert. The room was small, in a tower, with scarcely space for the Bishop's bed, that of his chamberlain and a small altar.[52] Nor did the power of Cantilupe's virtue cease with death. When his bones were returned to England for burial, it was reputed that they bled as they crossed land held by Gilbert de Clare, Earl of Gloucester, who had been a powerful rival for the temporal possessions of the church of Hereford.[53] From the moment the bones returned to Hereford it was generally believed that crops were more productive, fish more abundant and animals more fruitful.[54] It was common belief that he was a saint and that God worked miracles through him.[55] Even the Earl of Gloucester and his clergyman brother, Bogo de Clare, came to the tomb out of devotion. Others brought horses, oxen, flocks of sheep and other animals, all cured by the merits of the saint.[56]

In company with Cantilupe's virtues of devotion and sanctity, and the wide, positive impact that they had in sensory terms, the avoidance of personal, sensual stimulus formed an important theme in the Bishop's life. Witnesses spoke of his behaviour, customs and habits in terms of clothing and personal living arrangements, diet, and lack of pride in his own appearance and body. Former chamberlains and others described how Cantilupe wore a hair garment or a belt of hair under other clothing. Reports of these occur from both before and after he was elevated to the see of Hereford.[57] Nicholas de Warewik, a judge and royal councillor, who had been invited to dine with Cantilupe, came to the Bishop's chamber in the London house of the see. Cantilupe, however, was not completely dressed: he was without his cloak and with his other garments not yet gathered around his neck, and, to the distress of the Bishop, Nicholas glimpsed a knotted garment of horsehair under them.[58] Adam de Kyngesham had heard not only that he wore a hair garment next to the flesh, but that during Thomas' life, it had been thrown on the fire and it refused to burn – that is, the virtue attached to this item of clothing was such that it could not be destroyed.[59]

Hugh le Barber had been Cantilupe's barber in Paris and had subsequently served while he was at Oxford, keeping his chamber and making his bed. He was not, however, Cantilupe's closest personal servant; that was Roger de Kirketone, his *domicellus secretarius*, whom Cantilupe valued more than his other servants and who slept in Cantilupe's room. Hugh slept outside, except in the absence of Roger. He was nonetheless privy to some of Cantilupe's closest personal habits. Every day, Hugh had to clean large quantities of lice from Thomas' sheets and from his clothing and footwear – more lice than he had ever seen, either on paupers or the rich. If clothing were to be given away, he had to delouse it first.[60] He had asked Roger why Thomas had lice in this way, to which Roger replied that it happened naturally to some men more than others. Afterwards, when Thomas was chancellor of Oxford, Hugh, while making his bed, found between the sheets a very coarse belt of grey hair, about the size of a man's hand in breadth. It was also infested with lice. He cleaned the belt and put it under the foot of the bed. Roger, however, realised that he had seen it and he was made to swear not to reveal to anyone what he might find in Cantilupe's bed. He did not see the belt again, although after the Bishop's death, William de Montfort, the Dean of London and his executor, passed to him for safe keeping a hair garment that Thomas had used.[61]

While in the Bishop's chamber at Earley, Robert Deynte, also at this time a chamberlain, had given Richard de Kymberle, who served successively in Cantilupe's pantry, buttery and chamber, a small sack containing not only a coarse belt, such as friars wore externally, but also a rusty iron pauncer, or belt, with some thongs, about three fingers wide, that might be worn against the flesh.[62] Brother Robert of St Martin believed that Cantilupe wore a metal belt around his loins (*circa lumbos*) and also that he wore hair leggings.[63] William Cantilupe, the Bishop's nephew, who had been his servant and chamberlain during the period when Thomas was Archdeacon of Stafford and subsequently, had heard Robert Deynte refer to their master wearing a hair garment against the flesh, but he had not seen it. He believed the Bishop carried on his person the keys to a coffer which was thought to contain the hair garment and belts: he had seen protruding from the coffer some fastenings or buckles (*fibulas*) for the hair garments. Although he had been present on many occasions when Thomas retired to bed to sleep, when one would have expected a chamberlain to help him take off his clothing and footwear, William had not seen hair garments as Thomas wore them under a linen shirt which he did not take off, telling William and others present that priests ought to sleep in a shirt and breeches (*femoralibus*, leggings covering the thighs).[64] The chancellor of Hereford, Master Robert of Gloucester, had heard from Thomas's chamberlains that he wore a hair garment next to the skin and he also believed that he did not bath.[65] Robert Deynte noted that Thomas had the hair garment worn by his uncle, Walter Cantilupe, Bishop of Worcester, but that it had not seemed to him sufficiently abrasive. The new Bishop of Hereford had sent him to Oxford to have it made more coarse and he had brought it to Cantilupe at Prestbury about a month after he had been consecrated to the see. Thomas, however, had preferred to wear hair belts, as broad as a hand, rather than a hair garment. Robert had been responsible for washing these items, reporting that there were sometimes as many lice in them as a man might hold in his hand.[66] When, after his death, Robert washed Thomas' body, the impression of the hair belts remained. At this stage, the Bishop was not wearing hair breeches or leggings. Robert was not aware that the Bishop wore an iron breastplate either over, or under, the hair belt, or that he wore mail or iron belts.[67] To these servants, the Bishop's exemplar in denying sensory pleasure, in physically countering it through the mortification of the flesh, was outstanding, but it was also matched by modesty: it was not intended for show.

Asked if he had noticed any other especial austerity in the Bishop's life, Hugh le Barber answered that he had not, that Thomas was not a hypocrite and had no wish to show himself better than others. In his outer clothing and the arrangements for his bed (the soft furnishings of the bedchamber), he was similar to his peers. He used to wear a cloak (*mantellum*) within the house at *prandium*, and his clothing generally was like that worn by other prelates at university and while he was Archdeacon of Stafford. He used furs of vair (squirrel) on his clothing and one of his bedcovers was also furred with vair.[68] While his dress and gesture showed his humility, both he and his household used good, fine clothes and he had fine palfreys that carried him well.[69] Richard de Kymberle also confirmed that Cantilupe's external clothing, his riding habit, coverings of his bed and other accoutrements were like those of English prelates.[70]

Before Cantilupe's final journey abroad, William Jaudre, who had acted as caterer for Thomas' household in Hereford, buying fish and other necessities, had been given Cantilupe's clothing – a suit of four garments: a cloak, hood, tunic and *garnachiam*, possibly a supertunic, that is, it was composed like other suits of clothing worn by the upper classes. It was made of ruddy burnet (*brunetta rossa*), a woollen cloth of high quality. After Thomas' death and the return of his bones to Hereford, many people had asked William for parts of the clothing as relics. He had given away in this fashion small scraps of the *garnachiam* and tunic, but retained the cloak and hood (for which he had been offered £20 by a Welsh cleric) which were responsible for miracles. The cloak, in particular, was sought by women of Hereford for assistance in childbirth, and it had helped many. The garments, unaffected by age or moths, were shown to the commissioners.[71] Cantilupe's clothing thus became imbued with his saintly virtue: even though it was, as we shall see, similar to that used by other members of the episcopate, personal contact produced something that might have lasting distinction.

Cantilupe's ecclesiastical vestments were probably of considerable value. Master Robert of Gloucester had a vision on the night following Cantilupe's death, in which Thomas proceeded through Lyons towards the cathedral there (Cantilupe had attended both the First and Second Councils of Lyons, in 1245 and 1274). In the cathedral, he took off a cloak of perse (a prestigious dark blue cloth), vesting himself with the most beautiful white pontificals, before handling the host at an altar amid a throng of clerics in shining white vestments.[72]

Cantilupe also gave clothing away, to poor priests, both those serving his benefices and others; this probably came from the stock of the wardrobe or articles that he had previously worn. Hugh le Barber recalled two particular examples: a cloak of bluet, furred with miniver, which he gave to a vicar who was not serving one of his churches; and, when Cantilupe was at Paris, a cloak, supertunic and hood,

furred with miniver, and a tunic of good bluet of Ypres that he gave to William
Plantefolie on condition that the latter forswore taverns and gaming. Both Hugh
and Roger de Kirketone gave linen cloths and clothing to poor friars while
Cantilupe was at the universities.[73] William Cantilupe recalled that Thomas had
ordered clothing to be given to poor parishioners in those benefices he held before
he became a bishop, a practice he continued after he was appointed to Hereford.
He remembered this happening at Hampton, near Snitterfield and Bosbury. The
clothing was of coarse cloth (*grosso panno*) and of different colours.[74]

Further parts of the evidence show the arrangements of the Bishop's own
chamber. The bed – sometimes itself called 'a chamber' – was at this point the principal furnishing, often with elaborate textile hangings. Beds were in use throughout
the day, employed for seating, reading and for conducting business, as well as for
sleeping. Hugh le Barber noted that in Thomas's chamber there was no curtain or
screen, either of wood or stone, to prevent others lying in the chamber from seeing
Thomas's bed or him lying in it. The bed had linen sheets, but no mattress under
them. Instead there was a folded cloth or counterpane placed on the straw, upon
which the sheets were placed.[75]

William Cantilupe, the Bishop's nephew, also reported that he had linen sheets
and a mattress, just like other men. It was a fine and beautiful bed. Both he and his
predecessor, Thomas Bathewyng, used to sleep in Cantilupe's chamber and he
would call to them for a light and other things. He also believed that at times
Thomas did not sleep in the bed, but somewhere close to it – presumably on the
floor – as on one occasion he had been called and had come very promptly, slightly
disconcerting Thomas, whom William believed he had found getting into bed from
wherever he had been sleeping, but as it was dark he could not see where this was
or whether Thomas was dressed. He remembered, however, that this had taken
place at Earley, that is, during Thomas's episcopate, but he could not recall when
exactly it had happened.[76]

John Bute, too, believed that Thomas often slept on the ground, with a few cloths
between him and the ground, some more for his head and covered by a few more,
rather than in his bed. He recalled seeing this in Normandy, where Thomas spent
more than a year and a half and where Bute was his pantler, butler and sometimes
his chamberlain. He did not believe Thomas realised that he had noticed this: Bute
was in the habit of going into the chamber first thing in the morning, when the
chamberlain had gone to rouse the rest of Thomas's household, to speak about business relating to his domestic offices. A green curtain in the chamber obscured the
bed and the place in which he slept on the floor, and Bute did not enter this area.
There was, however, both summer and winter, a fire in the chimney opposite the
bed and its light shone through the curtain. Bute believed that Thomas was not
aware that he could perceive these arrangements, or that Thomas had arisen from
where he had been sleeping to get into bed to talk to Bute.[77] Robert Deynte reported

that Thomas did not often sleep in bed, but would lie dressed either on the bed or near it, on a rug. He would rarely lie down, but rather sit on the bed or near it, where, working with his books or writing, composing collations or sermons, he would fall asleep. Deynte used to sleep in the same room and frequently came to him to attend to the light in the lantern when he was working; Deynte became so tired – while Thomas did not sleep – that he sometimes thought he should secretly absent himself.[78]

Thomas did not sleep in the day, particularly after *prandium*, when others slept.[79] After one lunch, when two clerks, one Master Robert of Gloucester, had failed to settle a piece of business, Thomas called for a pen, ink and parchment in order to resolve the question himself; these were brought to him and placed on his bed. Robert returned to his room to sleep a little – it was after midday – only to be disturbed shortly afterwards by Cantilupe, with the pen, ink and parchment, so that they might do the work.[80]

Just as Edmund of Abingdon's servants had reported on the little time that he slept and his custom of not using his bed, Thomas Cantilupe similarly rejected the sensory pleasures of rest. In external appearances, however, the material furnishings of the chamber were like those of other bishops: the rejection of the sensory environment was a personal one.

Thomas' speech – an important indicator of virtue – was consistently described by the witnesses as being honest, truthful, beyond reproach and never deceitful; he never lied or uttered false words and always kept his word.[81] He was no gossip in childhood and his younger years; neither did he listen to jesting and idle tales.[82] His humility was apparent in his speech and gesture, with all ranks, and he spoke freely with both poor and rich.[83] Although he did not speak Welsh, the only oath he appears to have used was to swear by Dewy Sant, Saint David, that something would or would not be.[84] He rebuked others for vulgar language, including his future brother-in-law, Robert Tregoz, who had suggested crudely that Cantilupe's sister should be his wife.[85] Adam de Kyngsham found him affable and good company at table.[86] Brother Walter de Risebury noted that his speech was a little slow and dull, and that he was so occupied with things that he often had his confessor, Brother Henry de Belinton, a Franciscan, who was regarded as the best preacher in that region, preach before him when he visited his diocese.[87] But for all that, he had the voice of a saint. Nicholas de Warewik reported an incident that occurred while Cantilupe was at the Earl of Cornwall's castle at Wallingford, celebrating mass at Whitsun. When he started to sing 'Veni Creator', a great mass of small birds came and landed on the glass windows of the chapel. They went away when Thomas stopped singing and returned when he started again: Thomas' voice was a manifestation of his sanctity.[88]

Thomas' rejection of sensuality extended notably to his sense of taste. On the one hand, this led him to reject the consumption of meat and the carnality that was associated with it, to observe strict regimes of fasting, as St Edmund had done; and on the other, to note or even put forward for himself the temptations and stimulation of good food and wine, either in order to reject them, or to dissipate their effects.[89] Several witnesses reported that he was sober and that he had not eaten to satiety for many years, for the sake of virtue partaking of little food and drink.[90] He ate so little that Master Robert of Gloucester, while sitting beside him at a meal at Earley, was concerned that he would not survive and that he, Robert, would lose the promotion he hoped for: Cantilupe ignored him, then told him to eat and drink what he wanted but to be quiet and leave him in peace.[91]

His abstinence was tested by deliberate temptation. Thomas liked lampreys from the Severn above all other fish, as his household knew well. Nonetheless Nicholas de Warewik had seen him often abstain from these lampreys and other delicate and costly foods.[92] Cantilupe sometimes asked for them to be specially prepared, or for partridges or other birds; he would then smell them, or ask those around him whether they thought they were well prepared and smelled good, and send them away, perhaps tasting a single bowl. His wine was mixed with water and his ale was weak.[93] On Wednesdays and Fridays he drank water rather than wine, although on some Fridays he substituted weak ale for the wine.[94] William Cantilupe noted that Thomas was sober because he only ever drank a certain measure of liquor: he had never found in him, after drink or a meal, any alteration of the senses or morals, and he was in the same state after as he had been before.[95] John Bute, who had been his butler, noted that he had set before him at table one drink of white wine and another of water, or of water or weak ale if he was not drinking wine. His silver wine cup was small, holding no more than a medium-sized saucer, and he never drank his wine in one draught.[96] Hugh le Barber reported that it was common for him, when he took food, only ever to take things from the first dish and give the rest for alms.[97] William Cantilupe recalled that when delicate foods were set before him he rarely ate more than one or two dishes, but smelled the dishes and sent them away saying that as he had sensed them, he would not eat. When a variety of pottages and sauces was set before him, he often mixed them to diminish the flavour.[98]

The pattern of fasting among the devout, upper-class laity in the late thirteenth century resulted in abstinence from meat on about half of the days of the year; on these occasions fish usually replaced flesh as the main item of consumption.[99] The pattern of fasting that Thomas undertook similarly encompassed Wednesdays, Fridays and Saturdays, but in a more extreme fashion than many: he refused meat on Wednesdays, that is, he would eat fish on these occasions, but on Fridays he confined himself to bread and water, and on Saturdays abstained even from that, unless he was unwell – although at some periods in his life witnesses reported that he ate fish on this day and on Fridays. On the vigils of the Marian feasts he fasted on bread and water alone, again unusually for the secular clergy. He observed two or three advents in the year, following the habits of the exceptionally devout, abstaining as a prelude to the feasts of Christmas, Pentecost and St John the Baptist, even though other members of his household ate meat at these times. In the case of Christmas, his advent encompassed the six weeks preceding the feast rather than the more usual four; at Pentecost, he celebrated the advent of the apostles, running for nine days, from Ascension Day, rather than the more common pattern of abstinence for the three days preceding Ascension. Together these habits would have added more than two months of abstinence to the general expectation, making this a feature of his dietary regime on average for more than two days in every three. He did not drink after *prandium*, except on rare occasions when there were guests; neither did he usually have the second meal of the day, *cena*.[100] Bishop Swinfield noted that Thomas did not participate in the common drinking that usually happened in the afternoon between *prandium* and *cena*, unless there were guests of importance; even then, he often simply lifted the cup to his mouth and did not taste the contents.[101] Deliberately setting aside the pleasures of food and drink and the challenge these offered his senses, Cantilupe's pattern of consumption marked him out as a man of virtue. Unlike the mortification of the flesh by his garments, however, his abstinence was publicly known and clear for all to see.

Descriptions of Cantilupe's own body help us see how he managed the senses. The sacrist of St Bartholomew's in London, who had come to know Cantilupe while he was a canon of St Paul's, reported that he often sat below him in the choir and that he sang exceptionally well. In language which we have already seen has other resonances, he described Thomas as having the face of an angel; he was white and ruddy (*albus et rubicundus*), with a good beard and a long nose; the hair of his head and beard was partly white and part flecked with red.[102] Hugh le Barber recounted that Cantilupe was naturally discreet and prudent. He used all his senses for good, and not for any evil.[103] Another witness reported that he was sober and slim.[104] Bishop Swinfield noted that Thomas had a good complexion, despite his dietary rigours;[105] although Brother Eadmund de Aumari reported that abstinence, particularly on Fridays, made his face more pale than on other days.[106] William Cantilupe held that he was well proportioned and well formed. When he was on horseback, he rode with a stoop, looking at the ground, which William believed he did out of humility as he held his own physical appearance in contempt. He had two palfreys, one a beautiful morel, the other a grey. He often declined to ride the former, however, as he believed it would create pomp and attract the looks of bystanders.[107] These practices all point to him denying the pleasures of the senses of this world, or setting their experience at little value. The Bishop also had persistent illness, a suffering of his own, that added to his virtue. Three witnesses reported that Thomas often suffered from pains in his side, torsion or colic in his belly.[108]

Several witnesses recollected that they had never seen Cantilupe become angry, despite provocation. Gilbert de Clare, the Earl of Gloucester, had threatened him with a sword and had high words with him, but Thomas had told him that, say what he might, he would not become angry.[109] He was meek and mild, 'like a lamb', of gentle disposition (*deboneyre*) and long-suffering.[110] He would not be provoked to bad words: when faced with two members of his household, William de Neyvill and Peter de Ocle, who were both melancholic of disposition, impatient and garrulous, arguing between themselves and answering Thomas in a short and irreverent manner, he used to put up with their foolish answers benignly.[111]

The presence of women, lepers and the Jews provided challenges to the Bishop's virtue. In the first case, he sought to avoid contact and especially touch. Women might be a source of corruption: he was modest and chaste; he never freely had any consort with women, however old they were, even those serving in the kitchen or some other office and living in his household, although he sometimes found himself with them or they ate with him.[112] Discussing a nephew of Cantilupe's who was in the schools at Paris at the time he was there, Thomas lamented the attitude of youth to women. If any beautiful woman had looked at him, he would have blushed and averted his eyes, drawing his hood over his face so that he might not see her.[113] He hardly spoke to his own sisters when they came to see him,[114] and after a day in his household he courteously persuaded them to leave. The conversation of women he found vain and empty.[115] After he became Bishop of Hereford, he declined to kiss his sisters, even though he had done so before and it was the custom of English prelates to kiss family members. On one occasion, at Bosbury, he had declined to let Juliana Tregoz kiss him on the mouth, holding out his hand for her to kiss, which upset her greatly.[116]

He offered charity to lepers, although we are not told a great deal more than that he gave alms to them and to other unfortunates.[117] There was a noted exception to Cantilupe's mild words when it came to the Jews.[118] He had an abhorrence of them as enemies of God and rebels against the faith: if they did not wish to convert to Christianity, and it was to this end that he preached to them, he argued that they should leave England. Implicit in this was that they had no place in a Christian kingdom where they might derogate from the virtue of protection afforded by the faith.[119] Even conversion did not remove their taint: he argued successfully in the King's council that a converted Jew should not be accorded a position where he had power over the life and limb of a Christian.[120]

Like the descriptions of his clothing and the furnishings of his chamber, the arrangements for the management of his household were similar to those of other great ecclesiastics. Two witnesses, one a clerk in orders who had served in Thomas's chapel, pantry and chamber, the other his chamberlain, knew that Thomas had the virtues of prudence and discretion, through his good works and because he ruled his household well and wisely.[121] While he was at Oxford, he had entertained in London the future Edward I and many of the leading magnates. The household's silver had been brought to town especially for the event. Hugh le Barber had been incautious on this occasion. To his master's chagrin, Thomas's silver gilt, covered cup, weighing some seven and half marks, had been stolen by a servant who had been engaged to assist Hugh.[122] The expenditure on two visits of Edward I to Hereford during Thomas' episcopate was believed by one witness to have been £1,000,[123] an extraordinary sum, but a level of disbursement that might not implausibly have been committed to this entertainment.

The household was disciplined and we know that Thomas disliked disturbance or disruption.[124] Servants who erred were warned that if they did not amend their ways, they would be dismissed, but with their salary.[125] Hugh le Barber, Anselm the Marshal and another servant, wishing to go out at night, could not use the main household gate, but chose to climb over the wall of the church at Hampton. Brought before Thomas the following morning, they were told that he wished to dismiss them; when they sought his pardon, he warned them that if they ever offended again, they would be permanently excluded from his household.[126]

It was also indicative of Thomas' virtue that he had a well-regarded household (*bonam familiam*): the environment clearly had an aura of beneficence that went with it, derived from his virtue and reinforced at many points by his conduct and that of his servants.[127] It was only the younger members who survived to give testimony in 1307, but they provided strong evidence to its stability and merits. Many personnel served for long periods of time in a range of offices. Hugh le Barber had served Cantilupe first in Paris, as his barber for three years outside the household, before formally joining the establishment and coming to Oxford with Cantilupe. He served him continuously for six years before Cantilupe became Bishop of Hereford and for two years afterwards.[128] Richard de Kymberle, in 1307 the rector of Saham Monachorum in Norfolk, had spent a decade in the household, joining two years before Thomas became Bishop of Hereford and remaining until his death.[129]

Others served both Cantilupe and his successor, Richard de Swinfield, who had also been a chaplain to Cantilupe. William Cantilupe had spent many years in the service of his uncle in both England and France, going on to head the list of esquires in Swinfield's household in 1289–90.[130] John de Kemeseye had been with Cantilupe at Oxford, was in his household until his death, and then served in the household of Bishop Swinfield and remained closely associated with him until Swinfield's death in 1317.[131] Robert Deynte had been a junior servant in Cantilupe's pantry and buttery, before working as the Bishop's chamberlain, again first in a junior capacity. He was also an esquire in Swinfield's household in 1289–90 and his chamberlain at the time of a miracle that cured the new Bishop of the stone.[132] William, the watchman or wait, had been in the employ of Cantilupe's predecessor at Hereford, John Breton, and survived to meet the commissioners in 1307, when he was described as *simplex*, a layman, who gave his deposition in English and who knew little of the secrets of Cantilupe. He also served Swinfield as his household watchman until at least 1289–90.[133] Other servants, such as Harpin the Fowler, served both Cantilupe and Swinfield.[134] This continuity at Hereford marked a difference from the household of Peter of Aigueblanche (Bishop of Hereford 1240–68), who had employed many Savoyards as his domestic servants and who had spent much of his episcopate outside England.[135] Household service was an honourable employment and, in death, Cantilupe remembered his servants: Robert Deynte enjoyed the typical perquisite of his office, receiving Cantilupe's bed and its furnishings, along with 20 marks. Others received cloth, furs and horses; his cook, John, had 15 marks.[136]

Witnesses, responding to the second article of the commissioners, on what study and works Thomas had undertaken, what positions he had held and in what state he had left them before he became a bishop, noted that he built and rebuilt halls, chambers and barns on his estates, including a hall and chamber in the church house at Hampton Lucy, two barns and a chamber for the church estate at Dodderhill, further works at Bradwell, Sherborne St John and Kempsey, as well as rebuilding the chancels of Coleby and Winteringham.[137] At this period, one witness noted, Cantilupe was in possession of lands and ecclesiastical dignities worth about 1,000 marks per annum.[138] The see of Hereford, however, was among the least well endowed of English bishoprics and the number of residences used by the bishops in the late thirteenth century was limited. Although the bishopric had more than twenty manors, only a few were used extensively. The dating formulas in acts in Cantilupe's episcopal register show that Sugwas, Bosbury and Whitbourne were especially favoured as residences; the London house, Montalt, was used when the Bishop was in the capital, or when he stayed at Kensington, an Abingdon Abbey manor, and Tottenham, where the Priory of Holy Trinity within Aldgate held the advowson. The manor of Earley, close to Reading, was used as a resting place on the road between London and the bishopric, and may also have been a convenient place to retire to while remaining close to the capital. The twelfth-century palace at Hereford itself seems to have been little used, except for formal occasions.[139] Bishop Swinfield had a similar pattern of residence, favouring Sugwas, Bosbury, Whitbourne and Prestbury, but rarely Hereford itself.[140] There were other episcopal residences at Ross, Ledbury and Bishop's Castle. Little now remains of the fabric of these buildings, with the exception of the palace at Hereford and its clerestoried timber hall, a magnificent work of the late twelfth century. At Bosbury, an entrance range may date to the early fourteenth century; and cellars under the Royal Hotel at Ross may have been part of the episcopal house there.[141] This absence is a disappointment, given that Cantilupe had an interest in building. In addition, we know nothing of the decorative arrangements within these structures, although, if we conclude that Cantilupe lived outwardly in the same style as other late thirteenth-century bishops, as members of his household attested, we should expect that these houses were well ornamented and furnished with textiles.

If evidence is lacking for the built environment of the see of Hereford, there is much that could be written about the houses of bishops in general; and the architectural testimony of episcopal buildings is important if we are to consider the totality of the sensory environment that these men set out to create. Monastic authors took the war against sensuality to all the temptations of the world. Among those things that might distract the eye from its true purpose of inner contempla- tion, Aelred of Rievaulx highlighted the superfluous beauty of fine buildings. Although writing about the cloister, many of the things he mentioned were found in episcopal residences: he sought a construction without the distractions of dressed stone (that is, it should be of rough stone), ornament and decoration, be it rugs on marble floors or stories of peoples and battles of kings on the walls.[142] This senti- ment was echoed by a French Augustinian canon, Hugh of Fouilloy. Episcopal residences were built on a scale to rival churches; bishops delighted in painted bedchambers, the figures dressed in rich, coloured clothing – Trojans in purple and gold on the walls, Greeks with their arms and shields. The stone and timbers were clothed, but even old clothing and bread were denied to poor Christians at the gate.[143] While these tracts were addressed to a monastic audience, the buildings they described were not unfamiliar in secular society.

Every see in England had a series of residences: as the Cantilupe enquiry suggests, building could be a virtuous use of ecclesiastical funds. In default of residences which he may have built, the characteristics of medieval episcopal accommodation and some of the principal changes can be seen at other sees. In the eleventh and twelfth centuries, the Anglo-Norman episcopacy built a fine series of castles and palaces, often close to the primary church of the diocese; in the following centuries, there is more evidence for residences, less often fortified, on episcopal manors distributed around the dioceses, with, in addition, a London house. There were changes in this accommodation in the fourteenth and fifteenth centuries, in terms of numbers of chambers, with thought given to overall planning and quality of the residence. At all points, the developments paralleled those of secular lords.[144]

In the early twelfth century, Bishop Roger of Salisbury (1107–39) constructed four castles or residences within castles – the first at Kidwelly, c. 1106–15, and three more within his see, at Devizes, Sherborne and Old Sarum. His residence at Devizes was described by Henry of Huntingdon as among the finest in Europe.[145] At Sherborne, between 1122 and 1137, he built a fine house, perhaps on a pattern imitating a Benedictine cloister, around a courtyard, within an area surrounded by a curtain wall, itself largely encompassed by a moat and lake or marshy area. The north range of the inner courtyard had two chapels, one above the other, with the upper one, marked by its fine architectural detail, for the Bishop's own use. At Old Sarum, Roger built, for the bishops, a new palace arranged around a courtyard, again with a two-storeyed chapel, within the castle, reworking the former royal resi- dence there. In the later Middle Ages, Bishops of Salisbury made much less use of these castles and preferred to reside on episcopal manors. Bishop Mitford (1395–1407) spent much of the last year of his life at manor houses at Woodford and Potterne; other Bishops resided there and at Sonning, Ramsbury and

65 Sherborne Castle, built by Bishop Roger of Caen: the inner courtyard, looking towards the north from the south range, formerly the great hall. The Bishop's own chapel is decorated with fine stonework: the arcading is on the external face of the chapel.

Chardstock.[146] The plan of Sonning, which was recovered by excavation in 1912–13, shows a thirteenth-century hall over cellars with a two-storeyed chapel, adjacent to the Thames. In the mid-fourteenth century a new hall was built, along with a new kitchen and domestic offices. By the fifteenth century, the hall, Bishop's chambers, other chambers and domestic offices stood around a small courtyard on one side of the hall, with an outer court and gatehouse on the other.[147]

At Salisbury, a new palace was built with the transfer of the bishopric to New Sarum; at other sees, the investment in major palaces continued through the late thirteenth and early fourteenth centuries. The great hall of the Bishop's Palace at Lichfield was built between 1299 and 1314 by Bishop Walter Langton (Bishop of Coventry and Lichfield 1296–1321): in 1311–12 the walls were decorated with paintings of the life of Edward I, whose treasurer Langton had been, depicting not only the monarch's achievements as a Christian king, but also his martial successes.[148] The hall at Lichfield, with its attached private apartments, was similar in scale to the one at Wells, finished by 1292 under Robert Burnell (Bishop of Bath and Wells, 1275–92), another of Edward I's principal advisers. Bishop Burnell's work updated a major palace constructed by Bishop Jocelin (1206–42), who also built a substantial residence at Wookey.[149] The range of accommodation for this see can be seen from accounts for work done at Dogmersfield, the staging post in the journey between the see and London – much as Earley served Hereford – and a point for supplying provisions to the Bishop when in the capital.

66 Old Sarum Castle: parts of the palace of the Bishop of Salisbury, built by Bishop Roger. The range in the foreground contained chapels; immediately behind it were the Bishop's private chambers; on the embankment at the back, the kitchen tower and garderobes.

67 Old Sarum: the undercroft below the Bishop's private chambers.

68 Wells: the ruins of the great hall in the Bishop's Palace, facing east.

An episcopal house may have been established at Dogmersfield in the twelfth century. Henry III gave an additional seven acres of pasture for the park there in 1228.[150] In the first part of the fifteenth century it had a cloister or quadrangle, referred to as the 'lord's cloister', to distinguish it from a more general court, and it was in this area that the Bishop's chamber was located. That part of the cloister between the chapel and the hall was newly paved in 1426–7, early in the episcopate of John Stafford, and a new garden was also made, probably in the same area, between the hall and the Bishop's chamber. Just before Easter 1427, a carpenter made an Easter sepulchre and a chest for the Bishop's chamber. During the year, repairs were made to the guests' chamber next to the Bishop's chamber, in the same area of the cloister. There was a separate chamber for the steward; other rooms, as well as the kitchen, the larder and buttery, and all the Bishop's chambers, were cleaned against the Bishop's visit. The house also had a cowshed and a smithy, probably in the outer court.[151] In 1432–3, the straw from the manor's demesne lands was reserved for beds in the house and for litter in the stables; a carpenter was engaged to make new beds in the chambers; the walls of the hall and the chambers were whitewashed, the roofs were generally put in order, and the wall at the west side of the Bishop's garden was repaired. More than 46s. was spent on constructing a new lodge in the park and 62s. 3d. was spent assembling a woodpile for fuel for the hall and the chambers of the house against Christmas. Among the servants of the manor were a park keeper and a gardener for the Bishop's garden, the latter paid an annual stipend of 13s. 4d.[152] The property also had a substantial fishpond – as did nearby Frensham, the property of the Bishops of Winchester, and Fleet, the property of the Prior of St Swithun's, Winchester. Pike and bream were drawn from Dogmersfield for the table of Bishop Ralph of Shrewsbury when he stayed on the manor in March 1338, while most of his household had preserved marine fish; in 1366, pike from the pond were also chosen by John Barnet (Bishop 1363–6 and treasurer of England 1363–9) as a gift for his master, Edward III.[153] The house was used throughout the fifteenth century; it was here in November 1501 that Henry VII and his son, Prince Arthur, met for the first time Catherine of Aragon, who was being escorted to

69 Bishop Bekyngton of Bath and Wells in his chamber being presented with a book by Thomas Chaundler, later Dean of Hereford. The Bishop is seated on a cushion on his chair, below a celure; the wall is hung with textiles, cut around the window, which has lights with armorial glazing; and the floor is tiled. From Chaundler's *Liber apologeticus*, copied between 1459 and 1465.

London by the Bishop of Bath and Wells. Catherine, who had been resting, met them in her third chamber, that is, she had a suite of rooms assigned to her. Subsequently the parties confirmed in person the contract for the marriage between Arthur and Catherine, and the chamber of the Princess was used for dancing.[154] The house – of which no standing trace remains – was in part habitable as late as 1624–5, when it had a garden and orchard attached.[155] The Dogmersfield house was typical in its provision of accommodation for the Bishop set away from most of his household except his principal guests; in providing further areas where he might be apart from the world, as in his garden; and supplying virtuous food-stuffs, freshwater fish, that would have marked out both his rejection of carnality and his status. It was a comfortable and commodious residence, fitting for a prelate of high rank.

That the residence of a bishop might be suitable for a meeting between a king and his future daughter-in-law is a reminder of the scale of some of these properties. Henry de Gower, Bishop of St David's 1328–47, reconstructed two major residences in his see, at St David's and Lamphey, and he may have carried out substantial works at Swansea and possibly also at Llanddewi.[156] At St David's, the palace had two large halls, one for pilgrims and secular visitors, the other to accommodate the Bishop's household when it was in residence. Although on a smaller scale, two halls also featured in the Bishop's works at Lamphey, maintaining an important distinction

70 St David's Bishop's Palace, viewed from the cathedral. The Bishop's accommodation, closest to the cathedral, had its own hall, running across the south-east side of the courtyard. A further hall, with an impressive rose window, for the entertainment of pilgrims and guests, runs at a right angle to it.

between the episcopal establishment and general visitors, and the disturbance the latter might bring.[157] Perhaps slower than secular establishments to adopt some aspects of architectural fashion – even establishments such as the rebuilt palace of William Wykeham (Bishop of Winchester 1366–1404) at Bishop's Waltham maintained rooms that would be shared by servants in common, grouped by rank, for knights, esquires and clerks, rather than chambers for individuals and their immediate servants – episcopal builders might nonetheless create palaces that were employed for state occasions of great importance. Intimacy of accommodation was reached with a proliferation of chambers in the fifteenth century; for example, at Bishop's Waltham, in the creation of a range of lodgings late in the episcopate of Henry Beaufort (Bishop of Winchester 1404–47).[158] These were typical of the scale episcopal residences might achieve.

The architecture of these establishments linked directly to sensory arrangements in the separation of chambers, provision of gardens, fine buildings and landscaped settings. Episcopal inventories of the early fourteenth century give valuable evidence of the furnishings, the clothing and goods of those who lived in these establishments, which allows some further connections to be drawn out. Inventories compiled on the deaths of Thomas Bitton (Bishop of Exeter 1292–1307), Richard Gravesend (Bishop of London 1280–1303) and John de Sandale (Bishop of Winchester 1316–19) provide detailed descriptions of their material possessions.

71 The palace of the Bishop of St David's at Lamphey: the western hall, built in the thirteenth century. Henry de Gower built a second hall to the east, almost completely separate from the other accommodation of the palace.

Some of these we can tie directly to aspects of the senses and their management, as the references to St Edmund, Thomas Cantilupe and their households have shown. Winchester was the wealthiest see in medieval England. John de Sandale had a career as a royal administrator: he had been clerk and, from 1295, keeper of the great wardrobe; he managed the finances for the army in Gascony in 1296–8 and in Scotland in 1300; he was chamberlain of Scotland in 1305 and chancellor of the Exchequer in 1307, rising to become treasurer in 1310–11 and chancellor of England from 1314 to 1318. He was treasurer of Lichfield Cathedral in 1310 and by 1315 had gathered ecclesiastical preferment worth about £850 a year, including eight prebendal stalls and ten rectories. His tenure of the see of Winchester was brief, and the Crown acted swiftly on his death to secure the see and the Bishop's finances, to ensure that debts to the Crown were recovered. The Crown's administrators valued his estate at the very substantial sum of £5,948, including more than £755 in movable goods, jewels and cash at Southwark, £168 worth at Kirk Sandall, near Doncaster, and set a value of 200 marks on two carthorse-loads of silver and jewels which the Bishop's treasurer had removed from Wolvesey Castle; he and other servants had taken another horse-load of cash from Farnham to London to cover the expenses of the Bishop's funeral, and other members of the household had taken rings and further items worth £8 5s. 10d.[159]

The total receipts of the executors of Bishop Thomas Bitton, calculated in 1310, were of a similar order, £5,395, of which approximately £1,030 was in cash in the Bishop's treasury. Bitton came from a Gloucestershire family, some of whom held prominent places in the Church in the West Country; he had been Dean of Wells before his appointment to Exeter, and his uncle, William, had been Bishop of Bath and Wells.[160] Richard Gravesend was probably a relation of the Bishop of Lincoln who bore the same name. He had a university education, although it is not known where he studied. After his appointment to London, he also served Edward I abroad, in Gascony and on diplomatic missions to France and the Low Countries. On his death, his treasury was much less well endowed with cash – £148 – and the Bishop of London was much more heavily indebted than the Bishop of Exeter.[161] Gravesend's executors accounted for net receipts of approximately £1,580 and his overall estate was probably worth about £3,000.[162] The see of Exeter had twenty-three manors and that of London twenty-seven.

Besides cash, the treasuries of the three bishops contained jewels and stones, objects which we have seen were commonly believed to have a range of sensory consequences. Bishop Bitton was buried with a ring with a large sapphire; his pontifical ring, delivered to the King, was worth £1; and he bequeathed one ring with a diamond worth 26s. 8d. and another ring worth 10s. The other rings in his treasury were of small value, even though of gold: twenty-three were valued at 24s. in total, three at 2s. and a further seven were together worth 55s. He had three gold brooches, worth 18s., and two paternosters, worth 2s.[163] Gravesend's rings were equally modest: a pontifical ring bequeathed to Robert Winchelsey (Archbishop of Canterbury 1293–1313), and another pontifical ring worth 26s. 8d. There was in addition a gold ring with a crystal, worth 18d., and another, broken, with a ruby, worth 3s. 6d., five small rings worth 3s. 4d. and a brooch valued at 26s.[164] Sandale's jewels were of a completely different order. At Southwark he had a gold ring with a good ruby, valued at the exceptional sum of £100. Another ring set with a ruby was valued at 20 marks. While a group of nineteen rings set with rubies and emeralds were valued at 1s. each, he also had a further fourteen valued at more than £1, some of which were striking pieces, such as the ring set with a great oriental sapphire valued at 53s. 4d. Beyond these there were also a pontifical ring with a great sapphire surrounded by twelve small emeralds, valued at £5 6s. 8d., and a great ring of crystal, with a good sapphire – which, the inventory recorded, was believed to possess special virtues – surrounded by small rubies and emeralds and kept in a copper box, but valued at no more than 26s. 8d. The Bishop had three gold brooches: one set with three rubies and three emeralds was valued at £2.[165]

Besides these jewels and the benefits they conferred, all three bishops had small groups of stones and other items of virtue. Objects that may have possessed powers included, among Bitton's effects, an ox horn decorated with silver, valued at 10s.;[166] Sandale had a belt, held to be a relic, which had belonged to the former Archbishop, Robert Winchelsey.[167] Serpents' tongues, valued for their protection against poison, were present in all three establishments. Bitton had several beryls and two serpents' tongues, their mounts decorated with silver, besides some coral and another small serpent's tongue.[168] Gravesend had two 'trees', perhaps of coral, one with five serpents' tongues, the other with four.[169] Sandale had a gilt jewel, like a leaf or leaves, with a serpent's tongue, valued at 40s., nearly eight times the value of any of the serpents' tongues in the other two households.[170]

The investment in precious metals for use at meals and drinking is equally evident in all three households. Bitton had two substantial pitchers, probably of silver, one for wine, the other for water, and an impressive list of cups, headed by two of gold, worth £18. He also had a cup called *Hulle*, which was an heirloom and which was returned to the family by his executors.[171] All three bishops had sufficient silver plate to set at least a dozen places at table, and possibly between twenty and thirty depending on the number of courses in the meal. Bitton had forty-four silver spoons of the same set (besides six gold ones worth £7), forty-eight dishes and fifty saucers of silver, as well as another eight with each probably of a different size;[172] Gravesend had forty-four silver dishes, thirty saucers and forty-one spoons;[173] and Sandale had forty silver dishes and thirty-eight saucers.[174] The weight of silver in each case was similar, ranging from 52 lb. (Gravesend) and 56 lb. (Sandale) to 63 lb. (Bitton).

To the shining brilliance of this investment was added the piquancy of spices. All the inventories record spices and sweet confections in considerable quantity. Besides

77 lb. of sugar, 53 lb. of ginger, 29 lb. of cinnamon, 21 lb. of black pepper, 38 lb. of saffron, mace, cloves, fennel and liquorice, among Bishop Bitton's possessions were two gourds holding *gyngebrad*, further small boxes of the same and another confection made with pine-nuts.[175] Sandale had an 80-gallon barrel with sugar and other spices; some 38 lb. of saffron, 68 lb. of ginger, 40 lb. of galingale, 30 lb. each of cloves and mace, 30 lb. of pepper and 12 lb. of cinnamon, besides 6 lb. of sanders, often used for the red colour they gave to food.[176] Bitton's kitchen had a brass mortar, weighing 45 lb., for spices, along with an iron pestle and a stone, possibly a coarse one, for further grinding.[177] He also bequeathed to his successor at Exeter a hand-mill for grinding the ingredients for sauces.[178] His cooking equipment included a brass oven and cover, weighing 14 lb., probably for delicate work using charcoal, as well as a pair of basins for tarts and a frying pan.[179] Sandale had both brass and stone mortars.[180] These were typical of the cooking arrangements of the élite: the finest spices, the most advanced methods of food preparation, and, as we shall see, cooks who were rewarded appropriately. In matters of taste, these establishments sought the highest levels of fashion – and the bequest of the hand-mill for sauce ingredients suggests an expectation that this would continue. Other references show the common patterns of hygiene associated with the meal: silver ewers and basins for washing preparatory to the meals,[181] and Gravesend had in addition a pick, probably of silver, for cleaning teeth.[182]

The best garments used by the bishops employed textiles and furs of good quality and some value, although the fabrics were not of the highest cost. Among the first items in Bishop Bitton's possession were a supertunic of tripoli (a silk, known from its place of origin, probably now Trâblous in Lebanon), with a hood furred with miniver; a cloak of the same suit lined with black sindon (a fine linen), worth 40s.; a cloak of perse with a furred hood, worth 15s.; and a suit (*roba*) of burnet, comprising a lined supertunic and mantle furred with miniver, together worth 66s. 8d. He also had a supertunic of perse with a mantle furred with miniver, worth 60s, and a furred robe or suit of perse worth 20s., together with a range of garments of lesser value, for example a mantle furred with fox, and other items, some worn out, such as a furred robe of azure, which was valued at just over 13s.[183] Besides his own clothing, the wardrobe contained the remains of cloths of red, green, striped and mixed weave, as well as cloth that had been used for liveries for his clerks and for his grooms, and a large quantity of buckram for the poor – the distinctions of rank were clearly visible.[184]

The wardrobe of Bishop Gravesend was richer in its array and there were more garments, although some of the cloths were the same. Two supertunics, with a hood and mantle of burnet, furred with miniver, were valued at £6 12s. 4d. There were whole cloths of murrey (coloured like a mulberry, purple to red), cendrine (probably a white to grey, ashen colour) and perse, as well as garments of the last-mentioned – two supertunics with a hood, furred with miniver, worth 40s.; another two supertunics and hood, furred with miniver, were valued at 66s. 8d.; two more of russet of Douai, furred with *gris* and vair were worth 30s.; bluet was used for a corset (a close fitting, outer garment) and cloak; and there were three new furs for supertunics and one for a mantle, along with two caps of miniver worth £5.[185]

What is notable about Gravesend's wardrobe, however, is the absence of whole suits of clothes: there is no direct parallel to Bishop Bitton's suit of burnet, although the frequency with which supertunics occur in pairs suggests that Bishop Gravesend may have composed his clothing in a different way. Sandale's clothing, by contrast, was largely grouped into suits, although there were some groupings of pairs of supertunics that suggest a similar pattern to Gravesend. At Southwark, he had four complete suits of clothing and there were another two entries for the wardrobe at Sandall that could have been suits. One of the Southwark suits was of russet brustekin, possibly an embroidered fabric, consisting of five garments and two hoods, furred with *gris* with the exception of the mantle, which had lambskins from Lindsey instead, valued altogether at £6 13s. 4d. Another suit, of the same value, consisted of six garments of sanguine, furred with vair with the exception of the mantle, which had gross vair. A third suit, of tawny dyed with grain, contained six garments, one furred with vair, two with *gris*, the mantle with Lindsey lambskin and the cloak without fur, valued at £4 13s. 4d. There is no doubt that this was well-made, fashionable clothing, with choice furs, cloths and silks, an investment far outstripping that of the other two bishops. He did have, however, among the stock of his wardrobe, 5½ ells of burnet cloth, and a belt that he used was noted separately but was not appraised. The latter may perhaps be of more significance given the belts of hair worn by other bishops, as well as Sandale's ownership of Winchelsey's belt; he also had a fine belt of silver thread, with gilt fittings, worth 43s. 4d., for outward display, even if the others were worn discreetly.[186]

Clothing was both valuable and might transmit merits associated with the original owner, as we have seen with Thomas Cantilupe's garments. Both Bishops Bitton and Gravesend chose to bequeath their textiles probably within their families. Bitton left his best suit, of burnet, to Sibilla de Gosynton;[187] and Gravesend bequeathed most of his textiles to his sister-in-law and nieces.[188]

The textiles that constituted the beds were again a considerable investment. Bishop Bitton had two beds: the one he commonly used was valued at 60s. and left to his chamberlain, along with with his shoes, leggings and other bedding. It had three separate covers, two red (one valued at 25s.), the other white; a mattress lined with carde (a linen cloth); and fifteen sheets of a fine linen (*chalons*) worth 48s. The second bed was reported to be worn out and was valued only at 6s. 8d.: it had a cover of fox fur, two other old covers of fur, and one of silk. With this bed were seven pairs of sheets, five kerchiefs and pillows covered with sindon.[189] No separate bed was valued for Bishop Gravesend, although his bedding was worth considerably more than Bitton's: a bedcover of scarlet, furred with miniver, was valued at £5 alone, and there were other covers, blankets, *chalons* and sheets, coverlets and pillows. A kerchief of velvet furred with *gris* was valued at 13s. 4d. and the Bishop had a wardecorps – perhaps the equivalent of a dressing gown – of blanket.[190] The most expensive of Sandale's bedding was in Yorkshire, where a mattress, cover, curtains and sparver, or canopy, of blue samite, a rich silk, were valued at £6 13s. 4d. At Southwark there were a series of mattresses, counterpoints, curtains and sparver of sindon, along with a bedcover of burnet furred with *gris* which was valued at 40s.[191]

Beyond bedding, chests and bags, there is little evidence for furnishings in the bishops' chambers. Bishop Bitton possibly had two chairs, one white, of wood, the other of various colours. He also had a mortar (a small bowl for oil or wax, with a floating wick) to light the chamber.[192] It was probably here that all three kept their books. Gravesend would have been well acquainted with contemporary theories of sensory perception: he possessed a copy of Bartholomew the Englishman's encyclopaedia, a tract by Isidore of Seville, works of Avicenna and Aristotle, besides bibles and exegetical works.[193] Sandale's possessions included bibles, works on canon law, Priscian on grammar, the confessional treatise *Summa de viciis* by William Peyraut, and five books of romances as well as liturgical works, including the great pontifical of St Swithun's at Winchester, which the Prior succeeded in recovering.[194]

Rugs or tapets for the floor, covering chests or other furniture, along with bankers, cushions for benches in the hall, formed a moderately priced group of items. Bitton's included three tapets of miniver, along with long cushions of the same suit, presumably for his chamber; two tapets with yellow leopards, worth 60s.; three with his own arms, worth 20s; and others with assorted coats of arms.[195] There was a large tapet for the hall, worth 13s. 4d., probably for the area where the Bishop would have been seated and possibly to hang behind him as a dossal; a range of striped bankers, eighteen cushions, and a net for keeping out flies.[196] Bishop Gravesend had invested more in bankers. There was a dossal for the hall, worth £5, and a small dossal, worth £3 6s. 8d. There were also two curtains of green carde.[197] Bishop Sandale had in Yorkshire four new green tapets, worth 26s. 8d., but his best furnishings were at Southwark. Most remarkable among these was a set of two blue dossals, powdered with a design of three red leaves and a border of the arms of England. The set had ten matching hangings for the walls of the hall and chamber and was valued together at £30.[198]

The equipment of the chapels of the three bishops was in essential respects the same: incense boats, thuribles, containers for holy water and the chrism, and service books. Individual items lent an especial éclat to the liturgy. Bishop Gravesend had a tablet or tray, which had a silver leaf inserted, for giving communicants the pax, which his executors gave to the King's treasurer in order to smooth their path in negotiations and which was valued at £1.[199] The same object in Bishop Sandale's chapel was made of silver and enamel and valued at 15s.[200] The vestments and textiles for the chapel available to Sandale, however, were of much greater distinction. Some were indicative of his closer connection to the royal household: two albs at Sandall had the arms of England and France embroidered on the sleeves; a box at Southwark had ten orphreys, partly enamelled with the arms of England, given to the Bishop by the King.[201] Others were clearly of fine work: an altar frontal of green camaca (probably a fine silk), embroidered with the images of the Trinity and two angels censing, along with a super-frontal and a cloth for the lectern embroidered with an image of the Virgin and other images, were together valued at £10.[202] Sandale had vestments that were of similar colour to the altar hangings: red samite, green taffeta, black taffeta, white and of a colour that changed or shimmered as the light passed over it, which would have made it seem as if it were itself emitting light. The white set, of cloth of *Turky* – an oriental silk – comprising a chasuble, three choir copes, a tunic, a dalmatic and two frontals, was valued at £12 and was given to the Prior of St Mary Overy for the body of the Bishop on the day of his burial. Other sets were embroidered with the story of the passion and with images of the apostles.[203]

72 The Clare chasuble of 1272–94, possibly before 1284, typical of the prestigious, embroidered vestments that might be given to ecclesiastics of high rank. The chasuble has been cut away at the sides to suit later fashion, but was probably bell-shaped in origin.

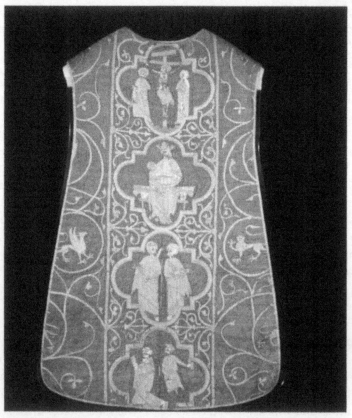

The Earl of Cornwall had given Bishop Bitton similarly striking vestments. A tunic and dalmatic of blue, with a white amice, stole and an embroidered maniple of blue sindon, were valued at £6 13s. 4d.; a chasuble, embroidered with the royal arms of England and France, a while alb with sleeves, stole and maniple of the same suit, possibly given by the King, were valued at £8 13s. 4d. Another set of vestments, embroidered on red samite, with altar hangings, was valued at £16 6s. 8d. Bitton also had vestments in black sindon. His mitre for the most solemn occasions, decorated with white pearls, was valued at £6 13s. 4d., although three mitres that were for daily use were valued at no more than 10s. together. Even the fly-whisk was of cloth of Tars, a silk, here valued at 1s.[204] Bishop Gravesend had an altar frontal and super-frontal, which were sold for £20; sets of green vestments, red vestments and blue vestments in cloth of Tars, others in black sindon, velvet and red *cangys*, probably another oriental cloth with a colour that appeared to change in different lights, a choir cope embroidered with images, valued at £13 6s. 4d., and a mitre, with white pearls, worth £4.[205] These vestments were the most highly valued clothing in each household and there was less disparity in cash value, or in ability for display, between the establishments than there was in non-liturgical clothing. Each bishop possessed vestments in a range of colours to match the liturgical pattern.

There was much to parallel secular establishments in the arrangements for travel and transport. The bishops had valuable palfreys: the best in the case of both Bitton (valued at £10 13s. 4d.) and Sandale (not valued, but the next in the inventory was worth £10) passed to the King, by custom. Although none is mentioned as passing in this way in Gravesend's accounts, it may have done so without appearing in the list as the most valuable of his horses was a palfrey, valued at £6 13s. 4d., for the Bishop's coach.[206] Sandale's horses were worth more than £64; his executors also valued two carts and a long cart among the household's accoutrements, but no coach.[207] Both Bitton and Gravesend had coaches, that of the former described as four-wheeled, the latter's as drawn by five horses; and Gravesend's household also had two carts and a long cart for moving its appurtenances.[208]

The pattern of servants was also common to the three households. The household was a male preserve – the exclusion of women was common in great households and especially so in those of the episcopacy, for many of the gender-based reasons that we have seen associated with sensory perception. Some members were probably related to the bishops. At the break-up of the establishment, Sandale's executors spent more than £75 paying off 108 individuals. Only one was female, a washerwoman; the steward of the household was his brother, Robert; and Alexander de Sandale and a clerk, John de Sandale, paid at the rate probably accorded to the esquires of the household, either had a family connection to the Bishop or came from his native town. There were probably eight clerks in the household and, although it is not possible from the list to establish who acted as chamberlain or personal servants, Sandale had three pages for his chamber.[209] Juliana de

Champaigne, probably a female relative and perhaps the daughter or wife of Sir John de Champaigne, a knight of the shire for Hampshire in 1316–19, came to Southwark for Sandale's funeral.[210]

Bitton's executors paid legacies to at least forty-five household servants and some also received perquisites customary to the performance of their duties, his chamberlain, Stephen de Tyting, for example, receiving his bed.[211] Gravesend left sums to be divided between his domestic servants, rather than enumerating them all. He did, however, single out for additional legacies three of his chamber servants, Jakett and young Jakett, who each received £1, and John de Tany, who received £2, and asked for the others to be rewarded additionally as a courtesy to reflect their periods of service.[212]

While there were many points in common between episcopal households at the end of the thirteenth and early fourteenth centuries, there were different emphases in different establishments which would have been readily apparent to the senses. In all cases, the greatest investment had been in the construction of buildings, itself an act of virtue – and one that created an environment that set the establishments apart from many in the locality. Within the buildings, the greatest investments were in textiles and jewellery for the person, either for liturgical purposes or for the chamber, rather than the hall (although some of this difference may have been because at this date in episcopal households the principal decoration of the hall was painted, rather than coming from soft furnishings and hangings). There was equally an investment in conspicuous consumption that went with the set pieces of medieval display: highly spiced foods eaten off silver. Decoration and ornament might make links showing the allegiance or career of the bishop. Within the household there were then further distinctions. Household servants were delineated by rank, matched by liveries, colours and patterns of clothing; and patterns of access within the household, to the Bishop's chamber or other rooms, were controlled to provide the minimum of disturbance to their master and his activities. The day-to-day clothing of the episcopacy, however, did have elements of restraint, particularly compared to secular society.

The evidence given in support of the canonisation of Edmund of Abingdon and to the Cantilupe commissioners demonstrates how an individual might seek to modify this overall environment, to deny sensory pleasure at a personal level, even if the overall arrangements for the household were similar to those of his peers. It is less clear whether anything similar can be deduced from the three episcopal inventories. Suggestions that the other bishops may also have been concerned at the potentially corrupting perception of sensory pleasure may be found in the presence of Archbishop Winchelsey's belt among Sandale's goods, and in the lower level of investment in luxury in Bitton's household, despite the substantial reserves of cash held at his death. All households had objects of virtue that might act for good, even if a comparatively small cash value was set upon them. Many items have resonances with the earlier discussions of sensory perception, such as the gleam of jewels, of plate and vestments; and the concentration of these items around the principal person in the establishment added greatly to that individual's sensory power, multiplying his already prominent and virtuous standing as a preacher of the word of God.

The households of the episcopacy were shaped by their spiritual purpose. Sensory perception might also be moulded to a particular end in other establishments. The Queens of England had a different purpose, at least in part, in managing their establishments, and an examination of their accommodation, households and activities in the following chapter shows both differences in emphasis and some similarities of intent.

Chapter 10

HOUSEHOLDS OF LATE MEDIEVAL QUEENS OF ENGLAND

In his *Jerarchie*, a work derived from the writings of pseudo-Dionysius on the celestial hierarchy and addressed to Queen Eleanor of Castile, John Pecham (Archbishop of Canterbury 1279–92) expounded the words of St Paul 'for our conversation is in Heaven', arguing that our life was not an earthly one, but one of the angels.[1] The analogy he used was the royal household. The three hierarchies of angels were likened to three categories of individual: those who were always at court, others who were usually away from it, in their bailiwicks or offices, and those who came and went between the two. The first, who were most loyally devoted to the King and most knowledgeable, resembled seraphim. The second were akin to cherubim: they were wise men, the clergy or good clerks, *philosophes* who oversaw those who governed. The third group were men whose judgement and great deeds were proven: these were the thrones, to whom was commited the government of the world. Each of the orders was itself divided into three. The marshal of the household (one of the dominations, who headed the second order) was the linchpin in the organisation: he received his commands from the thrones and gave them to others. Those instructions were of two natures, advancing good or rebutting evil. In contemplating the royal household and the divine powers invested in her husband, Eleanor would therefore have been encouraged to find the mirror of heavenly reality.[2] The connection was a strong one: the coronation order itself and the paintings of the Tree of Jesse, which had appeared at Windsor in the chamber of Queen Eleanor of Provence a year after her marriage to Henry III, strengthened the notion that the lineage of England was a dynasty indissolubly linked in its qualities to the House of David.[3]

In looking at the realities of the Queen's court, this chapter concentrates on the households of Eleanor of Castile and her fourteenth-century successors, Queen Isabella and Queen Philippa, the consorts of Edward II and Edward III, respectively, with some comparison with those of other queens. There are also useful parallels in the preparations made for three English princesses who went abroad to marry: Eleanor, the sister of Edward III, who became the wife of Reynald, Count of

Guelderland, in 1332; Joan, the daughter of Edward III, who was intended to marry Pedro of Castile in 1348; and Philippa, the daughter of Henry IV, who became Queen of Denmark in 1406.[4] The households of the late medieval Queens of England differed from those of their husbands – and the episcopacy – in a number of important ways. Most obviously, these were households led by women and, for the thirteenth, fourteenth and much of the fifteenth century, by princesses who had come from the Continent. They brought different expectations and new fashions to their domestic environments, making a sensory contrast that often struck contemporaries. When Eleanor of Castile arrived in London in 1255, her accommodation was decorated in Spanish fashion, with a plethora of textiles. There were silk hangings and tapets on the walls and even on the floors, a sight that Englishmen were more accustomed to find in churches, rather than the wall paintings that had traditionally adorned English royal apartments.[5] Her taste came to dominate fashion and the secular use of textiles in the royal and other households grew commensurately.[6] It was in this way that the daily realities of life – and general sensory expectations – might change.

The structure of the Queen's household reflected her status, different possibilities for display and her relationship with her children. Unlike both the King and the episcopacy, whose landed property remained relatively constant throughout the Middle Ages and in which the investments of one might be continued by a successor, the estates of English queens, although mainly from a recurrent group of holdings, often did not transfer directly to a successor. Widows held their dower after the demise of their husbands, and there were also periods when important estates remained in the hands of the King or reverted to the administration of the Duchy of Lancaster. There were, therefore, disjunctures in the development of these properties. While a queen might frequently reside with her husband and have suites of rooms in his palaces, the use of her own properties gives an indication of how her household was intended to function, her tastes and influences, and it was here that royal children were often to be found.

The assets of the queens were not as extensive as those of the monarch, but they were far from insubstantial. Their dower lands, almost by convention from about 1280, were worth about £4,500 p.a., although at times they might be considerably more valuable: Queen Isabella's estates were worth more than £13,000 in 1327 and the income from Queen Philippa's grew to about £7,000 a year by the end of her life.[7] There were, however, fluctuations: additional income might come from Queen's gold (10 per cent of voluntary fines made to the King, at times in the late thirteenth century worth more than £1,000 annually, but variable and an asset of little significance by the mid-fifteenth century) and direct grants from their husbands. In 1313–14, Queen Isabella received £5,559;[8] in 1358–9, John de Neubury, the controller of Queen Isabella's wardrobe, reckoning her assets at her death, including funds from the sale of some of her goods, as well as £311 found in gold in her closet and a gift of 1,000 marks from the King, accounted for a total receipt of £12,777.[9] The resources of other queens were closer to the size of their dower: in 1427–8, the treasurer of Queen Joan of Navarre, the widow of Henry IV, received £5,777;[10] in 1452–3, Queen Margaret of Anjou should have received more than £7,500, but some £2,808 in her account for that year was reckoned as arrears;[11] and in 1466–7, the income of Queen Elizabeth Woodville was, at £4,540, smaller than that of her predecessor but she had fewer bad debts.[12]

While much of this income went on the running expenses of the household and on display, especially in terms of clothing and jewellery, the principal residences of the Queen usually attracted a proportion. The records for the queens are far less complete than those of their husbands, but they nonetheless show a pattern of investment in favoured residences. Queen Eleanor of Castile started to build new accommodation at Haverfordwest, which she acquired in the year before her death and which was, in turn, held by Queen Isabella from 1331 until her death.[13] Eleanor's executors paid for carpentry there,[14] and for work on the King's baths at Leeds Castle after the estate had returned to his hands.[15] The establishment of permanent bathing arrangements marks a significant transition in royal provision: the sensory benefits of bathing have been outlined in earlier chapters, and were especially enjoyed by the female members of the royal household. At the same time, it is noteworthy that bishops, such as Edmund of Abingdon, avoided bathing and wore clothing all the time in a way that encouraged bodily parasites. Even peripatetic bathtubs are absent from their inventories.

Queen Eleanor had laid out £50 on works on the chapel at Leeds in 1288,[16] part of an ongoing campaign of works that followed the acquisition of the castle by the Crown in about 1278. With its own vineyard, fishponds and park, Leeds was used extensively by Edward I's two queens.[17] There were also major works under Eleanor at King's Langley, a site closely associated with royal children; within the royal household in the fourteenth century it was known as *Childre Langele*.[18] Here again there were gardens, one planted with vines, and there may have been more of Spanish taste: the Queen gave 40s. in May 1290 to gardeners from Aragon who had been employed at Langley and had licence to return home.[19]

Ludgershall – successively held by Eleanor of Provence; Edward II; his sister, Mary, who was a nun at Amesbury; and Queen Philippa, who had it from 1334 to 1355 – was also used for royal children, one of a network of manors in the south outside London to be employed for this purpose. A chamber had been built there for the future Edward I;[20] his mother, Eleanor of Provence, looking after his children while he was away on crusade, brought the young Prince Henry and his sister Eleanor there in 1274, prior to reuniting them with their parents.[21] The princes Thomas of Brotherton and Edmund of Woodstock, aged five and four, respectively, were there from July to October 1305;[22] and again, six years later.[23] Queen Philippa undertook extensive building works at Ludgershall in 1341–3. The chapel was rebuilt at this time, reglazed and a chamber block constructed: glazing has been

73 Leeds Castle, Kent. The thirteenth-century Gloriette, or keep, is at the left-hand side, linked to the main island by a bridge of stone corridors.

found with the arms of the Queen, and accounts describe the timbers of the chapel as painted with a beautiful blue.[24]

Philippa also carried out works at Odiham. The manor was granted to her in 1331 after the fall of Queen Isabella, who had held it from 1327; it had also been used by Eleanor of Castile and it was held by Edward I's second wife, Queen Margaret, from 1299 until 1318.[25] The castle still retained its impressive, octagonal keep, built by King John, who had turned to the site because it was convenient for hunting. Here some work had been carried out by Edward I, which had included constructing a chamber for the Queen. There were further works when Philippa acquired the property. The kitchen, which was built over the water, was remade, as were the bridges over the moat, which was also drained and cleaned; the cup-house – where the cups and other plate for the table were stored – was reconstructed, as was the King's garderobe, and a pale was built between the last and the kitchen. In the park, the Queen had a new garden, with seats that were roofed with turf and with its own garderobe, all enclosed with a timber palisade.[26]

Other properties favoured by the queens included Isabella's residence at Hertford, where she was to die in August 1358 – a property that, along with the nearby manor of Hertingfordbury, had been held by Aymer de Valence: his widow, Marie de St Pol, was a frequent visitor to Isabella in the last year of her life.[27] Isabella also held Leeds – which the Crown had exchanged, notwithstanding that she had been granted the

74 Ludgershall Castle: the new chapel and chamber block constructed for Queen Philippa in 1341–3, viewed from the north. The castle was set within a prehistoric hill-fort.

75 Ludgershall Castle: the works for Queen Philippa seen from the east, looking towards Henry III's great tower. The area to the right, to the north of the prehistoric bank, was a park.

reversion of the estate, and which she besieged in 1321 when she was refused admittance – and Castle Rising. At this last, a residence she used many times, there was considerable new building after she took possession in 1331, of a new chamber block, with garderobes, and domestic offices in the area to the south of the great keep.[28]

76 Odiham: King John's octagonal keep, now the only standing remains of this substantial castle.

Building accounts for the residences of the queens unfortunately do not tell us much about individual chambers. Where the King and Queen resided together, however, there is more information. The Queen's apartments in the palaces were distinctive in their schemes of decoration. Wall paintings in these areas in the thirteenth and fourteenth centuries followed themes with more iconographic prominence given to females and to contemplative themes, unlike those of the King's apartments, which had images connected to chivalry, royal genealogy and dynasty, or devotional interests closely associated with the sovereign. The Queen's apartments were also comparatively isolated from the main elements of the palace in terms of public living.[29]

These distinctions extended to the staffing and operation of the two households. The King's palaces typically had chambers available for allocation to individuals. At Westminster, a memorandum about repairs in 1307–11 noted various 'houses' and chambers which it had been the custom to allot to earls, countesses, barons and baronesses, as well as to the Queen's ladies: it was the practice for the King's household to assign accommodation to prominent individuals, including the Queen's principal servants.[30] This differentiation is unlikely to have applied to male servants commonly on the Queen's own properties. Structurally, the Queen's household differed from the King's, particularly in terms of its servants.

The great households of late medieval England – even those headed by women – were predominantly male institutions. In the household of Elizabeth de Burgh, listed in her livery roll of 1343–4, fewer than 10 per cent were women. Sometimes the proportion was smaller: Margaret of Clarence travelled to France in November 1419

77 Castle Rising: the garderobe block in the west range of the residential accommodation, built to the south of the great Norman keep, *c.* 1330, probably for Queen Isabella.

78 Clarendon: looking north-east, across the Antioch Chamber (with its two central supports) to the Queen's chambers and chapel (right-hand side) at the edge of the palace complex.

accompanied by 143 men and no women.[31] There were some changes in the fifteenth century as households at this time inhabited fewer major residences and stayed longer at them, which gave male servants an opportunity to marry, settle close by, and allowed their wives to join them in their domestic duties.[32] This pattern was also true of the households of the Queens of England: any enumeration of their household staff immediately shows the preponderance of men. The Pseudo-Aristotelian *Secreta secretorum*, with its counsel to princes, sumptuous copies of which were given to the future Edward III in 1326–7 by Philippa of Hainault and by Edward's tutor, Walter of Milemete, argued against the influence of women; the physical exclusion of women from great households mirrored this.[33] The Queen's household was always subordinate to that of the King – and when the two households were together, the Queen's usually ceased to function as a separate administrative entity.[34] The Queen was dependent on the King for her position: the sacral qualities of kingship were not extended to queens before Mary I, the first queen regnant since the Norman Conquest. Nonetheless the Queen was not without power, unofficial, but influential in selected circumstances.[35]

What were the consequences for a household led by a woman? From a purely functional point of view, the men in the Queen's household were responsible for much of the day-to-day work in the household departments: purchasing and preparing food and drink; serving it at table; maintaining security; and arranging transport. The greatest number of servants in the household was employed in the detail and labour of the kitchen, pantry and buttery, counting house, confectionary and stable. Equally there were others whose importance lay in their presence: simply being in the household, offering counsel, company and courteous behaviour, or, as clerks in the chapel, supporting the liturgy and devotion of the Queen. There were, however, some significant differences in terms of daily perceptions: the presence of a group of high-ranking women provided opportunities for display that was striking in its visual and general sensual impact.

This can be seen in the household of Philippa of Hainault, in the distribution of cloth or clothing at the summer livery in 1330, which was in many ways typical.[36] The livery was distributed to nine ladies; to eleven knights and bannerets of the Queen's household and that of the King (some of whom were employed principally outside the Queen's household); to twelve clerks, including some who belonged to the households of the King and his mother, Queen Isabella; to a further twelve clerks of Philippa's own household; to seven esquires of the King's household; to four esquires of Philippa's own household, who may have come with her from Flanders; to four esquires for serving her personal food and drink, with two for her son and two for her sister-in-law; to a further twenty-eight esquires of the Queen's household, some for the household departments; to seven yeomen (*valetti*) of her chamber; to two washerwomen and the under-clerk of the chapel (who would have been responsible for washing the vestments: women were usually not allowed to do this); to twenty-six other yeomen, for the household offices; and to two pages of her chamber.

For the issue of livery in the winter, there were in addition further categories of household staff: the Queen's minstrels; a group of outdoor servants, the carters and coachmen, the keepers of the coach-horses, the palfreymen and sumptermen (packhorse men), the *pre-equitatores* (those who rode before the household carts to make sure the way was passable); and a further group of junior servants, largely pages – junior in terms of hierarchy but not necessarily in terms of age. These included two pages and a porter for the Queen's chamber; the pages of her confessor (as these included a sumpterman and precursor – another servant whose role it was to precede others, this time on foot – these were probably adults); and eighteen pages for the household offices.[37]

The livery distributed to these servants differentiated them, with marked variations in the qualities, quantities and colours of cloth. The winter livery for the esquires was of ray cloths, from Ghent, of the colour of apple blossom; for the yeomen of the chamber there were also ray cloths from Ghent, but they were red, and in addition they received two short, brown cloths from Louvain. For the yeomen of the household offices there was ray cloth from Ghent of the colour of vetch flowers. The cloth for the yeomen of the stables, also ray of Ghent, was tawny in colour. The watchman had a checked cloth for his tunic and robe of chestnut colour; the pages of the offices had ray of Ghent in a brown and mixed colour; and the pages of the *pre-equitatores* had a similar ray, in brown.[38] The same differentiation occurred in 1332–3, when some, at least, of the colours were similar. The livery for the esquires, for example, was again of ray cloths, from Ghent, along with a yellow cloth (it had been apple blossom two years earlier); and the yeomen of the chamber again had red.[39] The visual distinctions here were common to the great household, but a close eye observing the court would have seen that they were not the whole picture, especially in terms of the Queen and her immediate entourage, and that there was here a significant difference from the King's household which focused on this group of women and their clothing, a point to which we shall return.

In 1452–3, the pattern of service in the household of Queen Margaret of Anjou was similar in many ways, but the documentation provided by her receiver-general allows us to study it from a slightly different perspective. Queen Margaret had three principal male servants within her household: her chamberlain, John Wenlok, who received £40 a year; and two knights, her stewards, who each received 40 marks (£26 6s. 8d.). Her principal female servant, Ismania, Lady Scales, was rewarded at the same level as her chamberlain; and three other ladies received an annual fee of £20 each. There was then a further group of noblewomen, some of whom had come to England with Margaret: Barbelina Herberquyne, who received 40 marks; ten other ladies who received £10 p.a.; and two female chamberlains who received £5 and 66s. 8d. each. There was only one other female servant, Mariona, a washerwoman, who was listed among the yeomen. There were twenty-three esquires, who were present in the household for varying periods of time: only one was present for the whole year, four were there for less than forty days, and the average attendance was 200 days. These were servants whose attendance was not expected to be continuous and who may have served on a rota, in and out of waiting. There were three clerks, for the Queen's closet, signet and jewels, and twenty-seven yeomen of the household, most of whom were present for well over 300 days in the year. There were a further four yeomen for household offices; a groom for the Queen's robes; nine for her chamber; and fourteen for the household offices; a page for the robes; two for the Queen's bed; two for her chamber; one for the baker of *painmain* – the finest white bread; and ten more pages for the offices. Those in junior roles – the yeomen, grooms and pages – served in the household almost without a break, unless given special leave of absence.[40]

The men in the household fell into three categories: first, the chief functionaries, such as the chamberlain; secondly, those who might be characterised as servants of honour, such as the esquires in waiting; and, thirdly, the menials, in the proper sense of the term, in constant service on the main tasks of the household. This was a reflection of the pattern of service generally at the Lancastrian and Yorkist courts, where the numbers were swollen for great occasions by those who might perform occasional service or who came in rotation.[41] The ranking accorded to Queen Margaret's female servants, although few in number, was high, equalling the standing of the chief domestic officers (but not those who served her outside the household, such as her chancellor and steward), and surpassing those in the most numerous male roles, including the servants of honour.

In the later Middle Ages, it is a feature of aristocratic domestic service that there was a growth in the number of servants of honour. This category included some of the pages, especially the pages of the chamber; and the footmen – servants whose job it was to attend aristocratic riders, running alongside, holding bridles, etc. In Queen Margaret's household two of the last, who were ranked with the esquires in the New Year's gifts, where they were given a bow apiece, each received an additional annuity of 40s.[42] A second group was associated with riding: henchmen, originally among the sumptermen in the royal household. Henchmen appeared first in the King's household in the 1340s. Their numbers, at least in the fourteenth century, were fewer in the Queen's household: Henry Naket served Queen Philippa in this guise in 1351–2, William Hengestman in 1357–8, on each occasion the only individual who can be identified in this role.[43] Their most striking appearance, however, was at the coronation of Richard III and Queen Anne, where five henchmen appeared in the Queen's train, riding on women's saddles, in company with her seven ladies-in-waiting.[44] The honorific establishment of the Queen's household was also boosted from time to time by the presence of members of the King's household, who might serve her as well and wear her livery. The numbers of women attending the Queen do not seem to have changed greatly in the fourteenth and fifteenth centuries; and their ranking, at a higher level than the male servants in waiting (many with aristocratic connections), is especially noteworthy. This pattern was reflected in conspicuous consumption and those aspects of the household that centred on display to a far greater extent than in the King's household.

In the 1330s and early 1340s, there were significant changes in fashionable clothing – which English moralists condemned and alleged could be traced to those from Hainault in the entourage of Queen Philippa, although in fact the change appears to have occurred quite widely and suddenly across Europe. The new fashions involved, for men, short, tighter-fitting garments, the use of new fastenings, especially buttons, and the use of 'frouncing', which was probably a way of gathering fabric with a thread. Women's attire was longer, but cut to fit the shape of the body in a way that contemporaries regarded as scandalous and sexually provocative. The sleeves of the garments of both sexes were closely cut, but those of the over-tunic ended at the elbow and continued with looser, hanging strips of fabric, a 'dagged' appearance that was also to decorate the edges of other garments.[45]

Queen Philippa's wardrobe accounts for the early 1330s show her household in the period immediately before these fashions took hold. What stands out is the quantity of her clothing and the variety in terms of cloth and its colours. Most of the cloth employed came from the Low Countries, particularly the south, with the addition of silks from the Mediterranean, linens from north-eastern France and furs from the Baltic.[46] The Queen's wardrobe prepared her new clothing from cloth that had been purchased on its behalf or that the Queen had received as gifts from others: the Queen dictated her own fashions. Most came in suits (roba), the numbers of garments in each varying between two and six, but iconographic evidence suggests that not all parts of the suit were intended to be worn at the same time.[47] The cloth was prepared for use by shearing it a second time, to make it shine, and, in the case of tirtain, an exceptionally fine textile, by pressing it or polishing it.[48] In 1330–1, the wardrobe made for the Queen no fewer than twenty-one suits of clothing; in 1332–3, between fourteen and sixteen, besides other garments.[49] Following the common pattern, the new clothing was prepared for the principal feasts of the year: Easter, Pentecost, Michaelmas, All Saints, with others for St Mary Magdalene and St Peter. In both years, the Queen had a child born in mid-June. In 1332, her clothing for the feast of St Mary Magdalene (as well as the summer livery of her household) was prepared to coincide with her formal up-rising after the birth; in 1330, the clothing for this feast may also have been designed to mark the same event.

The household was issued with clothes principally at the summer and winter liveries, at St Mary Magdalene, All Saints and Christmas, although the livery at All Saints was much reduced in quantity.[50] The Queen had new and striking clothes on these occasions, besides suits she was given. For her personal winter livery there were ten cloths of green silk, diapered with red and powdered with heads of griffins in gold; and eight cloths of white silk, diapered with yellow and powdered with gold crowns, which together cost £54.[51] The set of robes for her summer livery contained four garments, based on a green, long cloth from Louvain of 28 ells, at a cost of 110s., with a further 4½ ells for a lined corset, and 3 ells for seven pairs of hose against the same feast. At All Saints, the Queen had two sets of robes of three garments, a lined *cloch* (an upper garment, shorter than a corset and usually worn with a cap) taking 7¾ ells, and another six pairs of hose against the same feast. At Christmas she had two sets of robes of four garments, with additional items.[52]

Special grants of clothing were sometimes made to favoured ladies and maids of Philippa's chamber, such as Idona Lestrange, who was given 3 ells of sindon for her personal use, probably for underclothing, and, against Christmas 1330, 28 ells of sanguine cloth from Brussels, which cost £11 6s. 8d.[53] Generally, however, they received clothing only at the summer and winter liveries, and then in smaller amounts and in cloth of a lesser quality than that of the Queen, both diminishing with their rank. For the summer livery, against St Mary Magdalene, Joan de Carrue, the lady of the chamber, received short cloths woven at Malines, at 73s. 4d. a cloth (of 24 ells), and two pieces of sindon; the five maids received 15 ells and one piece of sindon each.[54] The knights and bannerets, members of the King's household, received 12 ells each of a parti-coloured cloth; Philippa's own principal household officials, for example her steward, received 10 ells, as did the steward of household of Queen Isabella. The livery for the clerks of the household was 12 ells of a different cloth of Malines, although her treasurer received 13¼ ells.[55] At Christmas, the Queen's ladies and maids had garments, with each additionally receiving corsets of 4 ells of sanguine tawny, with furs, a *cloch* of winter livery, at 3½ ells apiece, furs and a furred hood; at All Saints they had only received hoods.[56]

Some dress items the Queen renewed with an even greater frequency: in 1330–1, she had at least four dozen pairs of shoes and two dozen pairs of gloves; in 1332–3, there were at least fifty-four pairs of shoes and fifty-six pairs of gloves.[57] There was also a considerable turnover of linen. The purchases for the Queen of high-quality linen would have made very substantial quantities of *privata* – shifts, shawls or wrappers, possibly handkerchiefs, and undergarments – for the Queen's person. As well as having her linen washed very frequently, there was a great deal of it as well that was new.[58] The Queen's consumption of hose, however, was probably more restrained than that of kings. Edward II, for example, had been supplied for his personal use in the period 8 July 1315 to 31 December 1315 with no fewer than 110 pairs of hose of scarlet, dyed in various colours, at a cost of 4s. a pair, the equivalent of a pair almost every other day.[59] Queen Philippa's supplies were more modest and appear less frequently in her accounts: in December 1330, the red, mixed cloth, the sanguine of Louvain, and red, mixed cloth of Malines, bought for making the Queen's hose in preparation for Christmas, cost just under 22s. 6d.[60]

The Queen's clothing was impressive, for its quality, its colours and its shine, and for the frequency with which it was changed: the expectation was that the Queen would have new principal suits of clothes on average every two to three weeks, and new shoes and gloves every week. As well as the personnel of the great wardrobe who made many of the suits of clothing, the ladies and maids of her chamber and the seamstress of the chamber worked on the linen and kept the clothing at its most impressive. A ready, visual distinction within the household was drawn on the basis of rank: different cloths, different quantities and different qualities; and differences in the frequency of distribution.[61] In addition to the bold, bright colours of many of the cloths, there were others in the household that were more subtle in shade, in particular a range tending to add colour to white, from cendrine to the blush of apple blossom and the stronger colours of peach flowers; some cloths, the colour of vetch flowers, probably added to the range of hues between pink and violet, but may have been white or yellow. Additional decoration for the clothing might include vast quantities of spangles, as well as nets (probably for the hair) and head-cloths of fine materials.[62] This was clothing to which the Queen was particularly attached: in June 1331 she was herself obliged to enter the cellar at Eye to pay a forfeit of 40s. to recover one of her suits of clothing claimed by the butler of the Abbot of Peterborough.[63] There was almost certainly a prohibition on touching the Queen's clothing: arrangements for avoiding this are found in later household ordinances and this may be the implication of the purchase of 3½ ells of Brabant linen for the trousseau of Princess Philippa, in 1406, for wrapping her suits and garments to carry them into her chamber.[64] The Queen's household arranged special additional festivities at Epiphany and the Purification, particularly dancing: the costumes were a primary element in the spectacle.[65]

The contrast with the households of the episcopacy is marked. A very restricted range of garments, of middling-grade cloth, changing rarely in the year, formed the daily attire of a bishop; if there was a significant investment in clothing, it was in liturgical dress, which often had other connotations: connections with a benefactor as well as links to the ecclesiastical year. The Queen's household, and especially her person, on the other hand, presented an array of costume that was continually changing, of the highest grade, and which was kept at its most impressive. The households of other queens placed a similar emphasis on the visual effect of dress, although the way in which it was employed varied. It was a matter of honour that those of rank had dress commensurate to their station: Queen Isabella's wardrobe at Berwick on 28 June 1314 gave 40 marks to some German knights, who had come with the King from the battle of Bannockburn dressed in poor clothes, in order that they might buy new garments.[66] As well as the rich textiles, there was embroidered work, some probably carried out by the Queen's ladies, some by the wardrobe, but also by specialised workshops. The work appeared commonly on liturgical garments and on textile furnishings within chapels, as well as on some secular clothing. In addition it was sometimes employed on furnishings, notably on small cushions and on 'chambers' or bed hangings, usually to coincide with a special event.[67]

Embroidery was a common way of adding mottoes or badges to costume. In 1406, Princess Philippa had blue flowers embroidered on her gowns, probably forget-me-nots, a badge used by Henry IV. The green and red livery that was given to the household that accompanied her to Denmark had additions in both embroidery and pieces of cloth: crowns were applied in white cloth, the design varying according to rank in the establishment, as well as the motto *Soveraigne*.[68] The opulence of this wedding party culminated in Philippa's wedding dress, a tunic and mantle with a long train of white satin, worked with velvet and furred with miniver, the sleeves of the tunic furred with ermine.[69]

The messages conveyed by Philippa's dress are still apparent to us and the interpretation of the clothing of the widowed Queen Joan of Navarre is equally clear. Accounts dating from her imprisonment at Leeds Castle and subsequently show her dressed in black. This is most unlikely to have been the popular black of late medieval fashion, but long-term mourning, for she acquired few clothes for herself in any other colour, although her attendants had coloured cloth – and one of the Bretons among them may have owned the 'night cappe for a woman, rede after the gise of Bretaigne' inventoried among her goods in August 1419. In 1420–1, Joan had a black gown, black petticoats, as well as a black gown for night attire and black hose, while her socks (garments like slippers, but coming a short way up the leg) were made from a white cloth; in 1427–8, she had made for her seven gowns, ten kirtles, sixteen pairs of hose, all from black cloths, both wool and silk, and her socks this year were also made from black cloth. In the latter year, garnishes of fur, of miniver, ermine and sable, served to mark out her rank.[70] Another widow, Queen Isabella, had a very different wardrobe at her death in 1358: of the twenty suits of clothing and individual garments found, only one set was black, the others including fabrics in cendrine, red, violet, tawny, blue, murrey, marble, green and apple-bloom.[71] She chose to go to her grave wrapped in a tunic and mantle of red samite, a rich silk, lined with sindon, the clothing in which she had married Edward II fifty years earlier.[72] Occasionally, as well, routine items such as shoes could take on an exceptional cast. Both Queen Isabella and Queen Philippa rarely spent more than 6d. a pair. In France, on 4 May 1289, Eleanor of Castile may have spent more, buying from a merchant of Condom four gold purses and two pairs of gilt shoes for approximately 13s., although her purchases were normally priced at about the same level as those of the other queens.[73]

The visual impact of clothing was matched by that of other textiles in the household, particularly hangings for halls, chambers and beds. Not surprisingly, all again outclassed those found in contemporary episcopal households: but there was in addition variety, which few of the bishops could muster. Queen Philippa paid more than £12 to complete a new set of hangings for her hall, with a coster (side-hanging) and a banker (bench cushion) for the high table, worked with sirens and the arms of England and Hainault, the border decorated with leaves, birds, animals and *babewins* (grotesques), probably part of the preparations against the feast of her up-rising, in July 1332, after the birth of Isabella of Woodstock.[74] There may have been a dynastic message here, although heraldic decoration was very popular. The previous year she had received from the confiscated goods of the Mortimers a celure of jennet sindon, powdered with black roses, to hang over the high table in the hall (suspending a fine cloth in this way over the head of the Queen or other high-ranking individual was a visual mark of their honour and estate), along with a matching set of twenty-four tapets or rugs, dossals and bankers.[75] In July 1313, Isabella's tailor, John Faleyse, acquired cord in order to suspend a celure of Turkey cloth over the Queen in her hall at St Germain-des-Prés.[76] In the 1320s, the privy wardrobe in the Tower held three celures to hang over the Queen's table, made of gold cloth, with a blue field and flowers of gold and the arms of England and France in the border, along with three others of a slightly different design.[77] At Queen Isabella's death in 1358, she had dossals for her hall including an arras of the story of the cross, another with a red field powdered with leopards' heads, a set powdered with dolphins, and a further one which depicted scenes from the Apocalypse. There were four celures, three of sindon, respectively red, blue and black, and the last of red samite.[78] Princess Philippa took with her to Denmark in 1406 a cloth of gold, lined with buckram and with a silk fringe, to hang above her head as a cloth of estate when she sat at table.[79]

The furnishings of the queens' chambers were similarly rich, sometimes of astonishing price, particularly the beds. Textiles for the beds came in suits. On Isabella's death, her bed of blue camaca (a fine silk), embroided with owls in gold, silver and silk thread, was made up of a bedcover, a dossal, a celure, eight cushions, three curtains of sindon and twelve tapets or carpets of the same suit, valued together at 200 marks (£133 6s. 8d.); her next most valuable, an old one of blue silk, was worth £20, but was similarly composed, although it only had ten large carpets and a small one for the chamber door.[80] Additional carpets or tapets were made for the chamber. In 1330, eight white ones were added at a cost of more than £9 10s. for Queen Philippa, to match a white bed that had been made that January.[81]

The bed was made with sheets and covers, sometimes including other, finely embroidered counterpoints (the precursor of the counterpane). Queen Philippa's tailor, after recording the preparations he had made for the tapets for her chamber on the day of her up-rising after her first child, in July 1330, reported the costs of repairing an embroidered counterpoint that had been torn and chewed by dogs in the Queen's chamber.[82] In November 1332, Queen Philippa's bedcover was repaired with miniver; the curtains of purple sindon, worked with gold, silver and other colours, were repaired and painted about the time of her up-rising in July that year.[83] The cushions, filled with feathers, for her chamber were re-covered with sindon at the same time and needed other repairs during the accounting year.[84]

At the very close of her life, Eleanor of Castile had acquired both a feather mattress and 10 stone of feathers for cushions, possibly with the intention of making her more comfortable in her illness.[85] Featherbeds were not especially common in royal households of this period. Among the possessions of Queen Isabella, at her death, there was only one, while she had one mattress of samite and two of sindon, along with seven old ones – it was these that more usually formed the basis of the bed.[86] Equally there was only one featherbed in store among the vast array of textiles and furnishings in the privy wardrobe in the Tower in the 1320s, and that was old and worn out.[87]

The wardrobe made a further distinction for beds that were to be used for lying-in. In the 1320s, the privy wardrobe had a sparver for one that had presumably served for a royal birth, made of blue sindon, patterned with gold fleurs-de-lis and with a striped and checked silk cloth. A second sparver was striped with red, gold and blue on one side and had a striped and checked silk cloth on the other. A third sparver may have been intended for a bed used at an up-rising.[88] These items were readily identifiable to the wardrobe staff, even though we may not now be clear what features distinguished them for this purpose: the one prepared for Queen Philippa's up-rising after the birth of William of Windsor, in 1348, had a dossal, a quilt, a celure, three curtains, four woollen tapets, and used 21 ells of long, scarlet cloth, three cloths of gold, a fur bedcover made of 1,798 bellies of miniver, a piece of red velvet for a coverchief or headcloth, and a further fur cover for the coverchief, made of 348 bellies of miniver.[89] The implication of the bedding and associated textiles, with the accounts for its repair, is that the queens intended their chambers and beds for use, to be seen and to impress: we do not have references to queens avoiding the use of their beds as some at least of the episcopacy did. This was the highest level of luxury.

Queens were among the few individuals who had chairs, often kept in the chamber. Eleanor of Castile had at least six, which were regularly maintained;[90] and three pairs were delivered to Queen Isabella by the privy wardrobe in April 1327.[91] Princess Joan had two folding chairs among the goods she was to take with her to Bordeaux in 1348 and Princess Philippa took one, her private chair, covered with three ells of red cloth, to Denmark in 1406.[92] Other chamber goods included mirrors of ebony or steel, warming stoves, and combs, besides ewers and basins, coffers and trunks for clothing and linen.[93] It was in her chamber that Queen Isabella kept most of her jewels, her rings and cash, in her secret coffer, and distributed between here and her closet were crowns, liturgical books, many relics – including a head of one of the Eleven Thousand Virgins, with its own gold circlet set with pearls – and probably other books of romances.[94]

The relics indicated the Queen's personal preferences in devotion and they had a firm presence in the household. In the 1320s, the privy wardrobe held some in store: a small coffer, sealed by Brother John l'Armurer and Nicholas de la Beche, which Brother John said contained many relics; a small, enamelled reliquary with three containers in it, each with relics; a gilt and decorated tablet for use as a pax, to which relics were attached; and a great many belts for pregnant women, almost certainly endowed with powers to assist in childbirth. The store also contained three chemises that had come from Chartres, modelled on the Virgin's which was kept there, although one was old and worn out.[95] In 1337, Queen Philippa drew from that store a gold cross, with assorted relics.[96]

In 1330–1, Queen Philippa had six and possibly eight clerks for her chapel, and the number who served Isabella directly in that capacity at the end of her life was probably between four and six.[97] There is here a reminder that the sound of the liturgy was a constant presence in the household: the accounts of offerings show that the chapel of Queen Philippa operated on every day of the year from 1 April 1357, supplying a continuous benefit to the household.[98] The chapel vestments inventoried after Isabella's death included a rich sequence, with colours marking the different liturgical seasons. A white set, with a chasuble, two tunicles, three albs and amices, a cloth for the lectern with curtains, two altar frontals and three morses and orphreys, decorated with the arms of England and France, was valued at £12; a set in red velvet, powdered with red trefoils, the orphreys of blue cloth powdered with fleurs-de-lis, was valued at £26, which was also the value of a set of purple vestments. Others were in blue, tawny, cloth of gold, and green silk; there was a set in black for exequies; and at the Queen's death, a vestment and chapel furnishings in white camaca remained unfinished, with 480 gold roses not yet set in position.[99] Liturgical colours were echoed in the colouring given to the candles in the chapel of Eleanor of Castile.[100] The ornaments of Queen Isabella's chapel – chalices, cruets, thuribles, incense boat, bells, silver vessels for holy water, tablets for the pax, besides the liturgical books, one a gradual of the use of the French Franciscans, along with carpets, bankers and the vestments – were divided on her death between Edward III and Isabella's daughter, the wife of David II of Scotland.[101]

One textile deserves further attention: cloth of gold. Of considerable value, and having the merit of appearing to emit light, this was given by the queens as offerings, sometimes for vestments, and sometimes for draping on the bodies of the dead or to hang around their tombs. In 1311–12, Isabella offered one at the high altar at Rufford Abbey, gave another to a hermit in Windsor Forest and offered a third to the image of the Virgin in her chamber. Three cloths, however, were used for cushions for the coach that carried her ladies.[102] In September 1349, Queen Philippa bought one from the sacrist at Forde Abbey, then offered it there, to hang over the tomb of Hugh de Courtenay.[103] Two of Queen Isabella's were placed on her body on the day of her burial.[104]

Queens of England were anointed when they were crowned. Although their sacral power was not treated as the same as that of the King, it might nonetheless have a significant impact. The royal offerings made on Good Friday – in a ritual in which Queen Philippa participated – were subsequently melted down and transformed

into cramp-rings, employed against epilepsy.[105] The Queen's regalia – crown and ring – were blessed; although the Queen also received a sceptre, it was not treated in the same way as the King's. The floriate design of the sceptre likened the Queen to the Mother of God whose intercession was so valuable.[106] It was not surprising therefore that the Queen's jewels should attract special attention: the light radiating from their gleam carried an extra impetus when worn by the anointed.

In 1331, Queen Philippa had four crowns, three, including a large one, with ten florets each, and a further large one with eight florets, large rubies, emeralds and diamonds, that had been given to her by Queen Isabella.[107] All were held in the custody of one of the ladies of her chamber. In the same year four of the Queen's chaplets, in the custody of another of her ladies, were repaired.[108] Describing a jewel as a chaplet did not necessarily denote an object of lesser value. In 1358, Queen Isabella spent £150 on a chaplet of gold, with rubies, sapphires, emeralds, diamonds and pearls; another £87 on pearls and £80 on a crown;[109] but the series of crowns and chaplets given to Queen Philippa by the King in August 1337, from the stock of the privy wardrobe in the Tower of London, were mainly less valuable items, ranging up to £40, and included a chaplet of gold enamelled with the arms of England and France, valued at £8.[110] Items like this last were worn by the Queen's ladies: their jewels added to the impact and power of the display centred on the Queen.

There was a range of other jewels, some of considerable value and virtue. In November 1313, Queen Isabella gave Cardinal Arnold, returning to Rome, a gold ring with a great ruby, valued at £50, a reward for his efforts throughout the preceding year in the cause of peace.[111] Queen Philippa bought a series of brooches in 1331: two of gold, with diamonds and oriental pearls, which together cost £22 7s., were presented to her at Langley on the feast of the Purification.[112] There were many smaller items among the goods of Queen Joan of Navarre seized in August 1419, brooches that may have had heraldic connections, as well as others with mottoes or exhortations, and a pen-case, with a silver gilt inkhorn with the words 'God make us goode men'. The value of these items individually ranged from 6s. 8d. to £1.[113] If they were intended to produce an effect on their wearer through their words, other jewels had the same powers without text. Queen Isabella had, at her death, a silver serpent which had on its back a stone with a silver branch, on which were set five serpents' tongues, an assay against poison.[114] Margaret of Anjou had a salt in the shape of a serpent, which also had a tongue, repaired in 1452–3.[115] Other serpents' tongues may be linked with gold spoons and the taking of the assay of the Queen's food. In the 1320s, there were four gold spoons 'with four serpents' in the privy wardrobe.[116] Gold or gilt spoons often occur in pairs in the households of queens and princesses and it is to be suspected that they gave some special protection: two gilt spoons 'for the lady's mouth' were among the goods taken by Princess Joan to Bordeaux in 1348;[117] and two gold spoons headed the list of the goods of Queen Isabella, made at her death, passing to her daughter, Queen Joanna of Scotland.[118]

Although Isabella's gold spoons were valued at 72s., they formed a very small part of the plate in her household. The primary purpose of some of this was display, on cupboards, and so on; but other items had a functional purpose and we can see from this the scale on which the household operated. Isabella's scullery had plates and saucers of silver in a range of slightly different sizes, some groups containing as many as twenty-five items, but most in groups of eight (or multiples thereof), for example for fruit.[119] Her daughter-in-law probably had two services of silver in operation in 1331: one set, with the arms of England and Hainault, had thirty-six saucers of similar weights, twelve silver gilt spoons, and twenty-nine silver dishes, with a further one lost during the year. There were thirty-five silver cups with arms in an escutcheon – and Philippa pardoned her butler at Odiham that year for the loss of three further pieces of the set. Another set of sixty silver dishes had the arms of France and Hainault; there were also a further twenty-four spoons.[120] In 1348 Princess Joan took with her to Bordeaux thirty-six silver dishes along with the same number of silver saucers. She also had twenty-four spoons and twelve plain silver cups for wine – the lowest common denominator, suggesting a service which at its most magnificent could cope with three courses, with sauces, for twelve people, without recycling dishes through the scullery (and burnishing them after washing).[121] Princess Philippa in 1406 had a covered hanap (or cup), another fine hanap and twelve of silver; forty-eight silver dishes of two different sizes; twenty-four silver saucers and twelve silver spoons.[122] These silver dishes were undoubtedly reserved for those of the highest status: that Queen Philippa might make provision to entertain more people on a regular basis than the two princesses who went abroad to marry is perhaps unsurprising, but in all cases those who participated in meals in this way were a small and select group, and we should envisage a 'top table' of perhaps no more than eight or twelve.

What other things might affect the environment of this select group? Secular musical entertainment and dancing often marked the principal feasts or special occasions. In July 1331 Queen Philippa helped celebrate the marriage of one of the ladies of her chamber, Elena de Maule, and the King's minstrels were present. Cecily, a tumbler, visited in July, and the Queen also paid John le Nakerer (or drummer), one of the King's minstrels, the same month. On 25 July, women of Bilsthorpe, in Sherwood Forest, came to sing before her as she travelled to nearby Clipstone.[123] Young women danced for her when she approached Canterbury in June 1352 and a minstrel was rewarded for performing in front of the image of the Virgin in the crypt of the cathedral.[124] John Wayte, probably the watchman in her own household, performed with his colleagues for her at Christmas 1357.[125] Queen Isabella had visiting minstrels, some from the King and the Duke of Brabant, at Pentecost 1358, with others at Corpus Christi that year.[126] A cithara was repaired for Queen Joan of Navarre while she was imprisoned at Leeds Castle.[127]

The arrangements that the queens employed for travel marked them out from their contemporaries in much the same way as their jewels, textiles and high-quality furnishings did in their residences; and they attempted to provide the same magnificence of environment as they journeyed. They had the best palfreys when they rode and used the finest coaches, coach-horses and litters. Two of the six horses that pulled Queen Isabella's coach were valued at her death at £25 each; she had two lesser horses, as well as two mules, one fawn worth £3, the other white and valued at £10, that were used alternately as a pair for her litter.[128] Two other mules had been bought on the Queen's behalf at Avignon for £28 13s., but arrived after her death.[129] The saddles for ladies, in store in the privy wardrobe, included some with seats embroidered with silver and silk threads, adorned with black and white pearls, and there was equally fine harness to go with them.[130]

Expensive textiles covered their coaches. In 1330–1, Queen Philippa had a shining, red, mixed cloth of Malines; the openings to the coach were lined with green sindon.[131] The coach prepared for Princess Philippa in 1406, to take with her to Denmark, came with two covers of scarlet cloth; two of a red cloth, lined with Westvale, a linen, and waxed canvas, against the rain; and two of cloth of gold of Cyprus, lined with buckram and decorated with ribbon and silk fringe.[132] The arrangements made for her on ship included one 'cabin' – an area enclosed by textiles – for below-decks and one to stand on deck, both made of waxed canvas lined with red worsted and with a cloth of gold of Cyprus to hang over the Princess' head. There was also a cabin of worsted for her ladies.[133] If the Queen travelled by coach, so did the foremost of her attendants: the coach bought for Queen Isabella's ladies in 1313–14 cost just under £30.[134] In April 1327, Queen Isabella was given from the privy wardrobe a litter, covered with green velvet and embroided with the arms of France, its supporting arms decorated with a blue cloth of Tars, with gold flowers, along with two pillows of the same suit and a thick mattress with a red silk covering, striped with gold. There was also a rug and a red cover of scarlet cloth for the litter.[135]

This was an environment in which the exotic was not out of place. Queen Isabella was given a porcupine at Dover in 1314, which was fed on apples.[136] In 1332, Philip of Windsor brought back from Nijmegen for Edward III a bear given by the Count of Guelderland to mark his marriage to Edward's sister.[137] The royal menagerie was maintained in the Tower of London: while some beasts accompanied the King, it was more usual for the Queen and her ladies to have other exotica, especially birds. Eleanor of Castile was sent parrots by the Princess of Salerno in 1289[138] and Queen Joan of Navarre had a parrot in a cage in her chamber at Leeds in 1419–20.[139] Queen Isabella had two small cages bought for birds in her chamber in 1358, along with hemp-seed for their feed.[140] Others had nightingales;[141] and it was to Queen Eleanor of Provence that John of Howden addressed his poem, *Rossignos*, on the power of Christ's love through his incarnation.[142]

John concluded his poem with a prayer to the Virgin; her sweet odour was beyond the smell of cinnamon, of cloves and balsam.[143] This association between smell and the holy – or with objects with special powers – was present in various items in the royal household. In store in the privy wardrobe in the 1320s was a box made of nutmeg (*coffina de muge*) which contained a gold cross and other items, a comb and a mirror, perhaps for use by a priest; there was another small box of cedar, decorated with gold, which contained two rings with sapphires and a stone of chalcedony, all probably of special virtue.[144] The odour of sanctity might also be found about the person of the Queen: if the Queen of Heaven was redolent of spice, so earthly queens might not only imitate her but convey something of their own status through similar odours. If, as we have seen, the outer body reflected the condition of the inner soul, so the beautiful were (most likely to be) holy, and those that smelled good, equally so. Queenly virtue found its reflection in these traits.[145]

To ensure this virtue, the Queen's person was clean: arrangements for washing noted above indicate that there was regular bathing. Accounts of bathing equipment show repairs to Queen Isabella's bathtubs;[146] a bath 'tent' of linen was supplied by the privy wardrobe to Queen Isabella in April 1327;[147] in June 1332, Queen Philippa had two pairs of linen sheets made for her bath at Woodstock, possibly as relief given to her at the time of the birth of her daughter;[148] and at Sheen, in October 1349, a case was purchased for the basins that were used for washing the Queen's feet.[149] When Queen Joan of Navarre was imprisoned at Leeds Castle in 1419–20, 12 ells of linen were bought for her 'stewyng smokkes', the clothes she wore in the bath.[150] Among the goods found in Queen Isabella's chamber at Hertford after her death was a silver vessel for washing her head, along with arrangements for heating the water;[151] all this was in addition to the equipment for the regular washing of hands associated with the service of food and the office of the ewery.[152] As we saw in Chapter 7, evidence for aristocratic bathing is linked to the use of herbs and spices; and sometimes bathing may have taken place in a way that emphasised ritual cleansing.

Perfumes and cosmetics complemented the clean body. In January 1290, Eleanor of Castile bought four nutmegs and six pomanders.[153] These items were not common, even in the households of queens. The privy wardrobe had just one pomander in store in the 1320s.[154] Among Queen Isabella's effects was a pomander, set in gold.[155] Both Isabella and Philippa had an apothecary in their households, who was well remunerated for his task.[156] The apothecary prepared medicines as well as what we might now consider perfumes and cosmetics. In November and December 1313, an injury to Queen Isabella's hand and arm – almost certainly the result of a tent fire at Pontoise – kept her at Westminster, with the attentions of the surgeons of the King of France as well as those of two English surgeons. The treatments applied included ointments made from mace, rose-water, green wax, oils of hemp seed, laurel and other assorted oils, pomegranate juice, a plaster and leaves of roses of Orléans.[157] Her hand was still troubling her in September and October 1315, when similar remedies were applied.[158] The medicines with which she was supplied in 1311–12, although no less potent in their odours – fennel seed, grains of paradise, cardamom, long pepper, nutmeg – were connected with another ailment.[159] Many of these preparations would have been equally at home as constituents in cosmetics and some of their benefits were the same. Queen Joan of Navarre had two pots of citrinade and coinade while she was at Leeds Castle in 1419–20, which may have had a cosmetic use; but the little 'tastour' of silver for rose-water may have been for medicine as much as for any other purpose.[160] There are also occasional references to materials that may have been used to perfume rooms. The accounts for cleaning the rooms which the wardrobe of Queen Isabella was to use in 1313–14, both in England and abroad, note the purchase of fennel seed, which the household may then have cultivated for strewing on selected floors.[161]

Just as the Queen's body was clean, her clothes were exceptionally carefully treated. Laundresses were essential members of royal domestic establishments. The principal focus of their work was linen, usually for underclothes and napery – cloths, napkins, towels. Queen Philippa had a washerwoman for her chamber and one for the napery, and there was usually one for each of the royal children present with the Queen.[162] The washerwomen each received a set fee for the purchase of ashes and firewood, and there are references to the purchases of sulphur, soap and cauldrons.[163] The allowance for the washerwoman of the Queen's chamber was twice that for the washerwoman for the napery of her household or for the other washerwomen. It is probable that this related to the amount of washing she under-took, given that it was a fee for materials. On the analogy of later accounts, the implication is that the Queen's linen was washed more often, that is, that she wore her linen for a shorter period of time than others; and, as we have seen, she also had substantially more clothing than other members of her household.

Many of these distinctions accompanied a queen in death. Queen Eleanor's corpse was prepared with spices and a bushel of barley, probably to soak up fluids that remained after the embalming process. Incense was bought for censing the body, along with 6 ells of linen for wrapping the body.[164] The erection by Edward I of crosses at each point where his wife's body rested on its final journey reflected not only his devotion to her, but also the strong association derived from the sanctity of the touch of her anointed body, apparent also in the monuments that were then erected for each part of her body, for her viscera at Lincoln Cathedral, her heart at Blackfriars in London, and the corpse at Westminster Abbey.[165] The King also made this association with touch in a negative sense: Simon Fraser, a knight of Edward I's household, who had rebelled against him in Scotland in 1306, was hanged on Edward's orders, after which both his body and the gallows on which he was

80 The Eleanor Cross at Geddington. This is probably not the exact spot where Eleanor's body rested, which is much more likely to have been in the nearby church, but it is a mark of the sanctity that her anointed body brought to the place.

executed were burned, the latter almost certainly because it was contaminated by Fraser's treason.[166]

The overall environment of the Queen's household set out to create a number of different themes. First, the focus of the display was the Queen and her women. The male servants were deliberately subordinated to the overall purpose of queenship. They were not there to match the male display that featured in the King's household, with its direct associations with power and government. They came to the fore only in restricted pieces of protocol; they, like the overall position of the Queen, were subservient to the King and his household; and they were excluded or their role minimised on some of the principal occasions. That exclusion was explicit and extensive at the time of royal births, as we can see from the household ordinances of Henry VII of 1493.[167] The arrangements 'As for deliverance of a queene' included a formal meeting with the lords and ladies of high rank. The Queen then went to chapel and took communion, at which point the male nobility took leave of her. Her ladies and gentlewomen assumed all the functions normally undertaken by men in

81 The birth of
St Edmund, from the
Lives of St Edmund
and St Fremund,
written by John
Lydgate and presented
to Henry VI between
1434 and 1439. As the
male servants of the
household have been
excluded, the women
of the Queen's
household attend her,
acting in their roles, for
example as cup-bearer.
The Queen's chamber
is richly decorated,
with fine textiles for
the bed and a tapet or
rug in front of the fire.
A small cupboard, for
displaying plate, is in
the corner of the room;
and the Queen has a
chair beside her bed.
The floor to the
chamber is tiled.

the household, as butlers, pantlers, waiters, carvers and cup-bearers, taking the
goods from the men at the chamber door. This mirrored a common pattern of the
exclusion of males from births in the households of the upper classes. The men in
the Queen's household, however, were expected to take a lesser role on other occasions. On the first arrival of a new queen in the country and in the arrangements for
her coronation, it was envisaged that much higher status be given to the women
who accompanied her than to other members of her household. The ladies were to
be carried after the Queen on the way to her coronation; they had opportunities to
change their clothing that were not available to the male servants; their clothing was
to be sumptuous and even to mirror that of the Queen.[168] The Queen's household
was effectively missing a tier in its constitution: the males prominent in the King's
court were replaced in the Queen's household in terms of status by the Queen's
ladies, and there was an intimate connection between this female contingent and
the virtues they could promote in sensory terms. Men in the Queen's household

either had roles that gave the Queen honour – but without challenging the aspects of display at which she and her ladies excelled – or they performed menial service.

Medieval queens created an environment for their households that was distinctively feminine and different from other establishments in its sensory attainments. The arrangements for the presentation of the Queen relied heavily on showing her to the greatest effect, with the fine clothes, jewels and perfumes that surrounded monarchy. An impact on the senses was clearly intended, one that reflected on and related beauty and goodness, that made links to the powers of an anointed queen, yet her household was at the same time less in the public eye than that of the King. The environment was also foreign in its influences. Continental fashions and habits of living brought different expectations, which were in turn reflected in the domestic arrangements, in interior decoration and in gardens. Royal children, with their nurseries, toys and households, created an equally distinctive milieu. The Queen's residences and apartments were more intimate than those owned and developed by the King. Differences from the establishments of the episcopacy are also apparent: here the pleasures of the senses were not denied, but were made manifest in new and varied ways. Pecham's *Jerarchie* may have encouraged Queen Eleanor to contemplate ultimate realities through the medium of the royal court, but day-to-day living in her household was at the same time an experience that did not constrain sensory perception.

The households of the queens differed, in turn, from the households of the great nobles, where the competitive forms of public living and display were frequently combined in the later Middle Ages with elaborately choreographed routines that underscored social difference and hierarchy. In many ways the male aristocracy aped the practices of the King's court, if on a lesser scale – although monarchs sometimes felt that the difference was not as great as it should have been. It is these households that we shall consider as a concluding example of perception in the domestic environment.

Chapter 11

THE GREAT HOUSEHOLD AT THE END OF THE MIDDLE AGES

Sensory changes are conspicuous in the households of the aristocracy at the end of the Middle Ages. The fifteenth century saw an elaboration of ceremony in the great household. Descriptions of how it should function, prepared both by and for those with a particular interest in ceremony, especially its gentlemen ushers, servants of honour whose position first becomes clearly established in the second half of the fifteenth century, provide vignettes of the expectations of these establishments.[1] Changes to the physical environment can be seen in the layout of castles and great houses at this period. There were many, much smaller rooms; there was less use of the great hall by the lord, except for major festive occasions; the main arena for display was usually the great chamber of the house, a more intimate theatre. At the same time, there was more emphasis on the detail of the personal in terms of display, in things that could be sensed at close quarters, for example the wearing of fine clothing and elaborate jewellery; the use of perfumes, not only on the body, but within rooms; and the development of chamber music and the elaboration of music in chapels. In the wider environment, there were clearer patterns of arrangement and design in landscape and gardens. The great household was less peripatetic and investment was concentrated on fewer residences.[2] In sum, this was an environment that had changed significantly over the preceding two hundred years, and sensory perceptions and expectations had altered in like measure.

The Harleian household regulations of c.1460–c.1510 and the Second Northumberland Household Book both set out for the gentleman usher the daily routine of the household. The Harleian household regulations are not definitively associated with any specific establishment, although it has been argued that there is a connection to the Neville family; and the ceremonies of their households may well have been a model for its discussions.[3] The Second Northumberland Household Book[4] is different in content from the well-known First Northumberland Household Book, which was very closely tied to planning the domestic economy of the Earl of Northumberland and was based on model documents prepared over a number of years from 1512 onwards.[5] After a discussion of the daily routine to be overseen by the gentleman usher, the Second Household Book describes, from the same standpoint, a series of major events in the annual routine, including Maundy Thursday and Twelfth Night, and exceptional occasions such as birth and marriage, as well as events connected with the royal household at which a magnate might be present. Although it was written or copied by a member of the household of the Earl of Northumberland,[6] the Second Household Book is often framed with no particular household in view. Some of its contents are likely to have been drawn from the procedures of those of higher rank, especially those of the Crown and possibly a duke or marquess,[7] and it is most likely that it was intended to be a general model of practice for gentleman ushers probably in the period 1490–1530.[8]

The Harleian household regulations, in their description of the lord's rising, recorded that the gentleman usher carried out the 'chefe and principall' services.[9] The Second Household Book advised that 'all gentillmen uschars ought to be right well experte and parfite to know the rowmes of the yomen uschers, yomen of the chaumbre, groimes or childryn of the chaumbre, or ellis he is not able to occupy the said rowme as a gentillman ussher of the chaumbre . . .'[10] These officers were central to the smooth running of the ceremonial of the household, and through the prism of the gentleman usher we can follow the day's activities and its sensory connotations.

The Second Household Book starts by anticipating the arrival of the lord at a particular place. Much as a text from the first part of the fifteenth century had described the marshal of the hall as the 'gentleman harbinger',[11] the gentleman usher was to arrange the lodgings. He was first to be briefed by the lord and his closest servant, his chamberlain – a consultation that was encouraged throughout the text, which would save the gentleman usher much blame and trouble – and he was to understand how to prepare the lord's lodging, who was to reside there, what chambers were required for the lord, 'what arras, beddes and what stuf he shall prepayre the said chaumbres with all'.[12] At the same time he was to check the location for 'all maner of infirmities or outhir casualties that may be within the said plaice wheire he shall make and appointe the lodging', for anything that might annoy the lord, and he was to tell the lord as soon as he came to the lodging.[13]

The staff with him are outlined: a yeoman usher of the chamber, two or three yeomen of the chamber, a groom of the chamber, a groom porter for keeping the gates and one from the group who looked after the wardrobe of the beds. The gentleman usher was to gather all the keys to the place, giving the keys of the gates to the porter, 'theire shewing him the daunger of the said gaits if any suche be', and to repair any defects in the property that might trouble the lord. After that, he was to allocate the lord's lodgings and chamber, those for his wife, mother, sons and his chamberlain; and the rooms were to be fitted up with the lord's stuff: the soft furnishings that were usually carried with the household, in which a significant element of the sensory impact of late medieval lordship could be found.[14] Finally, he was to allocate lodgings for the head officers of the household, the household departments (houses of office) and lodgings for their staff; and then for the gentle servants 'as is ordourid daily to lye in the lodging of the said lord of dewty'.[15] These careful arrangements served to preserve the peace, essential for the household's honour. The harbingers were to be instructed to organise lodgings outside the precinct of the house – by which we may understand outside the main court, perhaps in a base court, blocks of lodgings or elsewhere in the locality, giving out 'billettes to the servauntes of all outhir wheire they shalbe loogid for knowlege that the herbigiours haith appointid theme theire for exchewing of variance that oone shuld not stryffe for the looging of an outhire'.[16] The lord was then to be met outside the gate and shown around the lodgings if he wished to see them.[17]

At the start of the day, in the Harleian regulations, the grooms of the bedchamber were to make a fire, and the gentleman usher was to enter with the grooms who were to help dress the lord. The yeoman of the wardrobe brought to the outer chamber the lord's clothing folded in a sheet, so that no one might touch it until it was passed to the gentleman usher to dress the lord.[18] In the Second Household Book, the gentleman usher was to arrive in the lord's chamber between eight and nine o'clock in the morning to oversee the yeomen ushers, yeomen of the chamber, grooms and children of the chamber, to make sure that they were performing their duties properly. He was then to keep to the door of the chamber where his master was, all day, while he was 'in waiting' – the First Household Book divided the day into shifts for gentle servants, giving their attendance in the morning, afternoon or evening – except when the lord was in council or in his secret chamber, an inner, private room. While the lord was engaged with others on business in this way, the usher was to wait outside the room, delivering the keeping of the door to a more confidential person.[19]

The next fixed points in the day were mass and breakfast. The gentleman usher was to establish where the lord was to hear mass, so that he might prepare what was necessary for one of his rank: a stool, carpet and cushions to kneel on, both the latter if he were in his chapel, cushions only if he were in the oratory in his closet.[20] The employment of textiles, particularly carpets for the lord and persons of rank to stand or sit on, or for wall or ceiling hangings in their presence, was a central part of the arrangements throughout both texts. Covering the ground in this way conveyed honour, much as covering cups or using canopies or celures overhead marked out this distinction. Special care was taken with managing the space around the lord and about handling anything the lord was to touch: the gesture of kissing the objects the lord was to handle or use assumes a good deal of prominence. In the Harleian regulations, as he put them in the traverse, the curtained compartment in which the lord was to hear mass, the gentleman usher was to kiss the cushions on which the lord was to kneel, and similar gestures were required in the Second

Household Book for the laying of the cushions for a countess going to hear mass in preparation for her confinement.[21]

In the Second Household Book, the lord's sword was delivered to him and was carried before him as he went to chapel, or when he went formally to any other place, at the great feasts of the year or if there were many guests in the household, by the gentleman usher if he were the most distinguished of those serving him, or by others of greater rank if they were there. The gentleman usher was also to establish important information about other fixed points in the day's routine, particularly where and when meals were required.[22] While the lord was at mass his bed was to be tidied, the linen folded away and the remains of the food and drink from 'all night' were to be taken away.[23] The lord's great chamber was where he usually dined as well as slept: the equipment for breakfast was therefore brought in at this point and a breakfast board set up on trestles on the carpet at the foot of the bed of estate. Other servants set up a ewery board, for food and drink delivered into the room as well as for water for washing. Both sets of regulations have descriptions of the elaborate routines for washing, before and after meals. In both, careful attention was given to taking assays of everything the lord was to touch, eat or drink. In the Harleian regulations, the lord sat down for breakfast, one of the gentleman ushers kneeling down, giving him his chair and cushion; the carver took the napkin from his shoulder and, unfolding it and kissing it, gave it to the lord, then uncovered the pottage and other foods; the cup-bearer brought the lord's cup whenever he wanted to drink, taking an assay of it, using a separate cup, kneeling or at least bending his knees.[24]

The meal in the great chamber was an exclusive affair for which only a small group of people were to remain in the room. The Second Household Book instructed the gentleman usher, who was to keep the chamber door, to send out of the room all except those who had been asked to eat there, the carver, sewer (the servant who is to serve the food), cup-bearer, henchmen and a gentleman waiter, the yeoman of the cellar or buttery, and the pantler for the lord's mouth.[25] Only the lord was to sit on a chair: the others were to sit on forms or 'buffit stooles'. No dishes were to be placed on the bed for fear of harming the counterpoint or the cloth on it, and the usher was to take great care that no one wiped or rubbed their hands on any of the arras or hangings of the chamber.[26] Further, at this meal and others in the chamber, the gentleman usher was to police the space: no man, of whatever degree, was to be so bold as to approach the lord's chair, stand under the cloth of estate should there be one, lean on the bed or stand on the carpet. All were to stand at the low end of the chamber, as far away from the lord as they could and especially to do so when the lord was speaking with anyone confidentially. 'And this to be callid boldely uppon yf the said parsounnes observe not the same.'[27] A similar restriction was enforced outside, when the lord was in procession on principal feasts or other significant occasions. Two gentlemen ushers were to go aside from the procession, preceding the lord by a little distance, one on each side, so that no man might approach him for any reason, perhaps to plead a case or to give him a petition.[28]

There is more specific information about the furnishings of the chamber in the directions contained in the Second Household Book for the arrangements to be made on the confinement of a countess. The chamber was to be hung with arras, counterfeit arras or similar, around all the walls, except for the windows. If there were window seats, these were to be covered with carpets and cushions, and curtains were to be hung so that they might be drawn across them – except in this special case, it was not usual to have curtains at windows. That this was a special case is confirmed by the directions for hanging a bridal chamber if the lady were a daughter of a peer of the rank of earl or above: 'the said windowis to have traverssis to draw befoir theim bicaus of keping of hir secreat from light upon the morrow bicaus sche cummeth not abroid'. For the confinement of the countess, hangings were also to be placed across the ceiling of the chamber. There was to be a bed of estate, with a celure and tester of embroidered work, arras, counterfeit arras, tapestry or a sparver with curtains. The framework of the bed was to be furnished with a large bed of down with a bolster, a counterpoint of arras or counterfeit arras, along with a pane of ermine and a sheet of fine linen for the pane to lie upon. Around the bed on the floor were two or three carpets, as needed.[29] There was also to be a second bed in the chamber, a low couch bed, next to the bed of estate, with a bed of down, a pair of fine sheets, two pillows and fine pillowcases, a pair of fustians (blankets), with a fine counterpoint of silk arras or counterfeit arras, a fine linen sheet to cast over the whole bed, a canopy with curtains, and carpets for the floor on each side. The whole was to be curtained off from the door to the chamber. Arrangements were also to be made for placing lights so that the bed area, as well as the altar and cupboard, could be illuminated when the curtains were closed.[30] The lady's room also had an altar, to be placed so that she could see it from wherever she lay in bed; a cradle, with a canopy, ready for the infant; and a cupboard, with three, four or five shelves according to the degree of the lady, illuminated by its own candlestick.[31] The chamber was furnished with a fireplace and chimney, a chair and further cushions.[32] It was one in a suite of at least three rooms: female servants, perhaps gentle, were accommodated in the second chamber; in the third were yeomen or grooms. It was the task of the last, when the lady went into labour, to instruct the chaplains to start their prayers – which they were to continue until she was delivered, at which point the chapel, if there was one, was to sing a Te Deum, or the chaplains were to say it, providing their mistress with continuous protection in sound at this dangerous moment.[33]

Just as breakfast was eaten by the lord in the great chamber, it was here that he ate lunch and supper, as well as participating in any formal drinking in the afternoon or in the food and drink of all night.[34] In the Harleian regulations, gentleman ushers and gentlewomen of presence (those that were to serve the lord and lady in person, directly in their presence) had their breakfast in the chamber immediately outside the great chamber, the yeomen ushers and chamberers eating at a table outside the great chamber door. If there were many important guests of rank, more than could be accommodated at the lord's own table, they were to be served their breakfast in the great chamber at a long table known as 'the knights' board'.[35] The head officers of the household were to eat in hall, along with the other members of the household and offer there and elsewhere in the household entertainment for visitors of less distinction or whose presence did not match the hours of the lord's meals. Meals in hall, were to take place simultaneously with those in the great chamber, but were to be completed in less time, not overlapping the chamber in any way: the lord and those visiting him in the great chamber, therefore, would not have seen the rest of the household at food.[36] The food for the great chamber was brought through the hall and it was the expectation that all there would bare their heads as a mark of reverence while the food passed.[37] Once the lord's board had been covered in the great chamber and lights set up, if it were winter, no grooms were to enter, except to renew the fire and lights, neither were any strangers under the degree of gentleman to be admitted. The keeper of the great chamber door was to enforce a similar injunction against the admission of all but the head officers and gentlemen at any time that the lord was within.[38]

At the end of the day, the great chamber was turned back into a bedroom. A fire was made there and the lord's bed was made with great deference. The Harleian regulations report that two grooms, with courtesies and bareheaded, made the bed, kissing their hands before unfolding the counterpoint, making the bed flat and soft. The chambermaid, if the lord was to sleep with his wife, was to bring into the room the fustians, pillows, head-sheet and sheets, all folded together in another sheet, and place them on a stool at the foot of the bed. The maid was to kiss her hands every time she delivered any of these items to the grooms making the bed.[39] The bread for all night was brought in wound in a napkin at both head and foot.[40]

In the Second Household Book, the gentleman usher was to signal the preparations for all night at eight o'clock by calling for a torch. All the servants who were to constitute that night's watch would then follow the groom with the torch to the ewery, pantry, buttery and cellar, collecting the lord's washing equipment and water, bread, wine and lights, which were all brought back for the gentleman usher to supervise, take the assay and place in the lord's secret chamber.[41]

In these day-to-day arrangements, one should note the privileging of access, and the creation of a special environment for the lord or lady. Other areas for general use were without the trappings that accompanied the lord: the rich, gleaming plate and textiles, the coloured arras and bed-hangings, and the soft furnishings, especially cushions and beds, or the carpets on which only the lord or the lady might tread. In that reserved area around the lord, gentle servants were continuously present. There were restrictions on touching materials within the great chamber or items that were to be touched by the lord or lady. When these had to be touched, that action was counterbalanced by the kissing of materials that the lord was to touch, or by the kissing of hands that were to touch them, or by reverence at the sight of them. The taking of assays protected the lord and lady from both poison and goods of substandard quality. Sound was carefully controlled around the lord: access to the space around him was restricted and people were expected to withdraw when he had private conversation. In his rooms resonance was dampened by the use of hangings and carpets. He and his wife had in addition the protection afforded by the continual round of the liturgy of the chapel. These sensory aspects were further developed in the arrangements for special occasions.

In the Maundy ceremony described in the Second Household Book, the lord is shown, both in chapel and in hall, performing a traditional series of actions – washing the feet of the poor, clothing and feeding them – in a fusion of liturgy with secular devotion and display in a carefully constructed sensory environment. The order for the day starts with the hallowing of the altars in the chapel and in the lord's closet, with the dean, subdean and whole chapel present, all in vestments. Two of the children of the chapel were to stand at either end of the altar, holding cruets with water and wine, while the remainder of the establishment processed there, led by those who were to read out the epistle and the gospel during the service. They were followed by the other children of the chapel and the gentlemen, with the dean and subdean ready to sweep the altars with brushes of box and yew. After this, they were to process to the hall, singing the responds for the day, which they were to continue until the gospeller began to read the gospel. The gentlemen and children of the chapel were then to go to the boards (tables) on the high dais and to stand in front of them for the remainder of the service. There was a lectern in the hall and here the gospeller stood to read, flanked by two children with candlesticks. Once the gospel was complete, the chapel then sang the Mandatum during the whole of the remainder of the Maundy ceremony.[42]

Before the choir arrived in the hall, a good deal of preparatory work had been done. The ewery board in the hall was covered by the yeoman of that office. Here he was to have the basins, all parcel-gilt, and two tubs, one with hot water, the other containing cold, both with flowers in them to scent the water. The yeoman also had an ash cup with which he was to put one measure of hot water and one of cold into every basin that was to be fetched for the lord by the gentlemen, the yeoman taking an assay of the water in every basin with two fingers.[43]

The servants from the wardrobe were placed in the middle of table on the right-hand side of the hall. They were to lay a linen cloth on the table, under the hoods and gowns that were to be given to the poor that day, along with the linen cloth that was to be given for their shirts. When the gentlemen ushers and yeomen ushers came to them for the cloth with which the lord was to serve, they were to kiss it before handing it over, and the ushers in turn were to kiss it before passing it to the almoner. The almoner was then to take an assay of the cloth, before taking it to the lord; one piece was to be used as an apron, the other as a towel around his neck. The almoner then received his own cloth for this purpose from a yeoman usher and the gentlemen went to the wardrobe to receive their cloths. After this, the washing of the feet of the poor took place. The final responsibility of the wardrobe was to deliver to each gentleman the gowns that were to be given away.[44]

On the left-hand side of the hall were stationed the servants of the pantry and larder. In front of them, on the table, on two linen cloths, were set out bread and other food. They were to hand over to the gentlemen as they came for it a wooden dish for each poor man, every one with a piece of saltfish and three herring; into the same dish the yeoman of the pantry was to place a cast of bread.[45]

The officers of the cellar dressed the cupboard in the hall with a cloth, pots of wine and as many ash cups as there were poor men. The cups were filled by the yeoman of the cellar, who was to take the assay before the cups were taken by the gentlemen to the lord for him to give to the poor to drink.[46]

The lord's cofferer or purse-bearer had in his hands as many halfpenny purses as there were poor men sat in the hall. Each purse contained as many pence as the lord was years old; but we are not told how many poor were to be present. In the household of Henry VII, there was a similar practice. In Henry's case, the number of poor present increased each year as well, to match the year of the King's life (for example, fifty, for his fiftieth year). The Second Household Book's cofferer was to kiss each purse as he delivered it to the gentleman who was to take it to the lord, who would hand it over to each poor man.[47]

After receiving the purse, each poor man was given clothing. The material for shirts, however, was worn as aprons, with the sleeves used as towels, by the lord and his gentlemen during the washing of the poor men's feet. The gentlemen were to take off their aprons and hand them over, to the lord, for him to give to the poor, along with the shirt sleeves. Additional shirts – there were more poor than there were gentlemen to wear shirts as aprons – were brought from the wardrobe table for the lord to hand over. The greatest of the gentlemen was to be last before the almoner and the lord to undo his apron.[48]

The lord may have been in the hall for the whole period, but he was to leave it to dress in a chamber adjoining the hall while the gospel was read. The reader was to pause at the words 'precinxit se' – 'He arose from supper and took off His garments, and when He had received the linen [towel], He girded himself.'[49] In this interlude, the gentlemen of the chapel were given drinks. When the lord returned, the reading from the gospel resumed. The lord was then given his apron by the almoner and the Maundy continued. At the end of the service, the lord was again to go into the chamber and change out of the gown that he had worn, which he then gave to the almoner, who was to carry it behind the lord through the hall until the lord selected the poor man to whom he wanted to give it. The almoner was then to kiss the gown and give it to the lord for him to hand out. After this, the lord was to go to stand behind the gospeller until he had finished the lesson. The members of the chapel left singing another responsory, returning to the chapel for vespers.[50]

The lord's gentlemen played an important role in the ceremony. Dressed in aprons, with towels, each gentleman fetched the basins and water from the ewery board for the lord to wash the feet of the poor men. As the basin was brought to the lord, they were to take the assay of the water with their fingers and to kiss it, then to kneel down and hold the basin under the feet of the poor man. They then went to the wardrobe table and fetched the gowns and hoods on outstretched arms: again, they were to kiss them and to hand them to the lord. The gentlemen helped the poor men dress in their new clothes. Next they fetched the dishes of food, kissing them before handing them to the lord to pass to the poor; they did similarly with the cups of wine, but taking the assay rather than kissing them. The purses were received from the cofferer and again, kissing them, they handed them to the lord. When they had finished these tasks, they undid their aprons and, kissing them, passed them to the lord to give to the poor men for shirts. The gentlemen were to take the assay of all things.[51]

During the ceremony, the feet of the poor were washed first by the almoner, making the sign of the cross on them; and subsequently, in the same manner, by the lord, attended by the gentleman holding the bowl, the dean and subdean of the chapel. The poor, however, had been prepared. The almoner had met them at the gate and brought them to the hall door, where hot water was ready. The under-almoner sat them on forms on either side of the hall, the oldest first and youngest last; and he then washed their feet. The poor men were to remain seated, with their hose rolled up to the knee: the boards from the trestle tables were placed under their feet to stop them becoming soiled again before the almoner and lord washed them once more.[52] These were not the only preliminaries. Storax or other fumigations were burned in the hall for the whole period of the service 'for voiding of contagious heires'; there was also a fire of charcoal in the hearth for the same purpose.[53]

This is useful evidence of the detailed and systematic thought that went into the ceremonies of the late medieval household. Once again, everything that was to be passed to the lord had a mark of reverence, usually a kiss, particularly from gentle servants, although from others as well. The lord's touch – and, indeed, that of his gentle servants – conveyed with it an intrinsic benefit as well as demonstrating humility. It was probably to this end that he and his gentle servants wore some of the clothes they were to hand over. The recreation of the Maundy itself, the words of Scripture – apart from the readings from the gospel and the liturgical singing of the choir, there was no other sound – brought all to the biblical sense of the event, in which they participated as if they were the original parties. The lord was no titular head of the household: his duty as head of this Christian establishment required him to play the role of Christ; in fact, to be Christ. It was through acts like this that his nobility was confirmed and made manifest to all around. His disciples – his gentlemen – benefited from the same association. Yet there were contradictions in this scene. The poor had to be cleaned before they were fit for the lord to touch; the water the lord used for washing the feet might be perfumed, but the first tub used by the under-almoner had no such odour. The fumigants burned in the hall were there to combat the disease that the poor might bring. The lord's protection was not compromised: assays were taken of all things he was to touch, even if he were then to pass them over directly.

Maundy Thursday was one of the few set-piece occasions on which the lord appeared in hall. In all of them, there was a considerable element of ceremony. It was not for nothing that the gentleman usher was to know if the lord wanted before him on any of the principal feasts his officers of arms, his minstrels or others.[54] The arrangements for the lord's appearance in hall at Twelfth Night, when a play and disguising constituted elements of the entertainment, were meticulously choreographed and furnished. Here were the textiles that accompanied him: a great carpet on the floor between the hearth and the screen; all the tables covered with carpets and to remain so until the lord left the hall for evensong; the walls hung with arras; the dais to have its tables covered with carpets, cushions for the lord and lady and a celure over their places; and the hall to have a great cupboard made for plate, to include seven-and-a-half tiers if the lord were a duke.[55] The members of the chapel were to be ready, too, on this occasion with carols, to accompany the lord and lady out of the hall to their own chamber.[56]

82 The keep at Warkworth was built by the Earls of Northumberland in the late fourteenth and early fifteenth century and was designed to support the elaborate ceremonial of the household. The castle was used by the family until 1537. Seen from the south, the chapel projects from the right-hand (east) end. Next to it are the double-height windows on the south side of the great hall.

83 The great hall in the keep at Warkworth. The dais area of this two-storey hall was well lit from the south (the right-hand side of this picture) and by windows behind the dais on the upper level. The wall immediately behind the dais remained blank for hangings. Late fourteenth or early fifteenth century.

The maintenance of the chapel and its choir, while not a new feature in great secular households – the chapel of Thomas, Duke of Clarence, in 1418–21, had twenty-one clerks, including four choristers[57] – is delineated in its full panoply in both the First and Second Household Books. The Second Household Book lists a similar number to Clarence's household: seventeen priests and gentlemen, besides the children. It allocates them in a specimen rota to the week's services – there were never to be fewer voices than would support four parts – along with a roster for the organ.[58] That the quality of the singing was just as important as the text of the liturgy is apparent from the instructions given to the dean of the chapel, as the lady of the house approached confinement, that the personnel of the chapel be well ordered and 'well agrad in the messe that they shall sing that daie so that it be doon and conveyd with as mooch melody as they cann'.[59]

If the descriptions of the Second Household Book and Harleian regulations are theoretical, they provide a dynamic element to our understanding of the everyday sensory environment. Wardrobe accounts are a much richer source for some aspects, for textiles and material culture in general, and can be linked to specific households. We are fortunate in having a substantial amount of documentation for the household of Edward Stafford, third Duke of Buckingham (d. 1521). The Duchy of Buckingham was one of only three non-royal, English dukedoms in the first two decades of the sixteenth century and Buckingham's possessions and expenditure marked him out as one of the wealthiest individuals in the country. Between 1514 and 1518 his annual expenditure was about £5,500, of which his household and wardrobe drew some £4,200. At the same time, he invested substantially in building projects, and there was in addition exceptional expenditure – about £1,500 was needed to pay for entertaining Henry VIII at Penshurst in 1519.[60]

The survey of Buckingham's estates made after his execution assessed their value and merits for the Crown as residences, hunting preserves and assets. At Thornbury, the rebuilding programme which began around 1511 was unfinished, suspended in 1519, but some parts were in use and had been lived in by Duke Edward. The south

84 Thornbury: the work of the third Duke of Buckingham, a print in the romantic tradition from T. Hudson Turner's *Some account of domestic architecture in England* (4 vols, Oxford, 1851–9).

side of the inner court had been completed 'with curious warkes and stately logginges'. The west and north sides were constructed to the height of a single storey, but had been roofed. The east side contained the hall and other domestic offices, but was of older construction. An outer court, with its lodgings, was unfinished. This was one of two sites where the survey described the gardens, with their orchards. Buckingham had invested in a new park immediately adjacent to the orchards by enclosing the land of his tenants, some of whom were left without compensation at his death.[61]

There were two residences in Kent. At Penshurst, there was 'a goodly maner and well buylded for the mooste parte of assheler stoone with a goodly hall, chambres and logginges and houses of offices accordingly'. The manor also had an orchard with fair alleys in it for walking. There was a base court, with a large stable and a good barn.[62] Tonbridge, four miles to the east, was an older Stafford residence, the strongest fortress that the Duke possessed. On three sides it was well protected, approached through a great thirteenth-century gate; the fourth side was covered by the river and was intended as a range of lodgings, but was unfinished. On the southeast, there was a square tower, the Stafford Tower, and close to it, on the river adjacent to the town bridge, an octagonal tower known as the Water Tower.[63]

85 Penshurst, Kent: the south front, with the Baron's Hall, built *c.* 1341, at the right-hand end, linked to the great chamber block (centre) built by John, Duke of Bedford, and subsequently in the possession of Edward Stafford, third Duke of Buckingham.

86 Tonbridge Castle, Kent, viewed from the town bridge. The Stafford Tower and the Water Tower, both fair towers, according to the survey of 1521, no longer survive; neither do the lodgings that were being built at that time along the river wall and which had been constructed to a height of 26 feet at the time of the Duke of Buckingham's execution.

Closer to London, at Bletchingley, there was a new-built manor, with many good lodgings and domestic offices. Here the walls of the hall, chapel, chambers, parlours, closets and oratories had all been wainscoted and ceiled, so that they could be used without hangings. This change in fashion would have enabled the property to be brought into use very quickly, without the need to transport arras, tapestries and other hangings. Although it would have created a very different environment acoustically, the main difference was in its visual impact. One of the primary concomitants of late medieval display was on the wane: the practices of the Second Household Book would have seemed just past the peak of fashion.[64] The same pattern can be seen in royal palaces, although the dating is less clear before the 1540s, partly because so little original wood panelling survives. In terms of scale, the wooden, panelled walls and ceilings predominated, some carved with linen folds or grotesques, some painted. From this point onwards, great houses began to employ decoration such as framed pictures, portraits and maps, all designed to hang on walls and to remain there, rather than employ textile hangings that had to be transported from site to site.[65]

Besides his London accommodation, the Duke had castles in a habitable state at Brecon, in the heart of his Welsh lordship, at Stafford and Maxstoke; substantial houses at Kimbolton and Writtle, which were both out of repair; and two further castles – Caus Castle in Shropshire and Newport, Gwent – which were ruinous.[66] Buckingham concentrated his investment on Thornbury, with improvements to his Kent and Surrey residences. Others, particularly the centres of his estates, were still usable, if not built in the latest fashion; a third group, left unrepaired, were super-numerary to his habits of life. The account of his wardrobe for 1517–18 shows more of how this was arranged: the buildings were a stage which could be dressed in order to show the Duke's magnificence, although the implication of the works at Bletchingley was that fashion was changing, and that 'less was more' in terms of its interior decoration.

In 1517–18, Buckingham's household servants were dressed in a livery of red and black: £113 was spent on cloth of these colours, probably against Christmas, with a further £25 on furs.[67] The five morris dancers who performed before him on Twelfth Night at Thornbury in 1518 had hose of red and black that was quartered; and the livery for the rest of the household may have been made in a similar way, although it would not have had the morris bells bought for the same performers.[68] It was only the Duke's servants who might wear his livery: these colours identified them with him particularly, and when they withdrew from his service they had to surrender their clothing, although they might be given other clothing in its stead.[69] The Duke's servants of honour included a footman, who was also bought hose of red and black kersey, and Buckingham's own hose were probably also of red and black, as sewing silk in these colours was bought for them at the end of March 1518.[70] At 12d., however, this was a tiny fraction of the Duke's expenditure on clothing. His tailors and hosiers worked by day and sometimes by night in his wardrobe at Thornbury in the months of November and December 1517.[71] In January 1518, a Florentine merchant was paid more than £180 for silks, satins, tinsels and velvets: these had probably been made into clothes for the Duke against the Christmas festivities.[72] Other dress and accoutrements included furred gloves; varnished stirrups; slippers, lined, unlined and furred, as well as night slippers (slippers were commonly worn in the household during the day, especially when one was to walk on textiles of considerable value); buskins; boots; palm or tennis shoes, with felt soles (used on an indoor court in January 1518); and pieces of embroidery that could be incorporated in garments, such as Stafford knots and swans (badges of the Duke), and his motto.[73]

The furnishings of the chamber were maintained by a group of arras-men, for whom dyed wools and needles were purchased.[74] John Baker was engaged for a total of six days in April 1518 in setting up arras and beds at Buckingham's London manor, Red Rose, as well as in taking them down again and packing them into wagons for their return journey to Thornbury.[75] In October 1517 the Duke bought a great carpet, 5 yards long by 2¼ yards broad, for £5, as well as three smaller carpets, just over 2 yards by 2 yards each, for £1 each.[76] The following August the household acquired four pieces of arras depicting 'the storey of Trooye', a suite probably intended for the Duke's great chamber or a room where his bed was on show, as he purchased a counterpoint with the same design to complement it. Together with another two counterpoints, all bought in Buckingham's presence, the total came to some £58.[77] There were also pieces of embroidery for window mantels.[78] Other bedding was purchased at much less cost: two large sparvers for 36s., twelve pallet coverlets for 3s. 4d. each, with twelve pillows of fustian, stuffed with 4 lb. of down each for 29s. 4d., are likely to have been for cadet members of the house or gentle servants.[79] A series of payments was made for repairs to the Duke's little clock, as well as to two large clocks.[80]

Three pairs of white pots of silver and some small pots of silver, bought for £63 10s. 8d., were the most substantial items of plate purchased in the year.[81] Others repaired plate: a goldsmith of Bristol made and gilded twelve hinges decorated or shaped with the ubiquitous Stafford knots.[82] A London goldsmith mended the lid of a great silver wine pot in September 1518.[83]

The Duke's chapel had impressive vestments. A series of twenty-eight embroidered scenes, as well as the Duke's arms, was purchased in London in November for £15 6s. 8d. for two copes and their hoods.[84] There was a similar investment in the liturgy and music. In November 1517, the Duke bought for 40s. from the widow of Henry Brewster, who had been master of the children of his chapel, his stock of music, some his own compositions, some the work of others in the chapel. The quantities suggest that some must have been part books. The stock consisted of a great book of four quires, with masses and anthems; seven quires of masses of five and six parts; four quires of masses of four parts on plainsong; five quires of five-part anthems to be sung after evensong; five quires of anthems to be sung after mass; four quires for services in Lent and five quires of carols for Christmas, all bound in forels (limp parchment covers), a *cantus firmus* mass based on *O mon coeur*, and a range of other masses and anthems in rolls and quires.[85] In early December 1517, Robert Ravynescroft, an organ-maker, was paid for repairing two pairs of the Duke's organs, gluing and new leathering the bellows of one and tuning the other over a period of four days.[86] Not all liturgical events took place in the Duke's chapel and closet. In March 1507, when the Duke's sister, Mary, died, he paid nineteen men of Thornbury for ringing the bells of the parish church from the time of her death until she had been buried. He also paid three local women for watching her body for two nights prior to her burial and for the offerings of three gentle-women lamenting at three masses on the day of her burial. The positive forces of sound and this genteel oversight were to protect her at this moment of transition.[87]

Besides the morris dancers, preparations for the Twelfth Night entertainments required seven carpenters, painting materials for the same and for the boar's head, and a miscellany of small items: paper, glue, resin, line, whipcord.[88] A grocer of Bristol supplied arcedine (a gold-coloured alloy of copper and zinc, often sold in sheets), white lead, vermilion, two dozen sheets of silver paper, three dozen sheets of gold foil and 26 lb. of chalk, all of which were probably associated with the

Christmas festivities and decorations, for food or for props or make-up for the disguisings.[89]

These purchases show the Duke managing his household at its most impressive: new accommodation, modelled to the highest quality, in the latest fashion, with decorative textiles, plate and servants, all marked with his badges or livery, a chapel with fine music, the splendour of the arrangements for the main festivals of the year and the entertainments associated with them. The sensory impact was intended to be considerable, and was intended to be observed at close quarters. Some aspects of these usages have a long pedigree, the protection afforded by liturgical sound, for example. Other elements in the accommodation suggest new usages on a day-to-day basis, such as finely appointed and much smaller chambers. These facilities offered new possibilities for the use of perfumes, but we have little evidence in this case of how these might have been deployed, although we have noted before that the Duke used perfumes on his own body and for washing his hair. More evidence for odour on the body and in residences in general comes from the household of Henry VIII.

In the King's household, smell was used in a number of ways. First, within the chapel, there was incense and images mounted with cinnamon, an aromatic whose associations with sanctity we have already seen.[90] Secondly, the inventory made on the King's death in 1547 lists some forty-five perfume-pans, for burning aromatics, which came from Greenwich, Hampton Court and Oatlands, although some were in store in the Jewel House at Westminster.[91] They included some very substantial pieces of silver, and these came in three broad sizes, with one group weighing between 40 and 60 ounces each; some much smaller items, of less then 10 ounces; and some very substantial pieces of metalwork, for example the perfume man at Westminster that was made with pillars and a clock of crystal, which weighed 194 ounces. Another one, also at Westminster and weighing 92 ounces, was embossed with signs of the planets and had on top a map of the country with two perfume-pans in it. Pieces like this would have been located in major state rooms.[92] Almost all the pans were of precious metal or marble, with the exception of four of iron at Oatlands.[93] Given the large numbers of rooms in the palaces, the perfume-pans can have been employed only in a limited range of venues at any one time.

In 1546–7, the King's apothecary spent 16d. on perfumes purchased on each occasion the household moved, probably in part to be used as fumigants against bad air or contagion on arrival at a new property. Occasionally there was a second purchase for a venue and sometimes this noted the rooms where the perfumes were used, including the King's closet and council chamber. Both the King and Princess Mary spent 16d. each with the apothecary on 24 September 1546 for perfumes for Windsor, Mary's specifically described as for her chambers. On 18 October 1546, perfumes were bought for where the King dined. Although the account does not tell us what these perfumes were, they were almost certainly aromatics with a charcoal base, to be burned in perfume pans.[94]

The 1547 inventory provides some further indications of the nature of the perfumes in the household. There were a number of perfume boxes: one containing 'burnynge perfumes' was taken in November 1549 to 'burne for the Kinges majestie'.[95] There were also two old pieces of ambergris; an ell of cinnamon stick, tipped with silver; and a white leather bag with a ball of balsam.[96] There is most evidence, however, for the use of perfumed waters on the person, in the silver bottles 'for swete waters', and the casting bottles for rose-water. Besides the last, of which there were some dozens, there were also pomanders, some from among the Queen's jewels, others enamelled, with gold.[97] There were two groups of perfumed dress accessories. In the Chair House at Westminster (where the King had his study) were eight pairs of perfumed gloves. There was also a series of walking sticks with a remarkable range of accessories, each with a perfume on top.[98]

Besides the perfumes for the rooms, in 1546–7 a second set of perfumes was purchased for use on the body, frequently in amounts of 5s., on an almost daily basis. While the apothecary's account does not specify that they were for personal use, they were almost certainly sweet powders. At one point they cease to be called perfumes and are described as Manus Christi powder or paste, purchased by the boxful at 5s. a time, or as dredge or paste, at 3s. 4d. Manus Christi was a sweet powder: in the eighteenth century it was made from sugar boiled with rose-water, violets or cinnamon, but there is no evidence that it took this exact form here. Like citrinade and the other sweet powders it could be used as a cosmetic, a perfume for the body and also to hang in sachets among clothes.

The third set of perfumes mentioned in the accounts were waters, particularly rose-water, but also marjoram-water with musk. A half-gallon flagon was refilled with rose-water at 4s. a time: in thirteen weeks, it was filled seventeen times, a rate of consumption of about 5½ pints a week. Rose-water was used continuously from the start of the account in August 1546 until the start of November, with smaller quantities used through November and December. Other plants followed: caprifoil-water (from honeysuckle or woodbine) was used from late September through to 16 January, and plantain-water, once in October and then three times between Christmas and New Year. Marjoram-water with musk was bought eleven times between late October and 20 December, although only once in November. Julep – in origin a rose-water (its Arabic and Persian meaning) – was probably used as a medicine or cordial, rather than as a perfume. It was probably intended as a gargle and sometimes appears with anise or unspecified roots. There was one purchase associated with the royal bath. On 12 November, 5s. 8d. was spent for the bath bag, with herbs, sponges, musk and civet. Two staves dressed with musk and ambergris were bought on 26 December for 26s. 8d. Other purchases of musk water, for the nose, between 22 December 1546 and 7 January 1547, followed for another fortnight by a mixture of nigella (fennelflower, with black seeds) and musk, had a medicinal purpose.

Together, these purchases show a notable use of perfume at the highest level, in considerable quantities, on a regular basis, with the aim of creating an atmosphere or aura around the King and the nobility, one that both protected the monarch from the ill-effects of bad odour and conveyed at the same time a good odour associated with his person or presence. The use of some smells, in a religious context and on the person, suggests that smelling good had important connotations, the beneficial qualities of the ecclesiastical context transferring readily to individuals of quality. The English aristocracy and monarchy may have been late into this field compared with their continental peers, but they participated in a way that would have marked them out by smell as an élite. Some of the growth in usage in England in the sixteenth century was the result of native production of scented waters, but the market for perfumed products of this quality remained limited. An apothecary's account for James I, for 1622, shows him using similar substances: scented waters (orange-water and rose-water); perfumes (probably aromatics) for chambers; with sweet powders used by the laundress and wardrobers. Sweet powders were used by the King's barber, too, along with rose-water; and the King also had orange flowers, rose leaves and clarified whey.[99]

If perfume and odour were essential adjuncts of the image of power, they were not the sole constituents of a sensory environment. By the mid-sixteenth century this environment was substantially different from that which had obtained in the mid-twelfth century or, indeed, in the mid-fourteenth. The studies of households have shown changes in fashion, such as the rise of textile hangings and their decline, or the move to smaller and more intimate wood-panelled rooms, equipped with fireplaces and perfumed. In a society that was acutely aware of hierarchy and its gradations there was a keen awareness of the quality of things, the refinement of perception and taste, a demand for the finest music or food. At the same time, there were substantial differences in mental outlook. Some might see allegory in their domestic establishment, and might attempt to deny all pleasure or temptation coming from the external senses. Others might create effects deliberately to overawe: the impressive jewels, textiles and odours of the households of queens and the upper classes made a direct impact through their virtues as well as carrying other, less direct messages. There were those who used their senses to see ultimate realities – or the mirror of them – in all things. There is no doubt that these attitudes were in place concurrently, but the balance between them changed during the later Middle Ages.

Chapter 12

CHANGING PERCEPTIONS

In late medieval England, the senses functioned within a framework that was very different to our present, Western, scientific model, and they did so in a plurality of ways. As well as vision, hearing, smell, taste and touch, there were significant additions to the sensorium, besides different modes and ambiguities of operation. For some, speech was a sense. A range of qualities – holiness, virtues and other, opposing attributes – had an effect that was conveyed by a process similar to touch, and they were also effective in general proximity, rather than requiring direct contact. Other characteristics had influence at a cosmic level.

Perception was a two-way process. The influences of the body flowed out through the senses, gateways which also brought to the individual impressions of the external world. Moral and spiritual qualities, as much as tangible ones, transferred in this way. In addition, these influences and qualities transferred to and from a range of objects and environments beyond those we would consider as implying animate life. Light shining from a jewel might bring benefit to the body; the virtue of a stone might be corrupted by the poor moral quality of the wearer; ghosts and angels could have an influence for good or bad. The boundaries of life, especially its end, were located at points beyond our expectation.

At times, these beliefs may appear chaotic; in practice, however, some groups had a more systematic appreciation of how the senses functioned. In its analysis of perception, medieval society fell broadly into five groups.[1] At one extreme lay the high culture of cognition, principally theologians and philosophers, whose daily task was to understand the senses and their operation, and who drew on Classical and Arabic learning in their construction of ordered models for its operation. Theirs was a culture of debate, interested in ultimate reality and how it might be perceived. A similarly learned approach underpinned the work of the foremost in the medical profession. The working of the body was described to them by earlier authorities, but gradually they developed a system of empiricism in which sensory perception played an important part. Thirdly, sensory perception was shaped into a framework for day-to-day life that was transmitted widely through ecclesiastical culture. This found its expression, often implicit rather than explicit, in monastic customaries, synodal statutes, manuals for confessors and similar didactic material that advocated particular routines of behaviour and the avoidance of others. The influences behind this documentation were principally those of a clerical, Latinate culture, but its texts moved from the theoretical to practical implementation. The sensory treatment of the dying, for example, shows how the sound of prayers and psalms, the light of candles and the use of incense, might create moral and spiritual influences, which continued to be effective after death, enduring to the resurrection, at which point the body and its senses were replaced by a resurrected entity and spiritual senses which mirrored those that had previously existed in corporeal form.

The episcopacy, the royal household and some aristocratic establishments came into regular contact with all three of the preceding groups and drew on elements of the practices of each in daily life. Their culture was Anglo-Norman or continental, vernacular rather than Latin, and this brought with it significant diversity. In 1354, Henry of Lancaster produced a list of the senses that differed from those his confessor might have recognised, but even so the treatise he composed to address the sins connected with those senses covered much ground that would have been familiar to his priest. The Anglo-Norman language itself approached perception in a different way to Latin. It is from this group, too, that we have some of the best evidence for individuals moulding their own sensory environments. Bishops who set out to deny sensory pleasure lived, nonetheless, in a world of sensation that their colleagues would have recognised.

The balance of the population, with its Middle English culture, constituted the fifth grouping. A great many elements were mixed up in its beliefs, some Christian, mitigated by a demotic presentation of its tenets from the pulpit; some of longer standing, but perhaps taking on a Christian hue. It is this group that looks both the most idiosyncratic and the most eclectic in its understanding of how the senses operated; at the same time, it is a section of society whose beliefs we can only glimpse obscurely through the homogenising effect of clerical culture. On the basis of modern anthropology, one would expect diversity of belief, and it is apparent here especially.

Between 1150 and 1550 there were significant changes in the ways it was believed the senses might operate. At the onset of the Middle Ages, the Church had to reach an accommodation with a range of practices that it deemed magical, both delete-rious and beneficial, many of which operated on the body through the senses. The task had largely been completed by the end of the twelfth century, or that is the impression given by the new Anglo-Norman clergy. The detail of the way it did this is no longer clear but, by accepting the efficacy of earlier traditions, it was able to meld ideas of holiness and sanctity into broad strands of popular religion.[2] In doing so, however, it did not eliminate popular ambiguity over the number and range of senses. If theologians debated the five senses and their operation, enumerating them for penitents, parish clerks in rural pulpits were not necessarily distant from their congregations in having a more blurred notion of the sensorium. The papacy, following its theologians, employed a restricted model of the senses, for example omitting from deliberations over canonisation mention of the odour that popularly evidenced sanctity: in the rigour of its pursuit of authoritative witness, commonly accepted notions of sensory perception were set aside.

Medicine also progressed from the transcendental notions that had traditionally constituted a part of perception. In the late thirteenth and early fourteenth centuries there was a move towards the practical use of observation by the senses and it is at this point that we find smell used as a tool for medical diagnosis rather than for moral characterisation.[3]

The popular model itself came under pressure. The views of heretics, especially the Lollards, and others who doubted the miraculous and the spiritual benefits accorded by touch or sanctity, presented a challenge to both popular religion and the ecclesiastical culture that kept religion enshrined in its sacred, Latin language. In 1430, after a period of imprisonment, John Reve of Beccles confessed that he had believed it was as meritorious for a Christian to be buried in a midden, meadow or field as to be buried in church or churchyard, that is, without the protection accorded by the touch of holy ground. This was a battle John had lost in the short term, but it was one that was to have profound effects on popular religion and the sensory framework in which it operated in following centuries. Doubts over religious practices, expressed more widely at the Reformation, impinged on many aspects of the generally accepted sensory model of the world.[4] Once belief in sacred powers and the transfer of virtues by touch was in doubt, why should one believe that other senses transferred anything of a moral or spiritual nature? Such uncertainty ultimately led to the development of scientific investigation in the seventeenth and eighteenth centuries, and to the 'closure' of the human body, a reduction of sensory perception to a definable, one-way process based on physiology and the nervous system.

Within these general frameworks, it is important to emphasise the variety in the way the senses functioned in late medieval England: they might bind groups together or they might create individual experience, placing the self in relation to ultimate reality in a way that is unlikely to have been understood before the eleventh century.[5] Christianity and devotional practice opened up a range of possibilities for sensory perception: meanings, virtues and influence might be implanted in the day-to-day environment, to be recalled as part of meditational practice. In monasteries, such as those of the Cistercian Order, there were common patterns of meditation, creating and calling to mind a common experience, perpetuated by the construction of claustral buildings to a common pattern.[6] Others were encouraged to find their own models for contemplation, a practice which might invest a transcendental significance in the perception of everyday things, or create biblical realities through the eye of faith. These realities were often made present through the allegorical and teleological meanings that were a prominent feature of medieval art.

At the end of the Middle Ages, visual culture was transformed by a different way of seeing, by realism and perspective. On the fringes of Europe, this development did not have its full impact on England until the sixteenth century, but it would have been apparent to those who visited Italy or the Low Countries in the fifteenth century.[7] It was indicative, however, of one way in which sensory perception could be modified – by the experience of new sensations. Fashion, discrimination and acquired knowledge brought external influences to England, particularly from the Continent, and sometimes from further afield, through education, trade, travel, pilgrimage and the exchange of ideas.

Examples from the great household show how this transition was made. New colours, new odours, new ways of living were all adopted, and adapted, in English environments. From 1505, Henry VII had a French cook, Pierrot, who had been the cook of the King of Castile, and one can imagine that royal cuisine was renewed and remodelled by his internationalism.[8] The Second Northumberland Household Book shared this interest in the quality of experience, valued for its own sake as well as for what that experience might represent in anagogical terms. Fine singing in the chapel was as important as the words of the liturgy; the poor were cleaned before the lord washed them, to preserve his gentle person from pollution, although he might still gain spiritual benefit from the Maundy in the conventional way; in exaggerated manners and etiquette, the lord's servants showed their obeisance to him, while the kisses they offered to things he was to touch saved the objects from contamination in a more traditional view of perception.

From the late medieval period comes considerable evidence for change in the day-to-day realities that intertwined with beliefs about perception. Even confining the discussion to food, hygiene and architecture, there can be little doubt about the scale of difference between the environment at the end of the Middle Ages and that of 1150 or 1300. While Henry VII's French cook doubtless brought to the King's table refinements, other, more general changes in diet – and taste – can be identified. In upper-class cookery the use of spices evolved to show a form of discrimination that was not found elsewhere, for example, in France; English cooking also made novel use of cooked fruits, flower blossoms and large quantities of sugar. Although the content of English recipes was slow to change, by the end of the sixteenth century many of the more exotic pottages typical of medieval cuisine had disappeared, and other dishes, such as fools and trifles, a feature of later English cookery, are first found in the early seventeenth century.[9] Aside from these changes in flavour and taste, there were also major changes in food in the later Middle Ages. Variety in diet expanded enormously in the mid-eleventh century, with an increase in the range and quantities of marine fish consumed, a greater range of birds, new patterns of hunting and the addition of spices. There was additional diversity in the late medieval period, especially in towns. Of outstanding significance, however, was the change in food availability, both in terms of range and quantity, from 1350 onwards. Copying the regimen of the upper classes, all those who could invested in meat, white bread and better-quality ales, and, where possible, in wine, beer and spices.[10] The upper classes then reshaped their diet to include new elements of exclusivity, restricting the consumption of, for example, wild birds and game, on a sumptuary basis.[11] The ordinary member of a great household in the early sixteenth century might aspire to these foods, but his expectations might also be frustrated:

> Thy meate in the court is neyther swanne nor heron,
> Curlewe nor crane, but course beefe and mutton,
> Fat porke or vele, and namely such as is bought
> For Easter price when they be leane and nought.
> Thy fleshe is restie [rancid] or leane, tough and olde,
> Or it come to borde unsavery and colde,
> Sometime twise sodden, and cleane without taste,
> Saused with coles and ashes all for haste,
> When thou it eatest it smelleth so of smoke
> That every morsell is able one to choke.[12]

Disappointment gives us a fair indication of what the pleasures should have been. These changes made eating in England at the end of the Middle Ages a different sensory experience to that of two hundred years previously: even if the understanding of the mechanisms of perception had not altered, possibilities and expectations had changed, along with the connotations, particularly of status, that went with them.

The possibilities changed in other areas too. In terms of hygiene, we have seen how the late medieval upper classes sought to mirror continental patterns of cleanliness and perfumery. In turn, these changes were copied more widely, although it was not until well into the sixteenth century that they made their impact in the provincial marketplace. The inventory of John Brodocke, an apothecary of Southampton who died in 1571, listed among the contents of his shop some thirty-two types of scented water, thirteen types of conserve, twenty varieties of confection and electuary, along with twenty-eight different powders, twenty-four assorted syrups and thirty-five varieties of oil, besides a long list of plasters, pills, ointments and simples.[13] These perfumes, additions to diet and valetudinarian preparations were available, as we have seen, some one hundred and fifty to two hundred years earlier around the north-west Mediterranean; and they were on sale in London at least fifty years earlier than in Southampton. Brodocke's shop is evidence for the widening availability of these possibilities and the changes that might come with them in terms of sensory perception. At the same time, we should expect both the connotations of perfume and practices of hygiene to change as this merchandise became accessible as well as desirable.

The buildings of the sixteenth century were very different from those at the start of the period we have been considering. Glazing, in particular, changed their character: the use of light in ecclesiastical buildings made the 'schoolroom' church interiors of the sixteenth century places where one might read a prayer book, whereas a Romanesque or Gothic church, with its restricted glazing, focused on the jewel-like qualities of light and its connotations of divinity, rather than providing illumination for all to read. In secular structures, glazing also brought new possibilities. Displaying the lord on his dais in hall had long been the intent of fenestration, but with the shift of aristocratic living to the lord's great chamber, measures were taken to create here an especially well lit environment. The lord spent much of his day in the great chamber, in study, at his meals, with his council and guests, and the greatest investments in display were made in this room, in textiles, beds and furnishings. In 1431–2, the twelfth Earl of Oxford spent 12s. on the repair of the glazing in a room at Castle Hedingham, strikingly known as the 'Glaschambre'.[14] The sixteenth century saw a further extension of glazing, not only in the houses of the highest ranks, but also in more modest dwellings, a change that was to alter daily life for many. These shifts in the sensory environment of the late medieval period, accelerating through the sixteenth century, speak of differences of perception: new tastes, new sensations and new realities for day-to-day life. Some sensations will have had different connotations, or ones that altered as exclusive practices became commonplace.

However much some medieval beliefs prefigure our post-Enlightenment understanding of the senses, other traditions were not discarded and continued to be employed and manipulated, sometimes by the very people who seemingly rejected these beliefs as criteria of significance in other circumstances. Even today, vestiges persist in ways that are now perhaps detached from general Western thought, such as the popular expressions cited at the start of this book. We are now subject to many more experiences of sensation from different cultures. To take one example, the media of mass communication have made us familiar with images of crowds clustering around the body of a martyr or holy man or woman, and it is not difficult to imagine those who pressed around the body of St Richard on its way from Dover to Chichester in 1253, attempting to touch it and thereby secure benefit. We should not assume, however, that the motivation of those we now see is the same as that of the inhabitants of Sussex seven hundred and fifty years ago or that they believe their senses operate in a similar way.[15]

Sensory perception may be cast, in scientific terms, as an experience that is common to all. It is shaped, however, by ideas that are distinct in time and place. It can be intensely personal and can differ from individual to individual; at the same time, it can be culturally determined and manipulated into forms of common experience. This book has outlined some of the commonalities of perception in late medieval England and considered some of the sensory connotations of life in the great household. There is much here still to explore and many other environments, such as the great cathedrals, where we need to understand more about perception in the Middle Ages in order to comprehend day-to-day experiences in the past.

ABBREVIATIONS

London is the place of publication, unless otherwise stated.

ABMA	British Academy, Auctores britannici medii aevi
Ailred	Walter Daniel *The life of Ailred of Rievaulx* ed. F.M. Powicke (1950)
Alphabet	*An alphabet of tales, an English 15th century translation of the Alphabetum narrationum of Etienne de Besançon from Additional MS. 25,719 of the British Museum* ed. M.M. Banks (2 vols, EETS, OS 126–7; 1904–5)
ANTS	Anglo-Norman Text Society
BA	*On the properties of things: John Trevisa's translation of Bartholomaeus Anglicus De proprietatibus rerum* ed. M.C. Seymour (3 vols, Oxford, 1975–88)
Barnwell	*The observances in use at the Augustinian priory of S. Giles and S. Andrew at Barnwell, Cambridgeshire* ed. J.W. Clark (Cambridge, 1897)
BAV	Biblioteca Apostolica Vaticana
BB	*The household of Edward IV: the Black Book and the ordinance of 1478* ed. A.R. Myers (Manchester, 1959)
Becket	*Materials for the history of Thomas Becket, Archbishop of Canterbury* ed. J.C. Robertson (7 vols, RS, 1875–85)
BL	British Library
Bodl.	Bodleian Library, Oxford
CCCM	Corpus christianorum, continuatio medievalis
CCSA	Corpus christianorum, series apocryphorum
CCSL	Corpus christianorum, series latina
CS 871–1204	*Councils and synods with other documents relating to the English Church. I.* AD *871–1204* ed. D. Whitelock, M. Brett and C.N.L. Brooke (2 vols, Oxford, 1981)

CS 1205–1313	*Councils and synods with other documents relating to the English Church. II AD 1205–1313* ed. F.M. Powicke and C.R. Cheney (2 vols, Oxford, 1964)
'DH'	Sister Mary Patrick Candon, 'The Doctrine of the Hert, edited from the manuscripts with introduction and notes' (Unpublished Ph.D. dissertation, Fordham University, 1963)
EC	*The court and household of Eleanor of Castile in 1290: an edition of British Library, Additional Manuscript 35294* ed. J.C. Parsons (Toronto, 1977)
Edmund	C.H. Lawrence *St Edmund of Abingdon: a study in hagiography and history* (Oxford, 1960)
EETS	Early English Text Society
EHR	*English Historical Review*
Erkenwald	*The saint of London: the life and miracles of St Erkenwald* ed. E.G. Whatley (Medieval and Renaissance Texts and Studies, 58; Binghamton, 1989)
ES	Extra Series
Etymologies	*Isidori Hispalensis episcopi etymologiarum sive originum libri XX* ed. W.M. Lindsay (2 vols, Oxford, 1911). This edition is not paginated and references are to books and sections in the text.
EWS	*English Wycliffite sermons* ed. A. Hudson and P. Gradon (5 vols, Oxford, 1983–96)
Executors	*Account of the executors of Richard, Bishop of London, 1303, and of the executors of Thomas, Bishop of Exeter, 1310* ed. W.H. Hale and H.T. Ellacombe (Camden Society, NS 10; 1874)
'Eynsham'	'Vision of the monk of Eynsham' in *Eynsham cartulary* ed. H.E. Salter (2 vols, Oxford Historical Society, 49, 51; 1907–8) ii, pp. 255–371
GH	C.M. Woolgar *The great household in late medieval England* (New Haven, 1999)
HAME	*Household accounts from medieval England* ed. C.M. Woolgar (2 vols, British Academy, Records of Social and Economic History, NS 17–18; 1992–3)
HBQI	*The household book of Queen Isabella of England for the fifth regnal year of Edward II 8th July 1311 to 7th July 1312* ed. F.D. Blackley and G. Hermansen (University of Alberta, Classical and Historical Studies, 1; 1971)

Henry VI	*Henrici VI Angliae regis miracula postuma ex codice Musei Britannici Regio 13. C. VIII* ed. P. Grosjean (Subsidia Hagiographica, 22; Brussels, Société des Bollandistes)
Hinton, *Gold and gilt*	D.A. Hinton *Gold and gilt, pots and pins: possessions and people in medieval Britain* (Oxford, 2005)
HKW	*The history of the King's works* ed. H.M. Colvin *et al.* (6 vols, 1963–82)
HO	*A collection of ordinances and regulations for the government of the royal household, made in divers reigns from King Edward III to King William and Queen Mary . . .* ed. Anon. for the Society of Antiquaries (London, 1790)
Hugh	*Magna vita sancti Hugonis: the life of St Hugh of Lincoln [by Adam of Eynsham]* ed. D.L. Douie and D.H. Farmer (2nd impression, 2 vols, Oxford, 1985)
Inventory of Henry VIII	*The Inventory of King Henry VIII: Society of Antiquaries MS 129 and British Library MS Harley 1419* ed. D. Starkey (3 vols, in progress, 1998–) i. References are to inventory numbers.
Jacob's Well	*Jacob's Well, an English treatise on the cleansing of man's conscience* ed. A. Brandeis (EETS, OS 115; 1900). The second part of the text is cited from Salisbury Cathedral MS 103.
JRULM	John Rylands University Library of Manchester
JWCI	*Journal of the Warburg and Courtauld Institutes*
LEJ	*Lower ecclesiastical jurisdiction in late-medieval England: the courts of the Dean and Chapter of Lincoln, 1336–1349, and the Deanery of Wisbech, 1458–1484* ed. L.R. Poos (British Academy, Records of Social and Economic History, NS 32; 2001)
Liber exemplorum	*Liber exemplorum ad usum praedicantium saeculo XIII compositus a quodam fratre minore anglico de provincia Hiberniae* ed. A.G. Little (British Society of Franciscan Studies, 1; 1908)
Macer	*A Middle English translation of Macer Floridus De viribus herbarum* ed. Gösta Frisk (The English Institute in the University of Upsala, Essays and studies on English language and literature, 3; 1949)
Macro	*The Macro Plays: The Castle of Perseverance; Wisdom; Mankind* ed. M. Eccles (EETS, OS 262; 1969)

Marbode	*Marbode of Rennes' (1035–1123) De lapidibus considered as a medical treatise with text, commentary and C.W. King's translation, together with text and translation of Marbode's minor works on stones* ed. J.M. Riddle (Wiesbaden, Sudhoffs Archiv Zeitschrift für Wissenschaftsgeschichte, Beihefte, 20, 1977)
MCL	*The monastic constitutions of Lanfranc* ed. D. Knowles, revised by C.N.L. Brooke (Oxford, 2002)
MK	*The book of Margery Kempe: the text from the unique MS. owned by Colonel W. Butler Bowdon* ed. S.B. Meech and H.E. Allen (EETS, OS 212; 1940)
NHB	*The regulations and establishment of the household of Henry Algernon Percy, the fifth Earl of Northumberland, at his castles of Wressle and Leconfield, in Yorkshire. Begun Anno Domini MDXII* ed. T. Percy (new edition, 1905)
NS	New Series
N-town	*The N-town play: Cotton MS Vespasian D. 8* ed. S. Spector (2 vols, EETS, SS 11–12; 1991)
OED	*Oxford English Dictionary* (2nd edition, Oxford, 1989)
OS	Old Series (Camden Society); Original Series (EETS)
Osmund	*The canonization of Saint Osmund from the manuscript records in the muniment room of Salisbury Cathedral* ed. A.R. Malden (Wiltshire Record Society, 1901)
Owst, *Literature*	G.R. Owst *Literature and pulpit in medieval England: a neglected chapter in the history of English letters and of the English people* (Cambridge, 1933)
Owst, *Preaching*	G.R. Owst *Preaching in medieval England: an introduction to sermon manuscripts of the period c. 1350–1450* (Cambridge, 1926)
PI	*Peter Idley's instructions to his son* ed. C. D'Evelyn (Modern Language Association of America, monograph series, 6; Boston and London, 1935)
PL	*Patrologiae cursus completus: series latina* ed. J.P. Migne (221 vols, Paris, 1844–1902). Cited by volume and column number
PRO	Public Record Office (the National Archives), Kew
PSQ	*The prose Salernitan questions edited from a Bodleian manuscript (Auct. F.3.10): an anonymous collection dealing with science and medicine written by an Englishman c. 1200 with an appendix of ten related collections* ed. B. Lawn (ABMA, 5; 1979)

RB	*RB 1980: the Rule of St Benedict in Latin and English* ed. T. Fry *et al.* (Collegeville, 1981)
RO	Record Office
Rot. Parl.	*Rotuli parliamentorum; ut et petitiones et placita in parliamento* ed. Anon. (6 vols, n.d., *c.*1783)
RS	Rolls Series
RWH, 1285–6	*Records of the wardrobe and household 1285–1286* ed. B.F. Byerly and C.R. Byerly (1977)
RWH, 1286–9	*Records of the wardrobe and household 1286–1289* ed. B.F. Byerly and C.R. Byerly (1986)
Seyntz medicines	*Le Livre de seyntz medicines: the unpublished devotional treatise of Henry of Lancaster* ed. E.J. Arnould (ANTS, 2; 1940)
SM	*The chronicle of William de Rishanger of the Barons' Wars. The miracles of Simon de Montfort* ed. J.O. Halliwell (Camden Society, OS 15; 1840)
SS	Supplementary Series
STC	*Short-title catalogue of books printed in England, Scotland, and Ireland, and of English books printed abroad 1475–1640* (2nd edition, 3 vols, 1976–91)
Swithun	M. Lapidge *The cult of St Swithun* (Winchester Studies 4.ii; Oxford, 2003)
Thurkill	*Visio Thurkilli relatore, ut videtur, Radulpho de Coggeshall* ed. P.G. Schmidt (Akademie der Wissenschaften der DDR Zentralinstitut für alte Geschichte und Archäologie, Bibliotheca Scriptorum Graecorum et Romanorum Teubneriana, Leipzig; 1978)
Towneley	*The Towneley plays* ed. M. Stevens and A.C. Cawley (2 vols, EETS, SS 13–14; 1994)
VV	*The book of vices and virtues: a fourteenth century English translation of the Somme le Roi of Lorens d'Orléans* ed. W. Nelson Francis (EETS, OS 217; 1942)
WN	*The life and miracles of St William of Norwich by Thomas of Monmouth . . .* ed. A. Jessopp and M.R. James (Cambridge, 1896)
Wulfstan	*The Vita Wulfstani of William of Malmesbury to which are added the extant abridgements of this work and the miracles and translation of St Wulfstan* ed. R.R. Darlington (Camden, 3rd series, 40; 1928)

BOOKS OF THE VULGATE

Act	Actus Apostolorum [Acts]
Apc	Apocalypsis [Revelation]
II Cor	Epistula ad Corinthios II [II Corinthians]
Ct	Canticum canticorum [Song of Solomon]
Eph	Epistula ad Ephesios [Ephesians]
Ex	Liber Exodi [Exodus]
Iac	Epistula Iacobi [James]
Io	Evangelium secundum Iohannem [St John's Gospel]
Lc	Evangelium secundum Lucam [St Luke's Gospel]
Mc	Evangelium secundum Marcum [St Mark's Gospel]
Mt	Evangelium secundum Mattheum [St Matthew's Gospel]
Phil	Epistula ad Philippenses [Philippians]
Prv	Liber proverbiorum Salomonis [Proverbs]
IV Rg	Liber Regum IV [II Kings]
Sap	Liber sapientiae Salomonis [Apocrypha: Wisdom of Solomon]
Sir	Liber Iesu filii Sirach
I Sm	Liber Samuehelis I [I Samuel]
Tb	Liber Tobiae [Apocrypha: Tobit]
I Tim	Epistula ad Timotheum II [II Timothy]

All biblical references have been taken from the Vulgate: *Biblia sacra secundum vulgatam versionem* ed. R.Weber *et al.* (4th edition, Stuttgart, 1994). Equivalent books in modern translations of the Bible, although noted here, differ at many points from their medieval Latin counterparts. For ease of reference, the Psalms have been given their numbers from the Authorised Version as well as those from the Vulgate.

NOTES

CHAPTER 1: INTRODUCTION

1. R.C. Finucane *Miracles and pilgrims: popular beliefs in medieval England* (1977) p. 22; *Hugh*, i, p. 70.
2. *The book of the knight of La Tour-Landry* . . . ed. T. Wright (EETS, OS 33; 1868) p. 82.
3. Compare the remarks about gesture, similar in many ways: K. Thomas, 'Introduction' in *A cultural history of gesture: from Antiquity to the present day* ed. J. Bremmer and H. Roodenburg (Cambridge, 1991) pp. 1–14, at 2–8.
4. Lc 8: 46; R. Kieckhefer *Magic in the Middle Ages* (Cambridge, Canto edition, 2000) pp. 34–5.
5. G. Casagrande and C. Kleinhenz, 'Literary and philosophical perspectives on the wheel of the five senses in Longthope Tower' *Traditio* 41 (1985) pp. 311–27, at 312–14.
6. *BA*, i, pp. 129, 147; below, Chapter 2.
7. R. Rivlin and K. Gravelle *Deciphering the senses: the expanding world of human perception* (New York, 1985); L. Watson *Jacobson's organ and the remarkable nature of smell* (1999); *The Oxford companion to the mind* ed. R. Gregory (2nd edition, Oxford, 2004).
8. *The varieties of sensory experience: a sourcebook in the anthropology of the senses* ed. D. Howes (Toronto, 1991), especially D. Howes, 'Sensorial anthropology', pp. 167–91, at 167–8; C. Classen *Worlds of sense: exploring the senses in history and across cultures* (1993); C. Classen, D. Howes and A. Synnott *Aroma: the cultural history of smell* (1994).
9. *The ancient mind: elements of cognitive archaeology* ed. C. Renfrew and E.B.W. Zubrow (Cambridge, 1994); Y. Hamilakis, 'The past as oral history: towards an archaeology of the senses' in *Thinking through the body: archaeologies of corporeality* ed. Y. Hamilakis, M. Pluciennik and S. Tarlow (New York, 2002) pp. 121–36; C. Jones, 'Interpreting the perceptions of past people' in *The archaeology of perception and the senses* ed. C. Jones and C. Hayden (special number of *Archaeological Review from Cambridge* 15: 1 (1998)) pp. 7–22.
10. A. Corbin *The foul and the fragrant* (1986), originally published as *Le Miasme et la jonquille* (Paris, 1982); L. Febvre *The problem of unbelief in the sixteenth century: the religion of Rabelais* trans. B. Gottlieb (Cambridge, Mass., 1982); R. Mandrou *Introduction to modern France 1500–1640: an essay in historical psychology* trans. R.E. Hallmark (1975); M.S.R. Jenner, 'Civilization and deodorization? Smell in early modern English culture' in *Civil histories: essays presented to Sir Keith Thomas* ed. P. Burke, B. Harrison and P. Slack (Oxford, 2000) pp. 127–44.
11. R. Jütte *A history of the senses: from Antiquity to cyberspace* trans. J. Lynn (Oxford, 2005); Classen *et al.*, *Aroma*, pp. 3–4.
12. T.M. Vinyoles i Vidal, 'Els sons, els colors i les olors de Barcelona pels volts del 1400' in *Història Urbana del Pla de Barcelona: Actes del II Congrès d'Història del Pla de Barcelona celebrat a l'Institut Municipal d'Història els dies 6 i 7 de desembre de 1985* ed. A.M. Adroer i Tasis

(2 vols, Barcelona, 1989) i, pp. 133–44; B.R. Smith *The acoustic world of early modern England* (Chicago, 1999).

13. E.g. M. Camille, 'Before the gaze: the internal senses and late medieval practices of seeing' in *Visuality before and beyond the Renaissance* ed. R.S. Nelson (Cambridge, 2000) pp. 197–223; E. Sears, 'Sensory perception and its metaphors in the time of Richard of Fournival' in *Medicine and the five senses* ed. W.F. Bynum and R. Porter (Cambridge, 1993) pp. 17–39, 276–83; C. Nordenfalk, 'Les Cinq Sens dans l'art du Moyen Âge' *Revue de l'Art* 34 (1976) pp. 17–28; L. Vinge *The five senses: studies in a literary tradition* (Lund, 1975); E.R. Harvey *The inward wits: psychological theory in the Middle Ages and the Renaissance* (Warburg Institute Surveys, 6; 1975).
14. D. Pearsall and E. Salter *Landscapes and seasons of the medieval world* (1973).
15. C. Morris *The discovery of the individual 1050–1200* (1972); D.F. Appleby, 'The priority of sight according to Peter the Venerable' *Mediaeval Studies* 60 (1998) pp. 123–57, especially p. 156.

CHAPTER 2: IDEAS ABOUT THE SENSES

1. This section is based on C.T. Lewis and C. Short *A Latin dictionary* (Oxford, 1879); *Dictionary of medieval Latin from British sources* ed. R.E. Latham, D.R. Howlett *et al.* (1975–); *Anglo-Norman dictionary* ed. L.W. Stone, W. Rothwell *et al.* (Publications of the Modern Humanities Research Association, 8; 1977–92); *Middle English dictionary* ed. H. Kurath and S.M. Kuhn *et al.* (Ann Arbor, Michigan, 1956–); and the *OED.*
2. *La Lumere as lais by Pierre d'Abernon of Fetcham* ed. G. Hesketh (3 vols, ANTS, 54–5, 56–7, 58; 1996–2000) ii, p. 174.
3. *Mirour de Seinte Eglyse (St Edmund of Abingdon's Speculum Ecclesiae)* ed. A.D. Wilshere (ANTS, 40; 1982) p. 6.
4. Ibid., p. 18.
5. C. Classen *Worlds of sense: exploring the senses in history and across cultures* (1993) pp. 50–76; J.M. Williams, 'Synaesthetic adjectives: a possible law of semantic change' *Language* 52 (1976) pp. 461–78.
6. *WN*, p. 218.
7. *Becket*, i, p. 403.
8. *Bedfordshire coroners' rolls* ed. R.F. Hunnisett (Bedfordshire Historical Record Society, 41; 1961) pp. 79–80.
9. *Becket*, i, p. 163.
10. B. Kemp, 'The miracles of the hand of St James' *Berkshire Archaeological Journal* 65 (1970) pp. 1–19, at 14–15.
11. *SM*, p. 93.
12. *Henry VI*, pp. 290–1.
13. *Osmund*, pp. 37–40.
14. H. Farmer, 'The canonization of St Hugh of Lincoln' *Lincolnshire Architectural and Archaeological Society Reports and Papers* 6 (1956) pp. 86–117, at 101.
15. BAV MS Lat. Vat. 4015, ff. 51v–54v.
16. *The Cyrurgie of Guy de Chauliac* ed. M.S. Ogden (EETS, OS 265; 1971) pp. 205–6; *Becket*, ii, pp. 252–3.
17. *MK*, p. 67.
18. *Thurkill*, pp. 6–7.
19. *EWS*, ii, pp. 117–18; v, pp. 172–3.
20. *Jacob's Well*, pp. 147–8.
21. *Becket*, i, p. 521.
22. *Ailred*, p. 59.

23. J.Y. Tilliette, 'Le Symbolisme des cinq sens dans la littérature morale et spirituelle des xie et xiie siècles' *Micrologus* 10 (2002) pp. 15–32, at 15–20.

24. Classen, *Worlds of sense*, pp. 2–3.

25. *Becket*, ii, pp. 226–7.

26. *Wulfstan*, p. 123.

27. *Osmund*, pp. 42–3.

28. *Edmund*, pp. 10–11.

29. IV Rg 4: 31. It was not only children who were described in this way: *Thurkill*, p. 7.

30. *Henry VI*, p. 296.

31. R.W. Ackerman 'The debate of the body and the soul and parochial Christianity' *Speculum* 37 (1962) pp. 541–65, at 545, 547–8; M.W. Bloomfield *The seven deadly sins: an introduction to the history of a religious concept, with special reference to medieval English literature* (Michigan, 1952, reprinted 1967) pp. 119–20, 171, 387–8.

32. C. Casagrande, 'Sistema dei sensi e classificazione dei peccati (secoli XII–XIII)' *Micrologus* 10 (2002) pp. 33–53.

33. J. Goering and P.J. Payer, 'The "Summa penitentie fratrum predicatorum": a thirteenth-century confessional formula' *Mediaeval Studies* 55 (1993) pp. 1–50, especially 18–20 and n. 73. The link between the greatest sins and the senses may be based on Judaic works from the early second century BC: Bloomfield, *Seven deadly sins*, pp. 44–5.

34. E.g. J. Goering and F.A.C. Mantello, 'The "Perambulauit Iudas . . ." (Speculum Confessionis) attributed to Robert Grosseteste' *Revue Bénédictine* 96 (1986) pp. 125–68, at 149; S. Wenzel, 'Robert Grosseteste's treatise on confession, "Deus est"' *Franciscan Studies* 30 (1970) pp. 218–93, at 291; J. Goering and F.A.C. Mantello, '"Notus in Iudea Deus": Robert Grosseteste's confessional formula in Lambeth Palace MS 499' *Viator* 18 (1987) pp. 753–73.

35. *Middle English sermons edited from British Museum MS. Royal 18 B. XXIII* ed. W.O. Ross (EETS, OS 209; 1940) p. 32.

36. *Instructions for parish priests by John Myrc* ed. E. Peacock (EETS, OS 31; 1868; 2nd revised edition, 1902) pp. 40–1.

37. For examples of ambiguity and the argument that this is linked to the sins of the tongue, A. Barratt, 'The five wits and their structural significance in Part II of *Ancrene Wisse*' *Medium Aevum* 56 (1987) pp. 12–24, at 15–20.

38. *VV*, pp. 153, 225.

39. *Seyntz medicines*, pp. 52–3, 107–8. For a continental, aristocratic parallel, the poem 'Les cinq sens' by the Burgundian, Olivier de la Marche, lists its third sense as La Bouche, whose speech is the messenger of the heart: L. Vinge *The five senses: studies in a literary tradition* (Lund, 1975) pp. 61–2.

40. *Macro*, p. 47.

41. *Heresy trials in the diocese of Norwich, 1428–31* ed. N.P. Tanner (Camden, 4th series, 20; 1977) pp. 72–3.

42. Salisbury Cathedral MS 103, ff. 122rb, 123va.

43. Salisbury Cathedral MS 103, f. 123va.

44. 'DH', p. 23; *Le Manuscrit Leyde Bibliothèque de l'Université BPL 2579, témoin principal des phases de rédaction du traité De doctrina cordis à attribuer au dominicain français Hugues de Saint-Cher (pseudo-Gérard de Liège)* ed. G. Hendrix (Ghent, 1980) p. 23; also D. Renevey, 'Household chores in *The doctrine of the hert*: affective spirituality and subjectivity' in *The medieval household in Christian Europe, c.850–c.1550* ed. C. Beattie, A. Maslakovic and S. Rees Jones (Turnhout, 2003) pp. 167–85, at 167–8.

45. C.S. Lewis *The discarded image: an introduction to medieval and Renaissance literature* (Cambridge, 1964) pp. 10–12, 18.

46. *Etymologies*, ii: lib. XI.i, 21–2.

47. F. Saxl, 'A spiritual encyclopaedia of the later Middle Ages' *JWCI* 5 (1942) pp. 82–142, at 117–18.

48. *BA*, ii, p. 1395.

49. *BA*, i, pp. 1–38.
50. *BA*, iii, p. 3, suggests multiplying by at least ten to give the original number of manuscripts that were in circulation.
51. I have used Trevisa's translation throughout. For the date, *BA*, ii, p. 1396; iii, pp. 7–8.
52. *BA*, i, pp. 108–23 (Book 3, cap. 17–22).
53. *BA*, i, p. 192.
54. *Etymologies*, ii: lib. XI.i, 47.
55. *Etymologies*, ii: lib. XI.i, 9–10.
56. *BA*, i, pp. 192–4.
57. S. Kemp, 'A medieval controversy about odor' *Journal of the History of the Behavioural Sciences* 33 (1997) pp. 211–19, at 211–12.
58. R. Palmer, 'In bad odour: smell and its significance in medicine from Antiquity to the seventeenth century' in *Medicine and the five senses* ed. W.F. Bynum and R. Porter (Cambridge, 1993) pp. 61–8, 285–7, at 62–3; B.S. Eastwood, 'Galen on the elements of olfactory sensation' *Rheinisches Museum für Philologie* 124 (1981) pp. 268–90.
59. K. Albala *Eating right in the Renaissance* (Berkeley, 2002) pp. 48–162; C. Rawcliffe *Medicine & society in later medieval England* (Far Thrupp, 1995) pp. 29–57.
60. *BA*, ii, pp. 1296–8.
61. Eastwood, 'Galen', pp. 280, 284.
62. E. Sears, 'The iconography of auditory perception in the early Middle Ages: on psalm illustration and psalm exegesis' in *The second sense: studies in hearing and musical judgement from Antiquity to the seventeenth century* ed. C. Burnett, M. Fend and P. Gouk (Warburg Institute Surveys and Texts, 22; 1991) pp. 19–42, at 25.
63. Robert Grosseteste *Hexaëmeron* ed. R.C. Dales and S. Gieben (ABMA, 6; 1982) pp. 255–7.
64. *EWS*, ii, pp. 270–1; v, p. 239.
65. Augustine *Confessiones* ed. L. Verheijen (CCSL 27; 1981) pp. 181–2.
66. *Memorials of St Anselm* ed. R.W. Southern and F.S. Schmitt (AMBA, 1; 1969) pp. 12, 41.
67. *Hugh*, i, p. 75.
68. *Hugh*, i, pp. 125–6.
69. *CS 1205–1313*, ii, p. 789.
70. *The cloud of unknowing and the book of privy counselling* ed. P. Hodgson (EETS, OS 218; 1944) pp. 38–9.
71. E.D. Craun, '"Inordinata locutio": blasphemy in pastoral literature, 1200–1500' *Traditio* 39 (1983) pp. 135–62, at 160.
72. Goering and Mantello, 'The "Perambulauit Iudas . . .", p. 136.
73. William Langland *The vision of Piers Plowman: a critical edition of the B-Text . . .* ed. A.V.C. Schmidt (new edition, 1987) p. 91.
74. *EWS*, ii, pp. 135–40.
75. G. Frank *The memory of the eyes: pilgrims to living saints in Christian late Antiquity* (Berkeley, 2000) pp. 12, 14–16, 29, 112–13.
76. Augustine, *Confessiones*, pp. 155–65.
77. Baldwin of Ford *Sermones de commendatione fidei* ed. D.N. Bell (CCCM 99; 1991) pp. 23–43.
78. *Jacob's Well*, pp. 222–3.
79. *Cloud of unknowing*, pp. 96–7.
80. C.V. Langlois *La Connaissance de la nature et du monde au Moyen Âge d'après quelques écrits français à l'usage des laïcs* (Paris, 1911); M. de Boüard, 'Encyclopédies médiévales sur la "connaissance de la nature et du monde" au Moyen Âge' *Revue des questions historiques* 112 (1930) pp. 258–304, at 272, 281–3, 295–6; D.A. Callus, 'Introduction of Aristotelian learning to Oxford' *Proceedings of the British Academy* 29 (1943) pp. 229–81; C.F. Lohr, 'Medieval Latin Aristotle commentaries: authors A–F' *Traditio* 23 (1967) pp. 313–413, at 313.
81. For an excellent summary, E. Sears, 'Sensory perception and its metaphors in the time of Richard of Fournival' in *Medicine and the five senses* ed. W.F. Bynum and R. Porter (Cambridge, 1993) pp. 17–39, 276–83, especially 23–4.

82. E.R. Harvey *The inward wits: psychological theory in the Middle Ages and the Renaissance* (Warburg Institute Surveys, 6; 1975) pp. 2–3.

83. For example, *The Cyrurgie of Guy de Chauliac*, p. 41.

84. Robert Kilwardby *On time and imagination: De tempore, De spiritu fantastico* ed. P. Osmund Lewry (ABMA, 9; 1987) pp. xvi, xx–xxii, 108–30. The conflict had been apparent in the Islamic world from the tenth century: P. Carusi, 'Les Cinq Sens entre philosophie et médecine (Islam x^e–xiie siècles)' *Micrologus* 10 (2002) pp. 87–98.

85. H.A. Wolfson, 'The internal senses in Latin, Arabic, and Hebrew philosophic texts' *Harvard Theological Review* 28 (1935) pp. 69–133; Harvey, *Inward wits*; M. Camille, 'Before the gaze: the internal senses and late medieval practices of seeing' in *Visuality before and beyond the Renaissance* ed. R.S. Nelson (Cambridge, 2000) pp. 197–223, at 197–201, largely following Harvey.

86. John Blund *Tractatus de anima* ed. D.A. Callus and R.W. Hunt (ABMA, 2; 1970) pp. vii–viii, 56–8.

87. BL MS Royal 12 B XIX, ff. 301v–304r. For the significance of after-images and the debate about sensory error, D.G. Denery II *Seeing and being seen in the later medieval world: optics, theology and religious life* (Cambridge, 2005) p. 117.

88. *BA*, i, pp. 110–11; C. Hahn, 'Visio Dei: changes in medieval visuality' in *Visuality before and beyond the Renaissance* ed. R.S. Nelson (Cambridge, 2000) pp. 169–96, at 174–5; Camille, 'Before the gaze', pp. 204–9.

89. This is an over-simplification of the debate, for which see K.H. Tachau *Vision and certitude in the age of Ockham: optics, epistemology and the foundations of semantics* (Leiden, 1988) pp. xvi, 3–4, 8–150; R. Pasnau *Theories of cognition in the later Middle Ages* (Cambridge, 1997) pp. 1–27, 161–219.

90. Denery, *Seeing and being seen*, p. 120.

91. *The Trotula: a medieval compendium of women's medicine* ed. M.H. Green (Philadelphia, 2001) pp. 9–14; B. Lawn *The Salernitan questions: an introduction to the history of medieval and Renaissance problem literature* (Oxford, 1963) pp. vii–xii, 1–72, 84; N. Orme, 'The cathedral school before the Reformation' in *Hereford Cathedral: a history* ed. G. Aylmer and J. Tiller (2000) pp. 565–78, at 565–7; and J. Barrow, 'Athelstan to Aigueblanche, 1056–1268' ibid., pp. 21–47, at 42–3; R.W. Hunt, 'English learning in the late twelfth century' *Transactions of the Royal Historical Society* 4th series, 19 (1936) pp. 19–42, at 23–4, 36–7.

92. *PSQ*, pp. 135–6 (*epare* for *epate* on 136), 207, 295.

93. *PSQ*, p. 106.

94. Lawn, *Salernitan questions*, p. 111.

95. Classen, *Worlds of sense*, p. 3.

96. P. Dronke, 'Les Cinq Sens chez Bernard Silvestre et Alain de Lille' *Micrologus* 10 (2002) pp. 1–14, at 1, 5; Sears, 'Iconography of auditory perception', pp. 23–4.

97. C. Burnett, 'The superiority of taste' *JWCI* 54 (1991) pp. 230–8, at 230–1.

98. Peter the Venerable *Contra Petrobrusianos hereticos* ed. J. Fearns (CCCM 10; 1968) pp. 122–3.

99. N. Campbell, 'Aquinas' reasons for the aesthetic irrelevance of tastes and smells' *British Journal of Aesthetics* 36 (1996) pp. 166–76, at 171–2.

100. F. Mütherich, 'An illustration of the five senses in mediaeval art' *JWCI* 18 (1955) pp. 140–1.

101. T. Borenius, 'The cycle of images in the palaces and castles of Henry III' *JWCI* 6 (1943) pp. 40–50; P. Binski *The Painted Chamber at Westminster* (Society of Antiquaries, Occasional Paper 9; 1986).

102. C. Nordenfalk, 'Les Cinq Sens dans l'art du Moyen Âge' *Revue de l'Art* 34 (1976) pp. 17–28; C. Nordenfalk, 'The five senses in late medieval and Renaissance art' *JWCI* 48 (1985) pp. 1–22.

103. R.L.S. Bruce-Mitford, 'The Fuller Brooch' *British Museum Quarterly* 17 (1952) pp. 75–6; R.L.S. Bruce-Mitford, 'Late Saxon disc-brooches' in *Dark-Age Britain: studies presented to E.T. Leeds* ed. D.B. Harden (1956) pp. 173–90; Hinton, *Gold and gilt*, pp. 110–12, 303 n. 8.

The senses may also appear individually on a sequence of five eighth-century *sceattas*: A. Gannon *The iconography of Anglo-Saxon coinage: sixth to eighth centuries* (Oxford, 2003) p. 167.

104. E. Bakka, 'The Alfred Jewel and sight' *Antiquaries' Journal* 46 (1966) pp. 277–82.

105. *The Harley Lyrics: the Middle English lyrics of Ms. Harley 2253* ed. G.L. Brook (4th edition, Manchester, 1968) pp. 31–2.

106. Ct 1: 4; *Macro*, pp. xxx, xxxiv, 119.

107. Alan of Lille *Anticlaudianus* ed. R. Bossuat (Paris, 1955) pp. 95–212; Mütherich, 'An illustration of the five senses', pp. 140–1.

108. 'Nos aper auditu, lynx visu, simia gustu,/ Vultur odoratu precedit, aranea tactu.' Thomas Cantimpratensis *Liber de natura rerum* ed. H. Boese (Berlin, 1973) p. 106; M. Pastoureau, 'Le Bestiaire des cinq sens (xiie–xvie siècle)' *Micrologus* 10 (2002) pp. 133–45.

109. E.C. Rouse and A. Baker, 'The wall-paintings at Longthorpe Tower near Peterborough, Northants' *Archaeologia* 96 (1955) pp. 1–57; E.C. Rouse *Longthorpe Tower, Cambridgeshire* (London, 1987). N. Coldstream *The decorated style: architecture and ornament 1240–1360* (1994) pp. 110–11 dates the work to the 1320s.

110. G. Casagrande and C. Kleinhenz, 'Literary and philosophical perspectives on the wheel of the five senses in Longthope Tower' *Traditio* 41 (1985) pp. 311–27; for the views of Albertus Magnus, doubting the sense of touch in the spider, F. Santi, 'Il senso del ragno. Sistemi a confronto' *Micrologus* 10 (2002) pp. 147–61.

111. *Rot. Parl.*, iv, pp. 213–42, at 230, 235.

112. Nordenfalk, 'Five senses in late medieval and Renaissance art', p. 7.

113. F. Joubert *La tapisserie médiévale au musée de Cluny* (Paris, 1987) pp. 66–92; A. Glaenzer, 'La Tenture de La Dame à la licorne, du *Bestiaires d'amours* à l'ordre des tapisseries' *Micrologus* 10 (2002) pp. 401–28.

114. *The 1542 inventory of Whitehall: the palace and its keeper* ed. M. Hayward (2 vols, 2004) i, pp. 194–7; ii, p. 90, no. 669; *Inventory of Henry VIII*, no. 10574.

CHAPTER 3: TOUCH, VIRTUES AND HOLINESS

1. *BA*, i, p. 92.

2. *BA*, i, pp. 108, 118–21; ii, pp. 1296–7.

3. *BA*, i, p. 222.

4. *BA*, i, pp. 152, 228, 237–8; *Etymologies*, ii: lib. XI.i, 90.

5. Mt 25: 31–46; G. Blangez, 'Destre et senestre, miséricorde et justice: un système de symboles' in *Mélanges de langue et littérature françaises du Moyen-Âge offerts à Pierre Jonin* (Aix-en-Provence, 1979) pp. 115–24.

6. *WN*, p. 171.

7. *Heresy trials in the diocese of Norwich, 1428–31* ed. N.P. Tanner (Camden, 4th series, 20; 1977) pp. 125, 128.

8. *Macro*, pp. xxxviii, 171, 223.

9. *Kent heresy proceedings 1511–12* ed. N. Tanner (Kent Records, 26; 1997) pp. 39–40, 73–4.

10. *English mediaeval lapidaries* ed. J. Evans and M.S. Serjeantson (EETS, OS 190; 1933) p. 34.

11. Ibid., pp. 30–2, 58–9; *Anglo-Norman lapidaries* ed. P. Studer and J. Evans (Paris, 1924) pp. 43–4.

12. Ö. Södergård, 'Un Art d'aimer anglo-normand' *Romania* 77 (1956) pp. 289–330, at 299, 309–10.

13. *The works of John Metham including the Romance of Amoryus and Cleopes . . .* ed. H. Craig (EETS, OS 132; 1916) p. 116.

14. *BA*, i, pp. 225–6; *Etymologies*, ii: lib. XI.i, 70–1.

15. *Becket*, i, pp. 193–4.

16. *Select cases from the coroners' rolls* AD *1265–1413* . . . ed. C. Gross (Selden Society, 9; 1896) pp. 2–3, 21–2; *Bedfordshire coroners' rolls* ed. R.F. Hunnisett (Bedfordshire Historical Record Society, 41; 1961) pp. 2, 18.

17. *Treatises of fistula in ano, haemorrhoids, and clysters by John Arderne* . . . ed. D'A. Power (EETS, OS 139; 1910) p. 22.

18. Owst, *Preaching*, p. 269; P. Fleming *Family and household in medieval England* (Basingstoke, 2001) p. 45; Hinton, *Gold and gilt*, pp. 340, n. 65; 367, n. 58.

19. J.-C. Schmitt *La Raison des gestes dans l'Occident médiéval* (Paris, 1990) pp. 33–55; J.-C. Schmitt, 'The rationale of gestures in the West: third to thirteenth centuries' in *A cultural history of gesture: from Antiquity to the present day* ed. J. Bremmer and H. Roodenburg (Cambridge, 1991) pp. 59–70; Augustine *De doctrina christiana* ed. I. Martin (CCSL 32; 1962) pp. 32–4; A. Bartosz, 'Fonction du geste dans un texte romanesque médiéval: remarques sur la gestualité dans le première partie d'*Erec*' *Romania* 111 (1990) pp. 346–60. For distinctions between types of signs, U. Eco, R. Lambertini, C. Marmo and A. Tabarroni, 'On animal language in the medieval classification of signs' in *On the medieval theory of signs* ed. U. Eco and C. Marmo (Amsterdam, 1989) pp. 3–41; K.M. Fredborg, L. Nielsen and J. Penborg, 'An unedited part of Roger Bacon's "Opus maius": "De signis"' *Traditio* 34 (1978) pp. 75–136.

20. *Swithun*, pp. 534–6, 672.

21. J.A. Burrow *Gestures and looks in medieval narrative* (Cambridge, 2002) p. 12; Fleming, *Family and household*, p. 49.

22. *The Anonimalle Chronicle 1333 to 1381* . . . ed. V.H. Galbraith (Manchester, 1927) pp. 144–8; Burrow, *Gestures and looks*, pp. 34–5.

23. *A contemporary narrative of the proceedings against Dame Alice Kyteler, prosecuted for sorcery in 1324, by Richard de Ledrede, Bishop of Ossory* ed. T. Wright (Camden Society, OS 24; 1843) pp. 5–6.

24. J. Goering and P.J. Payer, 'The "Summa penitentie fratrum predicatorum": a thirteenth-century confessional formula' *Mediaeval Studies* 55 (1993) pp. 1–50, at 29, 31–2, 37–8.

25. *PI*, p. 204.

26. *Alphabet*, ii, p. 355.

27. *Lanfrank's 'Science of cirurgie'* ed. R. von Fleischhacker (EETS, OS 102; 1894) pp. 8–9; *Treatises of fistula*, p. 5.

28. *BA*, i, pp. 263–4.

29. *VV*, p. 227.

30. *Manières de langage* ed. A.M. Kristol (ANTS, 53; 1995) pp. 24, 36.

31. *RB*, pp. 218–19, cap. 22; T.F. Kirby *Annals of Winchester College from its foundation in the year 1382 to the present time* (1892) pp. 509–10.

32. B.A. Hanawalt *The ties that bound: peasant families in medieval England* (Oxford, 1986) p. 48.

33. E.g. *LEJ*, pp. 50–1.

34. Robert de Avesbury *De gestis mirabilibus regis Edwardi tertii* ed. E. Maunde Thompson (RS, 1889) p. 456.

35. *The customary of the Benedictine abbey of Eynsham in Oxfordshire* ed. A. Gransden (Siegburg, 1963) pp. 51–2.

36. *Erkenwald*, pp. 158–61.

37. *WN*, p. 26.

38. *Calendar of coroners' rolls of the city of London* AD *1300–1378* ed. R.R. Sharpe (1913) p. 61.

39. *CS 1205–1313*, i, pp. 120–1, 465–6, 473, 560–1; *HKW*, ii, p. 674.

40. *Heresy trials . . . Norwich*, pp. 133–4, 136.

41. *RB*, pp. 244–5, cap. 43; V.I.J. Flint, 'Space and discipline in early medieval Europe' in *Medieval practices of space* ed. B.A. Hanawalt and M. Kobialka (Minneapolis, 2000) pp. 149–66, at 151–9.

42. *LEJ*, p. 144.

43. *Kent heresy proceedings*, pp. 102–5, 108–9.

44. *MCL*, pp. 146–55, 164–7, 170–3, 214–15; P.B. Griesser, 'Die "Ecclesiastica officia Cisterciensis ordinis" des Cod. 1711 von Trient' *Analecta Sacri Ordinis Cisterciensis* 12 (1956) pp. 153–288, at 234–7.

45. J.M. Bennett, 'Writing fornication: medieval leyrwite and its historians' *Transactions of the Royal Historical Society* 6th series, 13 (2003) pp. 131–62, at 136.

46. *LEJ*, p. 5.

47. *LEJ*, pp. 72–3, 75.

48. *Heresy trials . . . Norwich*, pp. 35–6.

49. BAV MS Lat. Vat. 4015, f. 90r.

50. *Wulfstan*, pp. 122–3.

51. *CS 871–1204*, ii, pp. 920–5.

52. *La Vie de Saint Laurent: an Anglo-Norman poem of the twelfth century* ed. D.W. Russell (ANTS, 34; 1976) pp. 11, 22–3, 55–7.

53. R. Bartlett *The hanged man: a story of miracle, memory, and colonialism in the Middle Ages* (Princeton, 2004); H. Summerson, 'Attitudes to capital punishment in England, 1200–1350' *Thirteenth-Century England VIII* ed. M. Prestwich, R Britnell and R. Frame (Woodbridge, 2001) pp. 123–33.

54. BAV MS Lat. Vat. 4015, ff. 230v–231r.

55. *Kent heresy proceedings*, pp. 76, 82, 86, 99.

56. *WN*, pp. 12–13.

57. *MCL*, pp. 58–65.

58. *MK*, p. 12.

59. L. Gougaud *Anciennes coutumes claustrales* (Abbaye Saint-Martin de Ligugé (Vienne); Moines et monastères, 8; 1930) pp. 24–36.

60. *Ailred*, pp. 30–1.

61. *Hugh*, i, pp. 23–4.

62. J. Goering and F.A.C. Mantello, 'The "Perambulauit Iudas . . ." (Speculum Confessionis) attributed to Robert Grosseteste' *Revue Bénédictine* 96 (1986) pp. 125–68, at 153–4; Owst, *Literature*, pp. 411, 413, 501; *EWS*, i, pp. 478–9.

63. *Barnwell*, p. 80.

64. *MCL*, p. 130; *CS 1205–1313*, i, p. 153.

65. *Jacob's Well*, pp. 104–5; the Devil's featherbed also features in *VV*, pp. 26–7.

66. Griesser, 'Ecclesiastica officia', p. 251.

67. *La Vie de Seint Auban: an Anglo-Norman poem of the thirteenth century* ed. A.R. Harden (ANTS, 19; 1968) pp. xvii, 19, 70.

68. *Ailred*, p. 39.

69. BAV MS Lat. Vat. 4015, ff. 173v–175v. Also *Becket*, ii, p. 122.

70. BAV MS Lat. Vat. 4015, ff. 165v–166r.

71. *Rot. Parl.*, iv, pp. 234, 237, 241.

72. *Inventory of Henry VIII*, 9061–73, 9076–7, 9251–5, 9258, 9365–9, 9384–6, 9440–1, 9763–77, etc.

73. Hanawalt, *Ties that bound*, p. 48; *LEJ*, pp. 436, 439.

74. Burrow, *Gestures and looks*, pp. 32–3, 51.

75. Aelred of Rievaulx *De spirituali amicitia*, in *Aelredi Rievallensis opera omnia, 1: Opera ascetica* ed. A. Hoste and C.H. Talbot (CCCM 1; 1971) pp. 306–8, Burrow, *Gestures and looks*, p. 50.

76. Griesser, 'Ecclesiastica officia', pp. 197, 200, 217–19.

77. E. Mikkers, 'Un "Speculum novitii" inédit d'Étienne de Salley' *Collectanea Cisterciensis Ordinis Reformatorum* 8 (1946) pp. 17–68, at 49.

78. *CS 871–1204*, ii, pp. 894–7.

79. *Dame Alice Kyteler*, p. 20.

80. H. Farmer, 'The canonization of St Hugh of Lincoln' *Lincolnshire Architectural and Archaeological Society Reports and Papers* 6 (1956) pp. 86–117, at 97.

81. *N-town*, i, p. 129.

82. *Towneley*, i, pp. 57–9, 63.

83. *Seyntz medicines*, pp. 178–9.
84. H. Johnstone, 'The wardrobe and household of Henry, son of Edward I' *Bulletin of the John Rylands Library* 7 (1922–3) pp. 384–420, at 419.
85. *La Vie Seint Richard evesque de Cycestre by Pierre d'Abernon of Fetcham* ed. D.W. Russell (ANTS, 51; 1995) p. 113.
86. *Heresy trials . . . Norwich*, pp. 89–90.
87. E.g. *EWS*, iii, p. 248.
88. R.C. Finucane *Miracles and pilgrims: popular beliefs in medieval England* (1977) p. 26; *EWS*, i, p. 268; G. Frank *The memory of the eyes: pilgrims to living saints in Christian late Antiquity* (Berkeley, 2000) pp. 118–25.
89. B. Kemp, 'The miracles of the hand of St James' *Berkshire Archaeological Journal* 65 (1970) pp. 1–19, at 6–7.
90. 'Liber miraculorum Beati Edmundi archiepiscopi' in *Thesaurus novus anecdotorum* ed. E. Martène and U. Durand (5 vols, Paris, 1717) iii, cols 1881–98, at cols 1888–9.
91. Robert Grosseteste *De decem mandatis* ed. R.C. Dales and E.B. King (ABMA, 10; 1987) pp. 9–10; A. Gurevich *Medieval popular culture: problems of belief and perception* trans. J.M. Bak and P.A. Hollingsworth (Cambridge, 1988) pp. 62, 96–7.
92. *Mirk's Festial* ed. T. Erbe (EETS, ES 96; 1905) pp. 18, 20.
93. *Jacob's Well*, p. 125.
94. *Erkenwald*, pp. 96–7.
95. Farmer, 'Canonization of St Hugh', p. 99.
96. *La Vie Seint Richard*, pp. 81–2.
97. *Osmund*, pp. 77–8.
98. J. Wickham Legg, 'On an inventory of the vestry in Westminster Abbey, taken in 1388' *Archaeologia* 52 part 1 (1890) pp. 195–286, at 200–1, 238.
99. *MCL*, pp. 122–5, 134–7.
100. *Instructions for parish priests by John Myrc* ed. E. Peacock (EETS, OS 31; 1868; 2nd revised edition, 1902) pp. 56–7.
101. *CS 871–1204*, ii, pp. 1060–1; *CS 1205–1313*, i, pp. 27–9, 372, 404; ii, p. 894.
102. Salisbury Cathedral MS 103, f. 165va.
103. *Liber exemplorum*, pp. 55–7, 143.
104. *Alphabet*, ii, pp. 460–2.
105. IV Rg 5: 1–19, cited, for example, by *Jacob's Well*: Salisbury Cathedral MS 103, ff. 123ra–b.
106. *MCL*, p. 132.
107. Kemp, 'Hand of St James', pp. 11–12.
108. *Alphabet*, ii, p. 353.
109. B.W. Spencer, 'Medieval pilgrim badges: some general observations illustrated mainly from English sources' in *Rotterdam papers: a contribution to medieval archaeology* ed. J.G.N. Renaud (Rotterdam, 1968) pp. 137–53, at 139.
110. *Three eleventh-century Anglo-Latin saints' lives: Vita S. Birini, Vita et miracula S. Kenelmi and Vita S. Rumwoldi* ed. and trans. R.C. Love (Oxford, 1996) p. 73 n. 3.
111. *Libri de nativitate Mariae: Pseudo-Matthaei evangelium textus et commentarius* ed. J. Gijsel (CCSA 9; 1997) pp. 67, 417–27; *N-town*, i, pp. 152–62; *Mirk's Festial*, p. 23.
112. *Lollard sermons* ed. G. Cigman (EETS, OS 294; 1989) pp. 59, 251.
113. *Alphabet*, i, p. 150; ii, pp. 466–7.
114. *Non-cycle plays and fragments* ed. N. Davis (EETS, SS 1; 1970) pp. lxx–lxxxv, 58–86.
115. J. Evans *Magical jewels of the Middle Ages and the Renaissance particularly in England* (Oxford, 1922) pp. 63–5.
116. R. Bartlett *England under the Norman and Angevin Kings 1075–1225* (Oxford, 2000) pp. 179–84.
117. C.T. Lewis and C. Short *A Latin dictionary* (Oxford, 1879), *sub* sacramentum.
118. E.g. *LEJ*, p. 385; *CS 871–1204*, ii, pp. 894–5, 897–8.
119. *Swithun*, p. 666.

120. *Liber regie capelle: a manuscript in the Biblioteca Publica, Evora* ed. W. Ulmann (Henry Bradshaw Society, 92; 1961) pp. 29–38. Richard II discovered this in the Tower after his coronation and sought to be re-anointed.

121. M. Prestwich, 'The piety of Edward I' in *England in the thirteenth century: proceedings of the 1984 Harlaxton Symposium* ed. W.M. Ormrod (Grantham, 1985) pp. 120–8, at 124–7; M. Bloch, *Les Rois thaumaturges: le caractère surnaturel attribué à la puissance royale, particulièrement en France et en Angleterre* (Strasbourg, 1924).

122. PRO E101/392/12, f. 34r; E101/393/11, f. 61r; E101/398/9, f. 34r; E101/398/9. f. 23r; E101/403/10, f. 34v; Hinton, *Gold and gilt*, pp. 361–2, n. 30.

123. *HO*, pp. 118, 120–2. For similar regulations in Henry VIII's ordinances of Eltham of 1526, *HO*, pp. 155–6. See also Chapters 10 and 11.

124. *CS 1205–1313*, i, pp. 31, 68–9.

125. Salisbury Cathedral MS 103, ff. 163va–164ra.

126. BAV MS Lat. Vat. 4015, ff. 64v–68v.

127. Griesser, 'Ecclesiastica officia', pp. 256–7.

128. Aelred of Rievaulx *De anima* ed. C.H. Talbot (*Mediaeval and Renaissance Studies*, Supplement 1; 1952) p. 23; *Ailred*, p. 63.

129. *Heresy trials . . . Norwich*, p. 141.

130. *CS 1205–1313*, i, p. 601.

131. *Mirk's Festial*, pp. 277, 280, 294–5, 297–8.

132. *CS 1205–1313*, i, pp. 128, 135, 172, 174, 273, 297, 601.

133. *CS 1205–1313*, i, p. 303.

134. J.R. Maddicott, 'Follower, leader, pilgrim, saint: Robert de Vere, Earl of Oxford, at the shrine of Simon de Montfort, 1273' *EHR* 109 (1994) pp. 641–53.

135. Ex 3: 1–6: *N-town*, i, pp. 59–60; *Towneley*, i, p. 75.

136. *SM*, p. 81.

137. *Macer*, pp. 113, 131, 158–9.

138. *Marbode*, pp. 7–8.

139. *Marbode*, pp. 35–6.

140. *La Vie Seint Richard*, pp. 56–7.

141. *Liber quotidianus contrarotulatoris garderobe anno regni regis Edwardi primi vicesimo octavo AD MCCXCIX & MCCC* ed. Anon. for Society of Antiquaries (1787) p. 280; PRO E101/393/4, f. 9r; Wickham Legg, 'Inventory of the vestry in Westminster Abbey', pp. 202, 251, 275. Dunstan's reputation as a craftsman was probably posthumous: Hinton, *Gold and gilt*, p. 329, n. 146.

142. *BA*, ii, p. 692; Hinton, *Gold and gilt*, pp. 187–8, 197, 200–1, 238, 258, 338–41, 360–1.

143. *English mediaeval lapidaries*, pp. 16, 22–3.

144. B. Nilson *Cathedral shrines of medieval England* (Woodbridge, 1998) pp. 36–8.

145. Kemp, 'Hand of St James', p. 16.

146. *Becket*, i, pp. 482–3.

147 BAV MS Lat. Vat. 4015, f. 74v.

148. Apc 21: 18–21. Compare *Erkenwald*, pp. 120–1; G. Constable, 'The vision of Gunthelm and other visions attributed to Peter the Venerable' *Revue Bénédictine* 66 (1956) pp. 92–114, at 108–9.

149. *Anglo-Norman lapidaries*, pp. 277, 285–6, 293–4.

150. P.D.A. Harvey and A. McGuinness *A guide to British medieval seals* (1996) pp. 58–9.

151. M. Henig, 'Archbishop Hubert Walter's gems' *Journal of the British Archaeological Association* 136 (1983) pp. 56–61.

152. *MCL*, pp. 42–5, 66–9; Wickham Legg, 'Inventory of the vestry in Westminster Abbey', p. 277.

153. *Three prose versions of the Secreta secretorum* ed. R. Steele (EETS, ES 74; 1898) pp. 87–8.

154. *The book of quinte essence or the fifth being; that is to say, man's heaven . . .* ed. F.J. Furnivall (EETS, OS 16; 1866) pp. 1–4; J. Hughes *Arthurian myths and alchemy: the kingship of Edward IV* (Far Thrupp, 2002) pp. 49–55.

155. PRO E101/414/16, f. 54r.

156. C. Rawcliffe *Medicine for the soul. The life, death and resurrection of an English medieval hospital: St Giles's, Norwich, c. 1249–1550* (Far Thrupp, 1999) p. 28; Evans, *Magical jewels of the Middle Ages and the Renaissance*, p. 144.

157. P.M. Jones and L.T. Olsan, 'Middleham Jewel: ritual, power, and devotion' *Viator* 31 (2000) pp. 249–90.

158. Spencer, 'Medieval pilgrim badges', pp. 137–53.

159. G.R. Owst, 'Sortilegium in English homiletic literature of the fourteenth century' in *Studies presented to Sir Hilary Jenkinson* ed. J.C. Davies (1957) pp. 272–303, at 302–3.

160. C.F. Bühler, 'Prayers and charms in certain Middle English scrolls' *Speculum* 39 (1964) pp. 270–8.

161. *The knowing of woman's kind in childing: a Middle English version of material derived from the Trotula and other sources* ed. A. Barratt (Turnhout, 2001) pp. 65–6.

162. *Treatises of fistula*, pp. 102–4.

163. Hinton, *Gold and gilt*, pp. 190–1, 341 n. 76.

164. *Lollards of Coventry 1486–1522* ed. S. McSheffrey and N. Tanner (Camden, 5th series, 23; 2003) pp. 40, n. 156; 238; *Statutes of the realm* ed. A. Luders, T.E. Tomlins, J. Raithby *et al.* (11 vols, 1810–28) i, p. 367; *CS 871–1204*, ii, pp. 920–5.

165. R. Kieckhefer *Magic in the Middle Ages* (Cambridge, Canto edition, 2000) pp. 89, 132.

166. IV Rg 25: 23, 'Maspha'; 25: 27, 'Evilmerodach'; J. Leclercq *The love of learning and the desire for God: a study of monastic culture* trans. Catharine Misrahi (New York, 1961) p. 96; Jerome *Liber interpretationis Hebraicorum nominum* ed. P. de Lagarde in *S. Hieronymi presbyteri opera: pars I Opera exegetica* ed. P. Antin (CCSL 72; 1959) pp. 57–161, at 116.

167. Gurevich, *Medieval popular culture*, p. 42.

168. BL Add. MS 60584, ff. 14r, 18v; *Geiriadur Prifysgol Cymru: A dictionary of the Welsh language* (in progress, Cardiff, 1950–), *sub* cariadog, caradoc; *The Harley Lyrics: the Middle English Lyrics of Ms. Harley 2253* ed. G.L. Brook (4th edition, Manchester, 1968) p. 76 n. 47. For other examples, *GH*, pp. 26, 154.

169. Hinton, *Gold and gilt*, p. 223.

170. P. Price *Bells and man* (Oxford, 1983) pp. 107–29.

171. Tony Hunt *Popular medicine in thirteenth-century England: introduction and texts* (Cambridge, 1990) pp. 74, 110; *Macer*, pp. 137–8.

172. *Macer*, p. 67.

173. *Seyntz medicines*, pp. 161–3.

174. *Macer*, pp. 65–6, 79, 81.

175. *Seyntz medicines*, pp. 207–11.

176. *PI*, p. 203.

177. *Lanfrank's 'Science of cirurgie'*, p. 191.

178. *Chronicles of the reigns of Stephen, Henry II and Richard I* ed. R. Howlett (4 vols, RS, 1884–9) ii, pp. 474–5.

179. Ibid., ii, p. 481.

180. *CS 1205–1313*, i, p. 70.

181. C.S. Lewis *The discarded image: an introduction to medieval and Renaissance literature* (Cambridge, 1964) pp. 103–10; Kieckhefer, *Magic in the Middle Ages*, pp. 126–8.

182. J.D. North, 'Some Norman horoscopes' in *Adelard of Bath: an English scientist and Arabist of the early twelfth century* ed. C. Burnett (Warburg Institute Surveys and Texts, 14; 1987) pp. 147–61.

183. J. Greatrex, 'Horoscopes and healing at Norwich Cathedral Priory in the later Middle Ages' in *The Church and learning in later medieval society: essays in honour of R.B. Dobson. Proceedings of the 1999 Harlaxton Symposium* ed. C.M. Barron and J. Stratford (Donington, 2002) pp. 170–7.

184. *CS 1205–1313*, i, p. 179.

185. *Manières de langage*, p. 3.

186. John of Salisbury *Policraticus I–IV* ed. K.S.B. Keats-Rohan (CCCM 118; 1993) pp. 57–61, 106–7, 147; Owst, 'Sortilegium', pp. 272–4; *VV*, p. 39; Gerald of Wales *Opera* ed. J.S. Brewer *et al.* (8 vols, RS, 1861–91) vi, pp. 87–8.
187. *LEJ*, pp. 339, 463.
188. PRO E101/414/16, f. 60v.
189. C. Burnett, 'The earliest chiromancy in the West' *JWCI* 50 (1987) pp. 189–95; L. Thorndike, 'Chiromancy in mediaeval Latin manuscripts' *Speculum* 40 (1965) pp. 674–706; *Works of John Metham*, pp. 85–116; J.F. Burke *Vision, the gaze, and the function of the senses in Celestina* (University Park, Pennsylvania, 2000) p. 39.

CHAPTER 4: SOUND AND HEARING

1. In the *Opus Maius*, summarised in the *Opus tertium*, in Roger Bacon *Opera quaedam hactenus inedita* ed. J.S. Brewer (RS, 1859) pp. 88–102; and K.M. Fredborg, L. Nielsen and J. Pinborg, 'An unedited part of Roger Bacon's "Opus maius": "De signis"' *Traditio* 34 (1978) pp. 75–136.
2. C. Burnett, 'Sound and its perception in the Middle Ages' in *The second sense: studies in hearing and musical judgement from Antiquity to the seventeenth century* ed. C. Burnett, M. Fend and P. Gouk (Warburg Institute Surveys and Texts, 22; 1991) pp. 43–69.
3. E.D. Craun *Lies, slander, and obscenity in medieval English literature: pastoral rhetoric and the deviant speaker* (Cambridge, 1997) pp. 23–6, 31–4.
4. 'Introduction' in *The second sense*, pp. 1–5, at 1.
5. F. de Saussure *Course in general linguistics* trans. R. Harris (1983), especially pp. 63–78.
6. Examples from anthropology are instructive: D. Howes 'Introduction: "To summon all the senses"' in *The varieties of sensory experience: a sourcebook in the anthropology of the senses* ed. D. Howes (Toronto, 1991) pp. 3–21, at 8–10, discussing the work of Paul Stoller, 'Sound in Songhay cultural experience' *American Ethnologist* 11 (1984) pp. 559–70, and *The taste of ethnographic things: the senses in anthropology* (Philadelphia, 1989).
7. *BA*, i, pp. 108, 113–15, 190–2; *Lanfrank's 'Science of cirurgie'* ed. R. von Fleischhacker (EETS, OS 102; 1894) p. 254; *The Cyrurgie of Guy de Chauliac* ed. M.S. Ogden (EETS, OS 265; 1971) p. 44.
8. Burnett, 'Sound and its perception', pp. 66–9; John Blund *Tractatus de anima* ed. D.A. Callus and R.W. Hunt (ABMA, 2; 1970) pp. 39–47; John of Salisbury *Policraticus I–IV* ed. K.S.B. Keats-Rohan (CCCM 118; 1993) pp. 46–51.
9. C.S. Lewis *The discarded image: an introduction to medieval and Renaissance literature* (Cambridge, 1964) pp. 111–12; Burnett, 'Sound and its perception', pp. 49–50.
10. *Memorials of St Dunstan, Archbishop of Canterbury* ed. W. Stubbs (RS, 1874) p. 170, and *Customary of the Benedictine monasteries of St Augustine, Canterbury, and St Peter, Westminster* ed. E.M. Thompson (Henry Bradshaw Society, 23, 28; 1902, 1904) i, pp. 329–30, both cited in C. Page, 'Music and medicine in the thirteenth century' in *Music as medicine: the history of music therapy since Antiquity* ed. P. Horden (Aldershot, 2000) pp. 109–19, at 109–11; P.M. Jones, 'Music therapy in the later Middle Ages: the case of Hugo van der Goes' in *Music as medicine*, pp. 120–44.
11. Lc 8: 8; *Lollard sermons* ed. G. Cigman (EETS, OS 294; 1989) pp. 93–4.
12. *EWS*, ii, p. 119.
13. M. Pastoureau, 'Le Bestiaire des cinq sens (xii*e*–xvi*e* siècle)' *Micrologus* 10 (2002) pp. 133–45, at 139–41.
14. *Seyntz medicines*, pp. 9–10.
15. B.R. Smith *The acoustic world of early modern England* (Chicago, 1999) pp. 49–52, 75–6.
16. *The political songs of England, from the reign of John to that of Edward II* ed. T. Wright (Camden Society, OS 6; 1839) pp. 231–6.
17. University of Southampton Library, MS 340/3, papers of Hope Bagenal.

18. G. Lawson, C. Scarre, I. Cross and C. Hills, 'Mounds, megaliths, music and mind: some thoughts on the acoustical properties *and purposes* of archaeological spaces' in *The archaeology of perception and the senses* ed. C. Jones and C. Hayden (special number of *Archaeological Review from Cambridge* 15: 1 (1998)) pp. 111–34, at 128; L'Abbé Cochet, 'Acoustic pottery' *The Gentleman's Magazine*, NS 15 (1863) pp. 540–3; M. Biddle, 'Acoustic pots: a proposed study' *Medieval Archaeology* 6–7 (1962–3) p. 304; A. Stock, 'A sounding vase at Fountains Abbey?' *Cistercian Studies* 23 (1988) pp. 190–1; D. Welander *The history, art and architecture of Gloucester Cathedral* (Far Thrupp, 1991) pp. 179, 181, 268–9, 386–8, 440.

19. Smith, *Acoustic world*, pp. 83–91.

20. *Inventory of Henry VIII*, 12183–95.

21. *GH*, p. 176; *A relation, or rather a true account, of the island of England; with sundry particulars of the customs of these people, and of the royal revenues under King Henry the Seventh, about the year 1500* ed. C.A. Sneyd (Camden Society, OS 37; 1847) p. 44.

22. *Le Jeu d'Adam: ordo representacionis Ade* ed. W. Noomen (Paris, 1971) p. 54; R.J. Dean and M.B.M. Boulton *Anglo-Norman literature: a guide to texts and manuscripts* (ANTS, Occasional Publications Series, 3; 1999) pp. 389–90.

23. The link is stronger than *OED*, *sub* noise *sb.*, suggests.

24. Owst, *Literature*, pp. 14–15, 37–8.

25. *Lollard sermons*, p. 134.

26. *BA*, i, p. 622.

27. 'The squire of low degree' in *Middle English metrical romances* ed. W.H. French and C.B. Hale (2 vols, New York, 1964) ii, pp. 721–55, at 722–3; *The book of the knight of La Tour-Landry . . .* ed. T. Wright (EETS, OS 33; 1868) p. 1.

28. *GH*, p. 114.

29. *La Vie Seint Richard evesque de Cycestre by Pierre d'Abernon of Fetcham* ed. D.W. Russell (ANTS, 51; 1995) pp. 106–9.

30. B.A. Hanawalt *The ties that bound: peasant families in medieval England* (Oxford, 1986) p. 44.

31. *Knight of La Tour-Landry*, p. 1; U. Eco, R. Lambertini, C. Marmo and A. Tabarroni, 'On animal language in the medieval classification of signs' in *On the medieval theory of signs* ed. U. Eco and C. Marmo (Amsterdam, 1989) pp. 3–41.

32. *SM*, p. 95.

33. *Etymologies*, ii: lib. XIII.viii, 1–2.

34. R. Keele, 'Richard Lavenham's *De causis naturalibus*: a critical edition' *Traditio* 56 (2001) pp. 113–47, at 124.

35. *BA*, i, pp. 590–1.

36. *English mediaeval lapidaries* ed. J. Evans and M.S. Serjeantson (EETS, OS 190; 1933) pp. 16, 33.

37. BAV MS Lat. Vat. 4015, ff. 142r–142v.

38. *Manières de langage* ed. A.M. Kristol (ANTS, 53; 1995) p. 8.

39. *PI*, pp. 121–2.

40. *MCL*, pp. xxviii–xxxix, 4–9.

41. *The customary of the Benedictine Abbey of Eynsham in Oxfordshire* ed. A. Gransden (Siegburg, 1963) pp. 16–17, 100–1.

42. P. Price *Bells and man* (Oxford, 1983) pp. 79–80, 83, 107–33.

43. *CS 1205–1313*, i, pp. 169, 175.

44. *CS 1205–1313*, i, p. 150.

45. BAV MS Lat. Vat. 4015, f. 123v.

46. *CS 871–1204*, ii, p. 801.

47. *Osmund*, pp. 37–8.

48. *York civic records, vol. III* ed. A. Raine (Yorkshire Archaeological Society, Record Series, 106; 1942) p. 70.

49. *HAME*, ii, p. 680.

50. A.F. Sutton, L. Visser-Fuchs and P.W. Hammond *The reburial of Richard, Duke of York, 21–30 July 1476* (1996) pp. 8, 14.
51. Price, *Bells and man*, pp. 111–14; *CS 1205–1313*, i, pp. 171, 268, 372.
52. *CS 1205–1313*, i, pp. 343, 361; see also Hinton, *Gold and gilt*, p. 343, n. 91.
53. *CS 1205–1313*, i, pp. 210–11. William I's crown may also have had small bells, possibly to ensure reverence and indicative of his status: Hinton, *Gold and gilt*, p. 142.
54. *CS 1205–1313*, ii, p. 894.
55. J.C. Cox *Churchwardens' accounts from the fourteenth century to the close of the seventeenth century* (1913) pp. 215–16, 226.
56. B.A. Hanawalt *Growing up in medieval London* (Oxford, 1993) p. 30.
57. *The early English versions of the Gesta Romanorum . . .* ed. S.J.H. Herrtage (EETS, ES 33; 1879) pp. 93, 95.
58. *BB*, p. 132.
59. *Manorial records of Cuxham, Oxfordshire, circa 1200–1359* ed. P.D.A. Harvey (Historical Manuscripts Commission, Joint Publications, 23; 1976) p. 637.
60. Owst, *Literature*, p. 13.
61. *BB*, p. 132.
62. *Macro* p. 159.
63. PRO E101/414/16, ff. 24r–24v.
64. *Becket*, i, p. 205.
65. 'Eynsham', p. 291.
66. *A contemporary narrative of the proceedings against Dame Alice Kyteler, prosecuted for sorcery in 1324, by Richard de Ledrede, Bishop of Ossory* ed. T. Wright (Camden Society, OS 24; 1843) pp. 5–6.
67. BAV MS Lat. Vat. 4015, f. 144v.
68. BAV MS Lat. Vat. 4015, f. 161v.
69. Not as implied by R.H. Hilton *The English peasantry in the later Middle Ages* (Oxford, 1975) p. 56, and Hanawalt, *Ties that bound*, p. 44.
70. M.R. James, 'Twelve medieval ghost-stories' *EHR* 37 (1922) pp. 413–22, at 418.
71. Gervase of Tilbury *Otia imperialia: recreation for an emperor* ed. S.E. Banks and J.W. Binns (Oxford, 2002) pp. 690–3.
72. *Ailred*, p. 51.
73. C.J. Holdsworth, 'Eleven visions connected with the Cistercian monastery of Stratford Langthorne' *Cîteaux* 13 (1962) pp. 185–204, at 203–4.
74. G. Constable, 'The vision of Gunthelm and other visions attributed to Peter the Venerable' *Revue Bénédictine* 66 (1956) pp. 92–114, at 109–11.
75. *Thurkill*, pp. 14, 17, 19–21, 24.
76. Holdsworth, 'Eleven visions', p. 202.
77. *Erkenwald*, pp. 162–3.
78. D. Stevens, 'A Somerset coroner's roll, 1315–1321' *Somerset and Dorset Notes and Queries* 31 (1985) pp. 451–72, at 461.
79. *Three eleventh-century Anglo-Latin saints' lives: Vita S. Birini, Vita et miracula S. Kenelmi and Vita S. Rumwoldi* ed. and trans. R.C. Love (Oxford, 1996) pp. 92–3.
80. *The late medieval religious plays of Bodleian MSS Digby 133 and E Museo 160* ed. D.C. Baker, J.L. Murphy and L.B. Hall Jr. (EETS, OS 283; 1982) p. 471.
81. See Chapter 7 for the association of its odour. R. Rastall, 'The sounds of Hell' in *The iconography of Hell* ed. C. Davidson and T.H. Seiler (Medieval Institute Publications, Kalamazoo, Early Drama, Art and Music Monograph Series, 17; 1992) pp. 102–31, at 110–11.
82. *N-town*, i, pp. 24, 32; ii, p. 419.
83. *Macro*, p. 1.
84. *Alphabet*, i, pp. 25–6. See also the exchange in *Mary Magdalen* in *Late medieval religious plays*, p. 63.
85. Smith, *Acoustic world*, p. 163.

86. BAV MS Lat. Vat. 4015, ff. 212v–219v.
87. BAV MS Lat. Vat. 4015, ff. 106v–107r.
88. *Henry VI*, pp. 19–20.
89. Act 2: 2; *EWS*, i, pp. 598–9.
90. Act 9; *Late medieval religious plays*, pp. 9, 13.
91. *WN*, p. 130.
92. *N-town*, i, pp. 391, 400–2; ii, p. 527.
93. Salisbury Cathedral MS 103, ff. 144va–145ra.
94. *MK*, pp. 11, 51–2, 86–8, 98, 185, 219, 283, 301–2.
95. *Thurkill*, pp. 39–41.
96. *Three Anglo-Latin saints' lives*, pp. lviii, 30–3.
97. *Becket*, ii, pp. 127–8: 'facta sicut homo non audiens'.
98. *The book of St Gilbert* ed. R. Foreville and G. Keir (Oxford, 1987) pp. 303, 315–16.
99. *Macer*, pp. 17, 98, 169.
100. *BA*, i, p. 367.
101. *Three receptaria from medieval England: the languages of medicine in the fourteenth century* ed. T. Hunt and M. Benskin (Oxford, 2001) p. 15.
102. *Henry VI*, pp. 224–6.
103. *La Estoire de Seint Aedward le rei attributed to Matthew Paris* ed. K. Young Wallace (ANTS, 41; 1983) pp. 123–4.
104. *MK*, pp. 28–9.
105. BAV MS Lat. Vat. 4015, f. 13r; 'DH', pp. 61–2.
106. *WN*, p. 41.
107. *Liber exemplorum*, pp. 110–11, 153–4.
108. *CS 1205–1313*, i, pp. 169, 174.
109. *CS 1205–1313*, i, p. 26.
110. Eph 5: 15; *EWS*, i, p. 683.
111. *Two Wycliffite texts: the sermon of William Taylor 1406; the testimony of William Thorpe 1407* ed. A. Hudson (EETS, OS 301; 1993) pp. 64–5.
112. J. Goering and F.A.C. Mantello, 'The "Perambulauit Iudas . . ." (Speculum Confessionis) attributed to Robert Grosseteste' *Revue Bénédictine* 96 (1986) pp. 125–68, at 138, 150–1.
113. *LEJ*, pp. 13, 29.
114. BAV MS Lat. Vat. 4015, ff. 140r–141v.
115. BAV MS Lat. Vat. 4015, f. 53r.
116. *Etymologies*, i: lib. III.
117. *BA*, ii, p. 1386.
118. I Sm 16: 23; *BA*, ii, p. 1386.
119. Innocent III *De sacro altaris mysterio. PL*, 217, col. 775.
120. *RB*, pp. 214–17, cap. 19.
121. M.B. Parkes *Pause and effect: an introduction to the history of punctuation in the West* (Aldershot, 1992) pp. 77, 103–5.
122. J. Leclercq *The love of learning and the desire for God: a study of monastic culture* trans. Catharine Misrahi (New York, 1961) pp. 301–2.
123. C. Waddell, 'The early Cistercian experience of the liturgy' in *Rule and life: an interdisciplinary symposium* ed. M.B. Pennington (Shannon: Cistercian studies series, 12; 1971) pp. 77–116, at 102–6; M. Cassidy-Welch *Monastic spaces and their meanings: thirteenth-century English Cistercian monasteries* (Turnhout, 2001) pp. 100–3.
124. E. Mikkers, 'Un Traité inédit d'Étienne de Salley sur la psalmodie' *Cîteaux* 23 (1972) pp. 245–88, at 253, 259–60.
125. *Barnwell*, pp. xi, 58–60.
126. *MCL*, p. 94.
127. W. Dugdale *Monasticon anglicanum* ed. J. Caley, H. Ellis and B. Bandinel (new edition, 6 vols in 8, 1846) vi, part 2, p. xlviii*.

128. *Two Wycliffite texts*, pp. 65–6.
129. *Mirk's Festial* ed. T. Erbe (EETS, ES 96; 1905) p. 1.
130. *The early English versions of the Gesta Romanorum . . .* ed. S.J.H. Herrtage (EETS, ES 33; 1879) pp. 136–9.
131. *BA*, i, p. 606.
132. C. Page *The owl and the nightingale: musical life and ideas in France, 1100–1300* (1989) pp. 8–41; C. Bullock-Davies *Register of royal and baronial domestic minstrels 1272–1327* (Woodbridge, 1986).
133. *Inventory of Henry VIII*, 2049–53; 2309; 11872–952.
134. E.g. P. Ellis *et al. Ludgershall Castle Wiltshire: a report on the excavations by Peter Addyman, 1964–1972* (Wiltshire Archaeological and Natural History Society Monograph Series, 2; 2000) pp. 163–7.
135. PRO E101/414/6, ff. 6r, 9r, 10v, 18r, 41r.
136. PRO E36/214, f. 12r.
137. BL Add. MS 21480, ff. 14v, 15r, 16r, 20r, 22v.
138. PRO E36/214, f. 123r.
139. *The York plays* ed. R. Beadle (1982) p. 212.

CHAPTER 5: THE SENSES OF THE MOUTH: SPEECH

1. *BA*, i, p. 200.
2. *BA*, i, pp. 206, 208.
3. *BA*, i , p. 213.
4. *Robert of Brunne's 'Handlyng Synne'* ed. F.J. Furnivall (EETS, OS 119, 123; 1901–3) i, p. 46; *PI*, pp. 129–30; E.D. Craun *Lies, slander, and obscenity in medieval English literature: pastoral rhetoric and the deviant speaker* (Cambridge, 1997) pp. 1–2.
5. *Memorials of St Anselm* ed. R.W. Southern and F.S. Schmitt (ABMA, 1; 1969) pp. 19–20, 22, 215–16.
6. J.F. Burke *Vision, the gaze, and the function of the senses in Celestina* (University Park, Pennsylvania, 2000) pp. 82–3; R. Horvath, 'Romancing the word: *fama* in the Middle English *Sir Launfal* and *Athelston*' in *Fama: the politics of talk and reputation in medieval Europe* ed. T. Fenster and D.L. Smail (Ithaca, 2003) pp. 165–86, at 168–9.
7. Salisbury Cathedral MS 103, ff. 146va–146vb; *Towneley*, i, pp. 92, 94; *Non-cycle plays and fragments* ed. N. Davis (EETS, SS 1; 1970) p. 71.
8. A. Hudson *The premature reformation: Wycliffite texts and Lollard history* (Oxford, 1988) p. 282.
9. *Mirk's Festial* ed. T. Erbe (EETS, ES 96; 1905) pp. 263–4.
10. H. Farmer, 'The canonization of St Hugh of Lincoln' *Lincolnshire Architectural and Archaeological Society Reports and Papers* 6 (1956) pp. 86–117, at 98.
11. *CS 871–1204*, ii, p. 1051; A. Gurevich *Medieval popular culture: problems of belief and perception* trans. J.M. Bak and P.A. Hollingsworth (Cambridge, 1988) pp. 211–15.
12. *Lollard sermons* ed. G. Cigman (EETS, OS 294; 1989) p. 93; D.G. Denery II *Seeing and being seen in the later medieval world: optics, theology and religious life* (Cambridge, 2005) pp. 23–4.
13. *EWS*, iii, p. 5; v, p. 283.
14. *CS 1205–1313*, i, p. 589; *Instructions for parish priests by John Myrc* ed. E. Peacock (EETS, OS 31; 1868; 2nd revised edition, 1902) p. 5; Salisbury Cathedral MS 103, ff. 164ra–164rb. For the association of names with virtues of their own, see Chapter 3.
15. *MCL*, pp. 180–93. Compare a secular death, that of Sir John Fastolf, which contained some similar elements: *GH*, p. 105.
16. Innocent III *De sacro altaris mysterio. PL*, 217, col. 775; *Visitations of religious houses in the diocese of Lincoln* ed. A.H. Thompson (3 vols, Lincoln Record Society, 7, 14, 21; 1914–29) i, pp. 18, 106; *Barnwell*, pp. 58–66.

17. *CS 871–1204*, ii, pp. 1048, 1060.
18. M.B. Parkes *Pause and effect: an introduction to the history of punctuation in the West* (Aldershot, 1992) pp. 15–16, 19, 35–8, 40.
19. *CS 1205–1313*, i, pp. 300, 488.
20. *Jacob's Well*, pp. 114–15, repeating a tale of Jacques de Vitry.
21. *Alphabet*, i, p. 86.
22. J. Leclercq *The love of learning and the desire for God: a study of monastic culture* trans. C. Misrahi (New York, 1961) pp. 18–19.
23. 'DH', pp. 2, 75–6.
24. *Hugh*, i, p. 76.
25. T.F. Kirby *Annals of Winchester College from its foundation in the year 1382 to the present time* (1892) p. 488.
26. *HO*, pp. *27–8.
27. *Hugh*, i, p. 75.
28. L.K. Smedick, 'Parallelism and pointing in Rolle's rhythmical style' *Mediaeval Studies* 41 (1979) pp. 404–67, at 404–5, 409, 463.
29. Robert de Avesbury *De gestis mirabilibus regis Edwardi tertii* ed. E. Maunde Thompson (RS, 1889) pp. 407–8.
30. Smedick, 'Parallelism and pointing', pp. 456–7, 463, 466; A. Hudson 'Middle English' in *Editing medieval texts: English, French and Latin written in England. Papers given at the twelfth annual Conference on Editorial Problems, University of Toronto 5–6 November 1976* ed. A.G. Rigg (New York, 1977) pp. 50–1.
31. JRULM Latin MS 235, f. 18r.
32. N. Denholm Young, 'The *cursus* in England' in idem, *Collected papers on medieval subjects* (Oxford, 1946) pp. 26–55.
33. PRO E101/414/6, ff. 9r, 30r, 47r, 50r, 59r, 67r, 87r.
34. *Eulogium historiarum sive temporis* ed. F.S. Haydon (3 vols, RS, 1858–63) iii, pp. 348–9.
35. *N-town*, i, p. 226.
36. S. Bardsley, 'Sin, speech, and scolding in late medieval England' in *Fama: the politics of talk and reputation in medieval Europe* ed. T. Fenster and D.L. Smail (Ithaca, 2003) pp. 145–64, at 162–3.
37 E. Sears, 'The iconography of auditory perception in the early Middle Ages: on psalm illustration and psalm exegesis' in *The second sense: studies in hearing and musical judgement from Antiquity to the seventeenth century* ed. C. Burnett, M. Fend and P. Gouk (Warburg Institute Surveys and Texts, 22; 1991) pp. 19–42, at 34–5.
38. *Jacob's Well*, pp. 110–11.
39. *EWS*, iii, pp. 312–14.
40. *Ailred*, pp. 6–7.
41. *Liber exemplorum*, p. 13.
42. E.g. *Heresy trials in the diocese of Norwich, 1428–31* ed. N.P. Tanner (Camden, 4th series, 20; 1977) p. 33.
43. R. Bartlett *England under the Norman and Angevin Kings 1075–1225* (Oxford, 2000) pp. 479–80; *Stubbs' Select charters from the beginning to 1307* ed. H.W.C. Davis (9th edition, Oxford, 1921) p. 173. Compare, for the transmission of heresy, M.G. Pegg *The corruption of angels: the great inquisition of 1245–1246* (Princeton, 2001) p. 78, which underplays the extent to which the contamination of heresy was a physical process.
44. *Lollards of Coventry 1486–1522* ed. S. McSheffrey and N. Tanner (Camden, 5th series, 23; 2003) p. 74; also pp. 89, 97–8.
45. *A contemporary narrative of the proceedings against Dame Alice Kyteler, prosecuted for sorcery in 1324, by Richard de Ledrede, Bishop of Ossory* ed. T. Wright (Camden Society, OS 24; 1843) pp. 27–9.
46. *CS 1205–1313*, i, p. 473.
47. *CS 1205–1313*, ii, p. 1045.

48. W. Holdsworth *A history of English law* (7th edition, 17 vols, 1956–72) i, p. 301; *Statutes of the realm* ed. A. Luders, T.E. Tomlins, J. Raithby *et al.* (11 vols, 1810–28) i, p. 64.

49. Gurevich, *Medieval popular culture*, p. 212.

50. J. Goering and P.J. Payer, 'The "Summa penitentie fratrum predicatorum": a thirteenth-century confessional formula' *Mediaeval Studies* 55 (1993) pp. 1–50, at 27.

51. R. Kieckhefer *Magic in the Middle Ages* (Cambridge, Canto edition, 2000) pp. 74–5, 82.

52. *The courts of the archdeaconry of Buckingham 1483–1523* ed. E.M. Elvey (Buckinghamshire Record Society, 19; 1975) pp. 257–8.

53. Tony Hunt *Popular medicine in thirteenth-century England: introduction and texts* (Cambridge, 1990) pp. 78–99.

54. D. Marner *St Cuthbert: his life and cult in medieval Durham* (2000) p. 22.

55. *CS 1205–1313*, i, p. 456; ii, pp. 1349–50.

56. S. Wenzel, 'Robert Grosseteste's treatise on confession, "Deus est"' *Franciscan Studies* 30 (1970) pp. 218–93, at 252.

57. Salisbury Cathedral MS 103, f. 112va.

58. J.A.F. Thomson *The later Lollards 1414–1520* (Oxford, 1965) p. 241.

59. Psalm 109 in the Authorised Version: *Three eleventh-century Anglo-Latin saints' lives: Vita S. Birini, Vita et miracula S. Kenelmi and Vita S. Rumwoldi* ed. and trans. R.C. Love (Oxford, 1996) pp. xci, 70–3.

60. *Swithun*, pp. 641, 676.

61. *Erkenwald*, pp. 142–3.

62. R. Rastall, 'The sounds of Hell' in *The iconography of Hell* ed. C. Davidson and T.H. Seiler (Medieval Institute Publications, Kalamazoo, Early Drama, Art and Music Monograph Series, 17; 1992) pp. 102–31, at 106–8.

63. *Jacob's Well*, p. 115; *PI*, p. 210; *VV*, p. 16; *LEJ*, pp. 275, 318, 331, 367; *Courts of . . . Buckingham*, p. 179.

64. A.J.M. Edwards, 'An early twelfth century account of the translation of St Milburga of Much Wenlock' *Transactions of the Shropshire Archaeological Society* 57 (1962) pp. 134–51, at 150; *Lanfrank's 'Science of cirurgie'* ed. R. von Fleischhacker (EETS, OS 102; 1894) p. 197; *The Cyrurgie of Guy de Chauliac* ed. M.S. Ogden (EETS, OS 265; 1971) p. 380.

65. 'DH', p. 89; but note an exception, a holy man with a speech impediment: M. Cassidy-Welch *Monastic spaces and their meanings: thirteenth-century English Cistercian monasteries* (Turnhout, 2001) pp. 213–14.

66. M.R. James, 'Twelve medieval ghost-stories' *EHR* 37 (1922) pp. 413–22, at 416, 418; J.-C. Schmitt *Les Revenants: les vivants et les morts dans la société médiévale* (Paris, 1994) pp. 227–30.

67. *The works of John Metham including the Romance of Amoryus and Cleopes . . .* ed. H. Craig (EETS, OS 132; 1916) pp. 142–3.

68. *Three Anglo-Latin saints' lives*, pp. cxli, clix, 91, 105.

69. *La Passiun de Seint Edmund* ed. J. Grant (ANTS, 36; 1978) pp. 1, 53, 97–100.

70. M.J. Carruthers *The book of memory: a study of memory in medieval culture* (Cambridge, 1990) p. 154; Schmitt, *Les Revenants*, pp. 147–8.

71. *Towneley*, i, pp. 84–5, 158.

72. *Chronicles of London* ed. C.L. Kingsford (Oxford, 1905), p. 31, discussed in P. Strohm *Hochon's arrow: the social imagination of fourteenth-century texts* (Princeton, 1992) p. 118.

73. *Lanfrank's 'Science of cirurgie'*, pp. 8–9.

74. *Treatises of fistula in ano, haemorrhoids, and clysters by John Arderne from an early fifteenth-century manuscript translation* ed. D'A. Power (EETS, OS 139; 1910) pp. 6–8.

75. *MCL*, p. 114.

76. *Barnwell*, pp. 212–14.

77. 'DH', pp. 70–2.

78. J. Goering and F.A.C. Mantello, 'The "Perambulauit Iudas . . ." (Speculum Confessionis) attributed to Robert Grosseteste' *Revue Bénédictine* 96 (1986) pp. 125–68, at 159–60.

79. Prv 15: 26, 'purus sermo pulcherrimus', rendered slightly differently in the text: *Jacob's Well*, p. 234; Salisbury Cathedral MS 103, f. 112va.

80. 'DH', pp. 90–4.

81. *RB*, pp. 190–1, 242–3, cap. 6 and 42; L. Gougaud *Anciennes coutumes claustrales* (Abbaye Saint-Martin de Ligugé (Vienne); Moines et monastères, 8; 1930) p. 14.

82. *MCL*, pp. 128–33.

83. *MCL*, pp. 146–7.

84. *Hugh*, i, p. 61.

85. *Barnwell*, pp. 136–8.

86. Leclercq, *Love of learning*, pp. 188–9.

87. *Statutes of the realm*, i, p. 29; *Fleta* ed. G.O. Sayles (3 vols, Selden Society, 72, 89, 99; 1953, 1972, 1984) i, p. 85; Holdsworth, *History of English law*, i, p. 327.

88. *Alphabet*, ii, pp. 304–5.

89. *The late medieval religious plays of Bodleian MSS Digby 133 and E Museo 160* ed. D.C. Baker, J.L. Murphy and L.B. Hall Jr. (EETS, OS 283; 1982) pp. xxx, xxxiv–xxxvi.

90. *N-town*, i, pp. xliii, 187–8.

91. *N-town*, i, pp. 267–70.

92. Rastall, 'Sounds of Hell', pp. 113, 122–4; G. Lester, 'Idle words: stereotyping by language in the English mystery plays' *Medieval English Theatre* 11 (1989) pp. 129–33, at 129–30; L. Forest-Hill *Transgressive language in medieval English drama: signs of challenge and change* (Aldershot, 2000) pp. 6–47, 85–135.

93. *Macro*, pp. xxxviii, 163, 165.

94. Craun, *Lies*, pp. 3–4, 13–18; Goering and Mantello, 'The "Perambulauit Iudas …"', pp. 125–68.

95. *VV*, pp. 54–5, 58–9; J.A. Burrow *Gestures and looks in medieval narrative* (Cambridge, 2002) pp. 46–7.

96. Gurevich, *Medieval popular culture*, pp. 189, 212–13; E.D. Craun, '"Inordinata locutio": blasphemy in pastoral literature, 1200–1500' *Traditio* 39 (1983) pp. 135–62, at 149–50; Robert Grosseteste *De decem mandatis* ed. R.C. Dales and E.B. King (ABMA, 10; 1987) pp. 26–7; Salisbury Cathedral MS 103, ff. 186rb–187va.

97. Sap 1: 11: *Middle English sermons edited from British Museum MS. Royal 18 B. XXIII* ed. W.O. Ross (EETS, OS 209; 1940) pp. xxxviii–xxxix, 24.

98. *Lollard sermons*, pp. 45–7.

99. *Two Wycliffite texts: the sermon of William Taylor 1406; the testimony of William Thorpe 1407* ed. A. Hudson (EETS, OS 301; 1993) p. 79. For the context of this case, Hudson *The premature reformation*, p. 371; H.G. Russell, 'Lollard opposition to oaths by creatures' *American Historical Review* 51 (1946) pp. 668–84, at 673.

100. P.R. Schofield, 'Peasants and the manor court: gossip and litigation in a Suffolk village at the close of the thirteenth century' *Past and Present* 159 (May 1998) pp. 3–42 at 35–6. For examples in ecclesiastical courts: *LEJ*, pp. 162, 277, 375.

101. *LEJ*, p. 316. For a case of 1328, in which a woman was accused of wearing a short jacket before Easter, M. Prestwich *Plantagenet England 1225–1360* (Oxford, 2005) p. 462.

102. *Courts of . . . Buckingham*, p. 149.

103. *Select cases on defamation to 1600* ed. R.H. Helmholz (Selden Society, 101; 1985) pp. xiv, xli, xliii–xlvii.

104. M. Carlin *Medieval Southwark* (1996) pp. 209–29.

105. Bardsley, 'Sin, speech, and scolding', pp. 145–64.

106. *Etymologies*, i: lib. IX.i, 3.

107. Gregory the Great *Moralia in Iob* ed. M. Adriaen (3 vols, CCSL 143, 143A, 143B; 1979–85) iii, p. 1346.

108. G. Hesketh, 'An unpublished Anglo-Norman life of Saint Katherine of Alexandria from MS. London, BL, Add. 40143' *Romania* 118 (2000) pp. 33–82, at 78.

109. P.O. Lewry, 'Thirteenth-century teaching on speech and accentuation: Robert Kilwardby's commentary on *De accentibus* of Pseudo-Priscian' *Mediaeval Studies* 50 (1988) pp. 96–185, at 96.
110. Hudson, *Premature reformation*, pp. 30–1.
111. Authorised Version, Psalm 51.
112. R.J. Dean and M.B.M. Boulton *Anglo-Norman literature: a guide to texts and manuscripts* (ANTS, Occasional Publications Series, 3; 1999) pp. 88–95, 278–322.
113. PRO SC 6/1261/9 (Aconbury); SC 6/1260/1–9, 15, and E 101/505/30 (Stamford).
114. S. Crane, 'Social aspects of bilingualism in the thirteenth century' in *Thirteenth Century England VI* ed. M. Prestwich, R.H. Britnell and R. Frame (Woodbridge, 1997) pp. 103–15.
115. Geoffrey Chaucer *The Riverside Chaucer* ed. L.D. Benson (3rd edition, Oxford, 1988) p. 25, l. 125.
116. Ibid., p. 27, ll. 264–5.
117. Walter Map *De nugis curialium. Courtiers' trifles* ed. M.R .James, revised by C.N.L. Brooke and R.A.B. Mynors (Oxford, 1983) pp. 496–7.
118. *Seyntz medicines*, p. 239.
119. A.C. Baugh and T. Cable *A history of the English language* (3rd edition, 1978) pp. 189–96; J. Milroy, 'Middle English dialectology' in *The Cambridge history of the English language* ed. R.M. Hogg *et al.* (6 vols, Cambridge, 1992–2001) ii, pp. 156–206.
120. *The chronicle of Jocelin of Brakelond concerning the acts of Samson, Abbot of the Monastery of St Edmund* ed. H.E. Butler (1949) p. 40.
121. *Caxton's Eneydos 1490 Englisht from the French Liure des Eneydes, 1483* ed. M.T. Culley and F.J. Furnivall (EETS, ES 57; 1890) pp. 2–3. I am grateful to Lynn Forest-Hill for this reference.
122. *Swithun*, pp. 641, 684–6.
123. *EWS*, i, pp. 598–600; v, pp. 61–2.
124. *Three Anglo-Latin saints' lives*, pp. 78–81.
125. *SM*, p. 89.
126. *Alphabet*, i, pp. 82–3.
127. *Macer*, pp. 78, 113–14.
128. *The Trotula: a medieval compendium of women's medicine* ed. M.H. Green (Philadelphia, 2001) p. 85.
129. Ibid., pp. 108–9.
130. *Marbode*, pp. 39–40, 44–5, 57–8.
131. *Gesta Henrici quinti: the deeds of Henry the Fifth* ed. F. Taylor and J.S. Roskell (Oxford, 1975) pp. 84–7.

CHAPTER 6: THE SENSES OF THE MOUTH: TASTE

1. *BA*, i, pp. 108, 117–18.
2. *The Cyrurgie of Guy de Chauliac* ed. M.S. Ogden (EETS, OS 265; 1971) pp. 45–6.
3. M. Pastoureau, 'Le Bestiaire des cinq sens (xiie–xvie siècle)' *Micrologus* 10 (2002) pp. 133–45, at 142–3.
4. *BA*, ii, pp. 1300–2.
5. *BA*, ii, pp. 1304–17; C. Burnett, '*Sapores sunt octo*: the medieval Latin terminology for the eight flavours' *Micrologus* 10 (2002) pp. 99–112; C. Burnett, 'The superiority of taste' *JWCI* 54 (1991) pp. 230–8.
6. *BA*, ii, pp. 1305–8; *PSQ*, p. 148.
7. *BA*, i, p. 118; ii, p. 1305; K. Albala *Eating right in the Renaissance* (Berkeley, 2002) pp. 82–4.
8. *Seyntz medicines*, p. 196.
9. *Cyrurgie of Guy de Chauliac*, pp. 477–8.

10. *Lanfrank's 'Science of cirurgie'* ed. R. von Fleischhacker (EETS, OS 102; 1894) pp. 9–10, 12–13.
11. *BA*, i, p. 614; ii, pp. 720–1.
12. *BA*, i, pp. 659–62; D. Serjeantson and C.M. Woolgar, 'Fish consumption in medieval England' in *Food in medieval England: diet and nutrition* ed. C.M. Woolgar, D. Serjeantson and T. Waldron (Oxford, 2006), pp. 102–30.
13. *BA*, ii, pp. 696–7.
14. *Etymologies*, ii: lib. XVI.ii, 6; *BA*, ii, pp. 875–6.
15. M. Prestwich *York civic ordinances, 1301* (Borthwick papers, 49; 1976) p. 12.
16. *Inventory of Henry VIII*, 427–9, 440–3, 1096–119, 2902; M.J. Carruthers *The book of memory: a study of memory in medieval culture* (Cambridge, 1990) p. 41.
17. C.M. Woolgar, 'Fasting and feasting: food and taste in the Middle Ages' in *Food: the history of taste* ed. P. Freedman (forthcoming); J.L. Flandrin, 'Internationalisme, nationalisme et régionalisme dans la cuisine des xive et xve siècles: le témoignage des livres de cuisine' in *Manger et boire au Moyen Âge* ed. D. Menjot (2 vols, Nice, 1984) ii, pp. 75–91.
18. *BA*, ii, pp. 1310–11.
19. *Seyntz medicines*, pp. 19–20.
20. *Manners and meals in olden time* ed. F.J. Furnivall (EETS, OS 32; 1868) pp. 151–3.
21. Ibid., pp. 172–5.
22. *Inventory of Henry VIII*, 1931,1936, 2001–3.
23. C.M. Woolgar, 'Diet and consumption in gentry and noble households: a case study from around the Wash' in *Rulers and ruled in late medieval England: essays presented to Gerald Harriss* ed. R.E. Archer and S. Walker (1995) pp. 17–32, at 29–30.
24. *HAME*, i, p. 227.
25. Cornwall RO AR12/25, mm. 2r, 3d, 5r.
26. *HAME*, i, pp. 229–34, 238, 240, 242, 245.
27. Hants RO, Winchester City Archives W/D1/121/48, W/D1/137/1.
28. Hants RO, Winchester City Archives W/D1/120/3, W/D1/124/24, W/D/143/5.
29. Hants RO, Winchester City Archives W/D1/137/38.
30. BAV MS Lat. Vat. 4015, f. 146v.
31. *The courts of the archdeaconry of Buckingham 1483–1523* ed. E.M. Elvey (Buckinghamshire Record Society, 19; 1975) p. 121.
32. *The Beauchamp cartulary: charters 1100–1268* ed. E. Mason (Pipe Roll Society, NS, 43; 1980) pp. 5–6.
33. *BA*, i, pp. 298–9.
34. *BA*, i, p. 304; N. Orme *Medieval children* (New Haven, 2001) p. 66.
35. *HO*, pp. 126–7.
36. J.L. Flandrin, 'Seasoning, cooking and dietetics in the late Middle Ages' in *Food: a culinary history* ed. J.L. Flandrin and M. Montanari (New York, 1999) pp. 313–27; M. Weiss Adamson, 'The role of medieval physicians in the diffusion of culinary recipes and cooking practices' in *Du Manuscrit à la table: essais sur la cuisine au Moyen Âge et répertoire des manuscrits médiévaux contenant des recettes culinaires* ed. C. Lambert (Montreal, 1992) pp. 69–80; Albala, *Eating right*, pp. 62, 159, 242–4.
37. Longleat MS Misc. IX, f. 15r; M.P. Cosman *Fabulous feasts: medieval cookery and ceremony* (New York, 1976) pp. 61–3; Albala, *Eating right*, pp. 166–7.
38. *Two fifteenth-century cookery-books* ed. T. Austin (EETS, OS 91; 1888) pp. 14–15.
39. Ibid., pp. 38–9.
40. Ibid., pp. 41–2.
41. Ibid., pp. 115–16.
42. A. Henry, 'Un Texte oenologique de Jofroi de Waterford et Servais Copale', *Romania* 107 (1986) pp. 1–37, at 12, 14, 17.
43. *Seyntz medicines*, p. 20.
44. J.E. Merceron, 'Cooks, social status, and stereotypes of violence in medieval French literature and society' *Romania* 116 (1998) pp. 170–87, at 174–5.

45. Serjeantson and Woolgar, 'Fish consumption in medieval England'; C.M. Woolgar, 'Group diets in late medieval England' in *Food in medieval England: diet and nutrition* ed. C.M. Woolgar, D. Serjeantson and T. Waldron (Oxford, 2006), pp. 191–200.

46. *HAME*, i, pp. 177–227.

47. *Hugh*, i, p. 78.

48. *Libellus de vita et miraculis S. Godrici, heremitae de Finchale auctore Reginaldo monacho Dunelmensi. Adjicitur appendix miraculorum* ed. J. Stevenson (Surtees Society, 20; 1845) pp. 71, 73, 79–83.

49. Owst, *Literature*, p. 501.

50. J. Goering and P.J. Payer, 'The "Summa penitentie fratrum predicatorum": a thirteenth-century confessional formula' *Mediaeval Studies* 55 (1993) pp. 1–50, at 5, 30–1, 37–8.

51. *Non-cycle plays and fragments* ed. N. Davis (EETS, SS 1; 1970) pp. lxxxv, 62–4.

52. For example, from its presence in mystery plays: *N-town*, i, p. 25.

53. P. Binski, 'Abbot Berkyng's tapestries and Matthew Paris's life of St Edward the Confessor' *Archaeologia* 109 (1991) pp. 85–100, at 91; *La Estoire de Seint Aedward le rei attributed to Matthew Paris* ed. K. Young Wallace (ANTS, 41; 1983) pp. xxiii, 93–4.

54. *CS 1205–1313*, ii, pp. 789–90.

55. *Hugh*, ii, p. 163.

56. Salisbury Cathedral MS 103, f. 104va.

57. *VV*, pp. 278–9.

58. *St Modwenna* ed. A.T. Baker and A. Bell (ANTS, 7; 1947) pp. xxi, 98–100.

59. Ibid., pp. 262–5.

60. *PI*, p. 132.

61. *PI*, p. 196.

62. *Thurkill*, p. 12.

63. See Chapter 2.

64. 'Gustate et videte quam suavis est Dominus'; Authorised Version, Psalm 34.

65. *Ailred*, pp. 22–3.

66. 'DH', pp. 144–5.

67. *The late medieval religious plays of Bodleian MSS Digby 133 and E Museo 160* ed. D.C. Baker, J.L. Murphy and L.B. Hall Jr. (EETS, OS 283; 1982) pp. xxx, 50, 92.

68. *Henry VI*, p. 1.

69. Salisbury Cathedral MS 103, f. 127va.

70. 'DH', pp. 18–19; Ct 5: 1.

71. *EWS*, ii, p. 135.

72. *MK*, p. 171.

73. *English mediaeval lapidaries* ed. J. Evans and M.S. Serjeantson (EETS, OS 190; 1933) pp. 63, 71, 91–3.

74. *BA*, i, p. 159.

75. C. Rawcliffe *Medicine & society in later medieval England* (Far Thrupp, 1995) p. 51.

76. *Macer*, pp. 113–14.

77. *Cyrurgie of Guy de Chauliac*, p. 477.

78. Tony Hunt *Popular medicine in thirteenth-century England: introduction and texts* (Cambridge, 1990) p. 207.

79. BL MS Sloane 3550, f. 31v, with related text in *The Trotula: a medieval compendium of women's medicine* ed. M.H. Green (Philadelphia, 2001) pp. 186–9.

CHAPTER 7: SMELL

1. C. Classen, D. Howes and A. Synnott *Aroma: the cultural history of smell* (London, 1994) pp. 3, 116.

2. B. Lawn *The Salernitan questions: an introduction to the history of medieval and Renaissance problem literature* (Oxford, 1963); *PSQ*, pp. 25, 165, 167, 331–4.

3. *BA*, ii, p. 1298.

4. Ex 30: 22–38.

5. Classen *et al.*, *Aroma*, pp. 52–4.

6. *Erkenwald*, pp. 22–3, 90 n. 16.

7. *Libellus de vita et miraculis S. Godrici, heremitae de Finchale auctore Reginaldo monacho Dunelmensi. Adjicitur appendix miraculorum* ed. J. Stevenson (Surtees Society, 20; 1845) p. 331.

8. *Becket*, ii, p. 37.

9. Mc 14: 3; Io 12: 3; *Wulfstan*, p. 182.

10. A.J.M. Edwards, 'An early twelfth century account of the translation of St Milburga of Much Wenlock' *Transactions of the Shropshire Archaeological Society* 57 (1962) pp. 134–51, at 138.

11. B. Nilson *Cathedral shrines of medieval England* (Woodbridge, 1998) pp. 28–9.

12. *Osmund*, p. 37.

13. II Cor 2: 15–16.

14. *Hugh*, ii, pp. 153–4 and n.

15. A.T. Baker, 'La Vie de Saint Edmond archevêque de Cantobéry' *Romania* 55 (1929) pp. 332–81, at 347–9.

16. *Three eleventh-century Anglo-Latin saints' lives: Vita S. Birini, Vita et miracula S. Kenelmi and Vita S. Rumwoldi* ed. and trans. R.C. Love (Oxford, 1996) p. 9.

17. *WN*, pp. 37, 39.

18. *Ailred*, pp. 76–7.

19. *The historians of the church of York and its archbishops* ed. J. Raine (3 vols, RS, 1879–94) ii, pp. 279–80.

20. *Mirk's Festial* ed. T. Erbe (EETS, ES 96; 1905) pp. 245–6. Compare also the description of the Virgin in L.W. Stone, 'Jean de Howden, poète anglo-normand du xiiiᵉ siècle' *Romania* 69 (1946–7) pp. 496–519, at 513–14.

21. P. Dronke, 'Tradition and innovation in medieval Western colour imagery' in *The realms of colour* ed. A. Portmann and R. Ritsema (Eranos 1972 Yearbook, 41: Leiden, 1974) pp. 51–107, at 77–88; D. Pearsall and E. Salter *Landscapes and seasons of the medieval world* (1973) pp. 56–75; *BA*, ii, pp. 788–90; 'Eynsham', p. 360; *The York plays* ed. R. Beadle (1982) pp. 62–3.

22. C.J. Holdsworth, 'Eleven visions connected with the Cistercian monastery of Stratford Langthorne' *Cîteaux* 13 (1962) pp. 185–204, at 191 n. 31, 197–201.

23. Following Psalm 140: 2 (V), 'dirigatur oratio mea sicut incensum in conspectu tuo elevatio manuum mearum sacrificium vespertinum'. *Lollard sermons* ed. G. Cigman (EETS, OS 294; 1989) pp. 77–8; *Jacob's Well*, p. 191; Baldwin of Ford *Sermones de commendatione fidei* ed. D.N. Bell (CCCM 99; 1991) pp. 39–40.

24. E.g. *MCL*; *Barnwell*; P.B. Griesser, 'Die "Ecclesiastica officia Cisterciensis ordinis" des Cod. 1711 von Trient' *Analecta Sacri Ordinis Cisterciensis* 12 (1956) pp. 153–288, at 260–1.

25. *Mirk's Festial*, pp. 49–50; *EWS*, ii, p. 238.

26. E.g. PRO E101/398/9, f. 23r; E101/401/2, f. 37r; BL Add. MS 35115, f. 33r; PRO E101/403/10, f. 35v.

27. M. Cassidy-Welch *Monastic spaces and their meanings: thirteenth-century English Cistercian monasteries* (Turnhout, 2001) pp. 65–7.

28. D. Robinson (ed.) *The Cistercian abbeys of Britain: far from the concourse of men* (1998) pp. 101–4, 176–9.

29. *Hugh*, ii, p. 163.

30. *N-town*, i, p. 263.

31. *Farming and gardening in late medieval Norfolk: Norwich Cathedral Priory gardeners' accounts, 1329–1530* ed. C. Noble and *Skayman's Book, 1516–1518* ed. C. Moreton and P. Rutledge (Norfolk Record Society, 61; 1997) p. 16.

32. *Liber exemplorum*, p. 30.

33. *Inventory of Henry VIII*, 35–41, 261.

34. W.L. Braekman, 'Fortune-telling by the casting of dice: a Middle English poem and its background' *Studia Neophilologica* 52 (1980) pp. 3–29, at 13.

35. T.H. Seiler, 'Filth and stench as aspects of the iconography of Hell' in *The iconography of Hell* ed. C. Davidson and T.H. Seiler (Medieval Institute Publications, Kalamazoo, Early Drama, Art and Music Monograph Series, 17; 1992) pp. 132–40; Owst, *Literature*, pp. 293–4, 413–14; *The early English versions of the Gesta Romanorum . . .* ed. S.J.H. Herrtage (EETS, ES 33; 1879) pp. 380–1.

36. *Becket*, i, pp. 416–17; ii, pp. 242–3.

37. 'Eynsham', pp. 305–23; *Thurkill*, pp. 28–9.

38. C.E. Woodruff, 'The will of Peter de Aqua Blanca Bishop of Hereford (1268)' in *Camden Miscellany XIV* (Camden, 3rd series, 37; 1926) p. vii; Matthew Paris *Chronica majora* ed. H.R. Luard (7 vols, RS, 1872–83) v, pp. 510–11.

39. *Libellus de vita et miraculis S. Godrici . . .*, pp. 356–8.

40. Geoffrey Chaucer, *The Riverside Chaucer* ed. L.D. Benson (3rd edition, Oxford, 1988) pp. 134–6, III (D) ll. 2089–294; P.B. Taylor, 'The Canon Yeoman's breath: emanations of a metaphor' *English Studies* 60 (1979) pp. 380–8, at 381.

41. *N-town*, i, pp. 39, 47.

42. *LEJ*, p. 427.

43. *BA*, i, pp. 369–70.

44. BAV MS Lat. Vat. 4015, f. 51r.

45. *Macro*, p. 43.

46. Salisbury Cathedral MS 103, ff. 153vb–154ra.

47. *Alphabet*, ii, p. 389.

48. E.g. *Macro*, p. 145.

49. G. Hesketh, 'An unpublished Anglo-Norman life of Saint Katherine of Alexandria from MS. London, BL, Add. 40143' *Romania* 118 (2000) pp. 33–82, at 74–6.

50. *Hugh*, ii, pp. 82–3.

51. *WN*, pp. 246–50.

52. *Becket*, ii, pp. 66–7.

53. J.A.F. Thomson *The later Lollards 1414–1520* (Oxford, 1965) pp. 148–50.

54. *BA*, ii, pp. 909–10, 1303–4.

55. *Two Wycliffite texts: the sermon of William Taylor 1406; the testimony of William Thorpe 1407* ed. A. Hudson (EETS, OS 301; 1993) p. 25.

56. BAV MS Lat. Vat. 4015, f. 118v.

57. *MK*, pp. 86–7, 301.

58. J. Goering and F.A.C. Mantello, 'The "Perambulauit Iudas . . ." (Speculum Confessionis) attributed to Robert Grosseteste' *Revue Bénédictine* 96 (1986) pp. 125–68, at 138, 151–2.

59. J. Goering and P.J. Payer, 'The "Summa penitentie fratrum predicatorum": a thirteenth-century confessional formula' *Mediaeval Studies* 55 (1993) pp. 1–50, at 37.

60. *Seyntz medicines*, pp. 13–14, 47.

61. R. Palmer, 'In bad odour: smell and its significance in medicine from Antiquity to the seventeenth century' in *Medicine and the five senses* ed. W.F. Bynum and R. Porter (Cambridge, 1993) pp. 61–8, 285–7, at 63–4.

62. C. Rawcliffe *Medicine & society in later medieval England* (Far Thrupp, 1995) pp. 60, 62, 149.

63. 'Eynsham', p. 291.

64. *Barnwell*, p. 202.

65. Palmer, 'In bad odour', p. 63.

66. *The knowing of woman's kind in childing: a Middle English version of material derived from the Trotula and other sources* ed. A. Barratt (Turnhout, 2001) pp. 76–9, 94–7; Rawcliffe, *Medicine & society*, p. 197; *The Trotula: a medieval compendium of women's medicine* ed. M.H. Green (Philadelphia, 2001) pp. 22–3, 31, 83–7, 104–5.

67. R. Kieckhefer *Magic in the Middle Ages* (Cambridge, Canto edition, 2000) pp. 132, 166.

68. Rawcliffe, *Medicine & society*, p. 40.

69. H. Johnstone, 'The Queen's household' in T.F. Tout *Chapters in the administrative history of mediaeval England: the wardrobe, the chamber and the small seals* (6 vols, Manchester, 1920–33) v, pp. 231–89, at 233, n. 3.

70. E.g. Tony Hunt *Popular medicine in thirteenth-century England: introduction and texts* (Cambridge, 1990) pp. 133, 207; P. Ruelle *L'Ornement des dames (ornatus mulierum): texte anglo-normand du xiii* siècle* (Brussels: Université Libre de Bruxelles, Travaux de la Faculté de Philosophie et Lettres, 36; 1967) pp. 68–9 (this text is not to be confused with the *De ornatu mulierum* of the Trotula ensemble, printed in *Trotula*, pp. 166–91); *Three receptaria from medieval England: the languages of medicine in the fourteenth century* ed. T. Hunt, with M. Benskin (Oxford, 2001) pp. 17–18, 105, 166–7; *The Cyrurgie of Guy de Chauliac* ed. M.S. Ogden (EETS, OS 265; 1971) pp. 477–8.

71. *Treatises of fistula in ano, haemorrhoids, and clysters by John Arderne from an early fifteenth-century manuscript translation* ed. D'A. Power (EETS, OS 139; 1910) p. 88.

72. M.R. McVaugh, 'Smells and the medieval surgeon' *Micrologus* 10 (2002) pp. 113–32.

73. Rawcliffe, *Medicine & society*, pp. 14–17.

74. M. Prestwich *York civic ordinances, 1301* (Borthwick papers, 49; 1976) pp. 12–13.

75. Walter Map *De nugis curialium. Courtiers' trifles* ed. M.R. James, revised by C.N.L. Brooke and R.A.B. Mynors (Oxford, 1983) pp. 462–5.

76. Pearsall and Salter, *Landscapes and seasons*, pp. 161–77, 198; *PI*, pp. 128–9; *The Harley Lyrics: the Middle English lyrics of Ms. Harley 2253* ed. G.L. Brook (4th edition, Manchester, 1968) pp. 43–4, 54–5; *Twenty-six political and other poems (including 'Petty Job') from the Oxford MSS Digby 102 and Douce 322* ed. J. Kail (EETS, OS 124; 1904) p. 143.

77. E.g. *Calendar of Nottinghamshire coroners' inquests 1485–1558* ed. R.F. Hunnisett (Thoroton Society, Record Series, 25; 1969) p. 28, exceptional only for its tragedy.

78. *Privy purse expenses of Elizabeth of York: wardrobe accounts of Edward the Fourth, with a memoir of Elizabeth of York, and notes* ed. N.H. Nicolas (1830) p. 4.

79. Pearsall and Salter, *Landscapes and seasons*, pp. 27, 56, 76–80.

80. *HKW*, iv, pp. 147–9, 223–4, 344–8; PRO E101/414/6, f. 61a, recto; E101/414/16, ff. 22v, 45r, 55v, 65v, 68v; BL Add. MS 21480, ff. 17r–v; PRO E36/214, ff. 4r, 52r, 53r, 71v, 134v, 147v, 153v, 158v, 159v–160r.

81. *Six ecclesiastical satires* ed. J.M. Dean (Kalamazoo, Michigan, 1991) pp. 232–4.

82. Dronke, 'Tradition and innovation', p. 80.

83. *Middle English sermons edited from British Museum MS. Royal 18 B. XXIII* ed. W.O. Ross (EETS, OS 209; 1940) p. 75.

84. M. Kowaleski *Local markets and regional trade in medieval Exeter* (Cambridge, 1995) p. 190.

85. *Calendar of coroners' rolls of the city of London AD 1300–1378* ed. R.R. Sharpe (1913) pp. 198–9.

86. *Nottinghamshire coroners' inquests*, p. 141.

87. Kitchens: *HKW*, ii, p. 652; prisons: *The life of St Catherine by Clemence of Barking* ed. W. MacBain (ANTS, 18; 1964) p. 47; latrines, rubbish pits and dirt: Rawcliffe, *Medicine & society*, p. 40; E.L. Sabine, 'Latrines and cesspools of mediaeval London' *Speculum* 9 (1934) pp. 303–21; L. Thorndike, 'Sanitation, baths, and street-cleaning in the Middle Ages and Renaissance' *Speculum* 3 (1928) pp. 192–203; Hinton, *Gold and gilt*, p. 233; M.S.R. Jenner, 'Civilization and deodorization? Smell in early modern English culture' in *Civil histories: essays presented to Sir Keith Thomas* ed. P. Burke, B. Harrison and P. Slack (Oxford, 2000) pp. 127–44; Kowaleski, *Local markets*, p. 190; C. Platt *Medieval Southampton: the port and trading community, AD 1000–1600* (1973) p. 171; J. Steane *The archaeology of the medieval English monarchy* (1993) p. 116. For dunghills: *Manières de langage* ed. A.M. Kristol (ANTS, 53; 1995) p. 30; *The cloud of unknowing and the book of privy counselling* ed. P. Hodgson (EETS, OS 218; 1944) pp. 46, 191; *PI*, p. 148.

88. *PSQ*, pp. 190–1; *Manners and meals in olden time* ed. F.J. Furnivall (EETS, OS 32; 1868) pp. 179–80.

89. *Liber exemplorum*, p. 121.

90. *Calendar of coroners' rolls of . . . London*, p. 221.

91. E.g. J.E. Merceron, 'Cooks, social status, and stereotypes of violence in medieval French literature and society' *Romania* 116 (1998) pp. 170–87, at 172–4.

92. E. O'Connor, 'Hell's pit and Heaven's rose: the typology of female sights and smells in Panormita's *Hermaphroditus*' *Medievalia et Humanistica* NS, 23 (1996) pp. 25–51.

93. *The late medieval religious plays of Bodleian MSS Digby 133 and E Museo 160* ed. D.C. Baker, J.L. Murphy and L.B. Hall Jr. (EETS, OS 283; 1982) pp. xxx, 69, 212.

94. *Riverside Chaucer*, pp. 103–4, ll. II (B¹) 1163–77.

95. *Towneley*, i, pp. 57–8.

96. G.P. Largey and D.R. Watson, 'The sociology of odors' *American Journal of Sociology* 77 (1972) pp. 1021–34; L. Golding *The Jewish problem* (London, 1938) pp. 59–60; Classen, Howes and Synnott, *Aroma*, pp. 161–79; C. Roth *A history of the Jews in England* (3rd edition, Oxford, 1964) pp. 9, 13.

97. *EWS*, ii, p. 137.

98. Classen, Howes and Synnott, *Aroma*, p. 31.

99. *Seyntz medicines*, p. 25.

100. *Macro*, pp. 28–9.

101. William Langland *The vision of Piers Plowman: a critical edition of the B text* ed. A.V.C. Schmidt (new edition, London, 1987) p. 197; *N-town*, i, p. 242; *Towneley*, i, p. 429.

102. *Hugh*, ii, p. 219.

103. 'The squire of low degree' in *Middle English metrical romances* ed. W.H. French and C.B. Hale (2 vols, New York, 1964) ii, pp. 721–55, at 742.

104. *Lollard sermons*, pp. 90–1, 167–8.

105. *Jacob's Well*, p. 118.

106. *Seyntz medicines*, pp. 104, 107–8.

107. *Towneley*, i, p. 149.

108. *Early English versions of the Gesta Romanorum . . .*, p. 280.

109. M. Pastoureau, 'Le Bestiaire des cinq sens (xiiᵉ–xviᵉ siècle)' *Micrologus* 10 (2002) pp. 133–45, at 141.

110. B.A. Hanawalt *The ties that bound: peasant families in medieval England* (Oxford, 1986) pp. 50–1.

111. Ibid., pp. 41, 43, 61.

112. *Hugh*, ii, p. 49.

113. *Inquests and indictments from late fourteenth century Buckinghamshire: the superior eyre of Michaelmas 1389 at High Wycombe* ed. L. Boatwright (Buckinghamshire Record Society, 29; 1994) p. 21; *Nottinghamshire coroners' inquests*, p. 128.

114. A. Collet, 'Traité d'hygiène de Thomas le Bourguignon (1286)' *Romania* 112 (1991) pp. 450–87, at 477.

115. *EWS*, iii, p. 115.

116. A.G. Vince, 'The Saxon and medieval pottery of London: a review' *Medieval Archaeology* 29 (1985) pp. 25–93, at 46–7, 50, 72; D. Gaimster and B. Nenk, 'English households in transition c. 1450–1550: the ceramic evidence' in *The age of transition: the archaeology of English culture 1400–1600* ed. D. Gaimster and P. Stamper (Society for Medieval Archaeology Monograph Series 15; 1997) pp. 171–95, at 177.

117. S. Vandenberghe, 'Une Aquamanile ornitomorphe trouvée à Gistel (Flandre occidentale) (an ornithomorphic aquamanile from Gistel)' in *Everyday and exotic pottery in Europe c. 600–1900: studies in honour of J.G. Hurst* ed. D.R.M. Gaimster and M. Redknap (Oxford, 1992) pp. 336–9.

118. *GH*, p. 167; *Calendar of coroners' rolls . . . of London*, pp. 147–8.

119. *VV*, pp. 179–80.

120. *The book of St Gilbert* ed. R. Foreville and G. Keir (Oxford, 1987) pp. 332–5.

121. *GH*, p. 167, citing PRO E101/546/18, f. 67r.

122. *Thurkill*, pp. 5–6.

123. T.F. Kirby *Annals of Winchester College from its foundation in the year 1382 to the present time* (1892) p. 510.

124. *Manières de langage*, pp. 43–4.

125. *Hugh*, i, p. 39.

126. G.R. Owst, 'Sortilegium in English homiletic literature of the fourteenth century' in *Studies presented to Sir Hilary Jenkinson* ed. J.C. Davies (London, 1957) pp. 272–303, at 287 n. 5.

127. J. Larmat, 'Les Bains dans la littérature française du Moyen Âge' in *Les Soins de beauté* (Nice, 1987) pp. 195–210; M. Bloch *Feudal society* (2 vols, 1965) ii, p. 316; *The receyt of the Lady Kateryne* ed. G. Kipling (EETS, OS 296; 1990) pp. 50–1, for the appointment of Knights of the Bath, going to their baths and the dubbing of some that did not.

128. A.J. Taylor, 'Count Amadeus of Savoy's visit to England in 1292' *Archaeologia* 106 (1979) pp. 123–32, at 125, 129.

129. BL Add. MS 38006, f. 5r: bath on 21 May; the date of marriage is given as 22 May in E.W. Safford, 'An account of the expenses of Eleanor, sister of Edward III, on the occasion of her marriage to Reynald, Count of Guelders' *Archaeologia* 77 (1927) pp. 111–40, at 113 (which edits the duplicate of the account, PRO E101/386/7).

130. John Russell's *Boke of Nurture*, in *Manners and meals*, pp. 183–5. The tent-like arrangement was common, e.g. in the 1320s, BL Add. MS 60584, f. 24r.

131. Arrangements for bathing do not feature in J.B. Post, 'A fifteenth-century customary of the Southwark stews' *Journal of the Society of Archivists* 5 (1974–7) pp. 418–28; for the Continent, Paulino Iradiel, 'Cuidar el cuerpo, cuidar la imagen: los paradigmas de la belleza femenina en la Valencia bajomedieval' in *Les Soins de beauté*, pp. 61–86, at 72–3.

132. *Three prose versions of the Secreta secretorum* ed. R. Steel (EETS, ES 74; 1898) pp. 82–3.

133. *HKW*, ii, pp. 926, 934, 974.

134. *HKW*, ii, pp. 697, 998.

135. *HKW*, i, p. 246.

136. Gaimster and Nenk, 'English households in transition', pp. 181–2; S. Thurley *Whitehall Palace: an architectural history of the royal apartments, 1240–1698* (New Haven, 1999) p. 49.

137. *Wulfstan*, p. 120.

138. *St Modwenna* ed. A.T. Baker and A. Bell (ANTS, 7; 1947) pp. 50–1.

139. R.D. Berryman *Use of the woodlands in the late Anglo-Saxon period* (British Archaeological Reports, British Series, 271; 1998) pp. 35–7.

140. H. Johnstone, 'The wardrobe and household of Henry, son of Edward I' *Bulletin of the John Rylands Library* 7 (1922–3) pp. 384–420, at 411; Tony Hunt *Popular medicine in thirteenth-century England*, p. 206.

141. *RWH, 1286–9*, p. 393, no. 3238.

142. Longleat House MS Misc. IX, ff. 15r–16r.

143. *HAME*, ii, p. 578.

144. L. Bolens, 'Les Parfums et la beauté en Andalousie médiévale (xième–xiiième siècles)' in *Les Soins de beauté*, pp. 145–69, especially 152, 154, 157.

145. R. Staccini, 'L'inventario di una spezieria del Quattrocento' *Studi Medievali* 22 (1981) pp. 377–420, at 387–409.

146. *BA*, ii, pp. 1029–31; *Seyntz medicines*, pp. 149–50; *Treatises of fistula*, pp. 92–5; I. Leveque-Agret, 'Les Parfums à la fin du Moyen Âge: les différentes formes de fabrication et d'utilisation' in *Les Soins de beauté*, pp. 135–44, at 138–40.

147. *Farming and gardening in late medieval Norfolk*, p. 9.

148. H. Braunschweig *Liber de arte distillandi, de simplicibus* (Strasbourg, 1500); and *Liber de arte distillandi de compositis* (Strasbourg, 1512); translations and other editions listed in *Short-title catalogue of books printed in the German-speaking countries and German books printed in other countries from 1455 to 1600 now in the British Museum* (1962) pp. 148–9; English editions, *The vertuouse boke of distyllacyon* (1527), see STC 13435–7.

149. Ruelle, *L'Ornement des dames*, p. 68.

150. Ibid., pp. 70–2.

151. J.R. Gilleland, 'Eight Anglo-Norman cosmetic recipes: MS. Cambridge, Trinity College 1044' *Romania* 109 (1988) pp. 50–69, at 57, 63–4: nutmeg rather than musk.
152. Ibid., p. 58.
153. B. Laurioux *Le Règne de Taillevent: livres et pratiques culinaires à la fin du Moyen Âge* (Paris, 1997) pp. 175–9.
154. T.H. Turner, 'The will of Humphrey de Bohun, Earl of Hereford and Essex, with extracts from the inventory of his effects, 1319–1322' *Archaeological Journal* 2 (1846) pp. 339–49, at 344, 348.
155. PRO E101/411/9, m. 26r.
156. *Rot. Parl.*, iv, pp. 219, 221, 224; *The Bedford inventories: the worldly goods of John, Duke of Bedford, Regent of France (1389–1435)* ed. J. Stratford (Reports of the Research Committee of the Society of Antiquaries of London, 49; 1993) pp. 171, 231, 254.
157. PRO E36/214, f. 111v.
158. E.g. *Les Collections du trésor royal sous le règne de Charles VI (1380–1422): l'inventaire de 1400* ed. P. Henwood (Paris, 2004) pp. 158–9.
159. BL Add. MS 60584, ff. 12r, 59v; and a branch or root of a balsam tree with a spine or thorn of the same: f. 23r, described as both at different points, f. 51v, as a root.
160. *Seyntz medicines*, p. 13.
161. *Rot. Parl.*, iv, p. 218.
162. PRO E101/411/9, m. 24r.
163. *HAME*, ii, p. 667.
164. A.A. Ruddock *Italian merchants and shipping in Southampton 1270–1600* (Southampton Records Series, 1; 1951) p. 72.
165. *Alphabet*, i, p. 170.
166. *Privy purse expenses of Elizabeth of York . . .*, p. 4.
167. *NHB*, pp. 371, 430.
168. Longleat MS Misc. XIII, ff. 73v, 86v.
169. Leveque-Agret, 'Parfums à la fin du Moyen Âge', pp. 136–8.
170. D.A. Bidon and F. Piponnier, 'Gestes et object de la toilette au xivème et xvème siècles' in *Les Soins de beauté*, pp. 211–44, at 221–4.
171. *Inventory of Henry VIII*, 3282–3.
172. *MCL*, pp. 136–9.
173. *Barnwell*, p. 194.
174. BL MS Sloane 3550, f. 32r; *Women's writing in Middle English* ed. A. Barratt (London, 1992) pp. 35–7; cf. also the use of scented ointments, J.R. McGuire, 'L'Onguent et l'initiative féminine dans *Yvain*' *Romania* 112 (1991) pp. 65–82.
175. *Macro*, p. 371.
176. *PSQ*, pp. 354–5.
177. *Calendar of coroners' rolls of . . . London*, p. 100.
178. *Bedfordshire coroners' rolls* ed. R.F. Hunnisett (Bedfordshire Historical Record Society, 41; 1961) p. 15.
179. Hanawalt, *Ties that bound*, p. 147.
180. *Middle English sermons*, pp. 274, 373.
181. E.g. PRO E101/393/4, f. 12v; BL Add. MS 35115, f. 43r; *Privy purse expenses of Elizabeth of York . . .*, pp. 121–2, 124.
182. *HBQI*, p. 150.
183. PRO E101/375/9, f. 25r.
184. S. Wenzel, 'Robert Grosseteste's treatise on confession, "Deus est"' *Franciscan Studies* 30 (1970) pp. 218–93, at 291.
185. Johnstone, 'The wardrobe and household of Henry, son of Edward I', p. 408.
186. *BA*, ii, pp. 1255, 1265–6.
187. *BB*, p. 118.
188. *BB*, p. 120.

189. *HO*, p. 215.
190. *Manners and meals*, p. 179.
191. *Rot. Parl.*, iv, p. 239.
192. *Alphabet*, i, p. 171.
193. Rawcliffe, *Medicine & society*, p. 43.
194. BAV MS Lat. Vat. 4015, f. 140v; f. 142r for the date.
195. *Middle English sermons*, p. 279; Owst, *Literature*, p. 35.
196. 'DH', pp. 6–8.
197. *MCL*, p. 82.
198. *Barnwell*, pp. 72, 154.
199. *Mirk's Festial*, pp. 129–30.
200. Winchester College Muniments, 1.
201. *Towneley*, i, p. 237.
202. *Hugh*, ii, pp. 121–3.
203. 'Eynsham', p. 323.
204. *The book of quinte essence or the fifth being; that is to say, man's heaven . . .* ed. F.J. Furnivall (EETS, OS 16; 1866) p. 2.
205. Gervase of Tilbury *Otia imperialia: recreation for an emperor* ed. S.E. Banks and J.W. Binns (Oxford, 2002) pp. 574–5.
206. *Manières de langage*, p. 10.
207. *BA*, i, pp. 561–2.
208. Gaimster and Nenk, 'English households in transition', p. 179.
209. PRO E101/414/16, f. 56r.
210. *Privy purse expenses of Elizabeth of York . . .*, p. 21.
211. Gaimster and Nenk, 'English households in transition', p. 177.
212. *Middle English metrical romances*, ii, p. 747.
213. *Mémoires d'Olivier de la Marche* ed. H. Beaune and J. d'Arbaumont (4 vols, Paris, 1883–8) ii, p. 351.
214. G. Kipling *Enter the King: theatre, liturgy, and ritual in the medieval civic triumph* (Oxford, 1998) pp. 17, 94, 163.
215. *Non-cycle plays and fragments* ed. N. Davis (EETS, SS 1; 1970) pp. xxix, xxxi–xxxv, 8, 15.
216. Rawcliffe, *Medicine & society*, p. 42.
217. PRO E101/414/16, ff. 28v, 60v.
218. *Farming and gardening in late medieval Norfolk*, p. 9.
219. *English mediaeval lapidaries* ed. J. Evans and M.S. Serjeantson (EETS, OS 190; 1933) p. 32.
220. *Marbode*, pp. 57–8; *English mediaeval lapidaries*, p. 33.

CHAPTER 8: VISION

1. *BA*, i, pp. 108–13; *Etymologies*, ii: lib. XI.i, 21.
2. I have benefited particularly from G. Frank *The memory of the eyes: pilgrims to living saints in Christian late Antiquity* (Berkeley, 2000) and J.F. Burke *Vision, the gaze, and the function of the senses in Celestina* (University Park, Pennsylvania, 2000).
3. *BA*, i, pp. 108, 172–3, 178–85; *The Cyrurgie of Guy de Chauliac* ed. M.S. Ogden (EETS, OS 265; 1971) pp. 42–3.
4. *The early English versions of the Gesta Romanorum . . .* ed. S.J.H. Herrtage (EETS, ES 33; 1879) p. 240.
5. *BA*, ii, pp. 1153–4.
6. JRULM Latin MS 235, f. 28r.
7. *Anglo-Norman lapidaries* ed. P. Studer and J. Evans (Paris, 1924) pp. 277, 284.
8. *BA*, i, pp. 90–1.

9. C.S. Lewis *The discarded image: an introduction to medieval and Renaissance literature* (Cambridge, 1964) pp. 49, 55.

10. C. Hahn, 'Visio Dei: changes in medieval visuality' in *Visuality before and beyond the Renaissance* ed. R.S. Nelson (Cambridge, 2000) pp. 169–96, at 171.

11. P. Dronke, 'Les Cinq Sens chez Bernard Silvestre et Alain de Lille' *Micrologus* 10 (2002) pp. 1–14, at 6–7.

12. Frank, *Memory of the eyes*, pp. 137, 171, 174, 176–80.

13. A. Paravicini Bagliani, 'En Guise d'introduction: le pouvoir pontifical a-t-il besoin des cinq sens?' *Micrologus* 10 (2002) pp. ix–xiv, at xi–xii.

14. *WN*, pp. 31–3.

15. *Etymologies*, ii: lib. XI.i, 36; reiterated in *BA*, i, p. 178.

16. A.J.M. Edwards, 'An early twelfth century account of the translation of St Milburga of Much Wenlock' *Transactions of the Shropshire Archaeological Society* 57 (1962) pp. 134–51, at 150.

17. *The works of John Metham including the Romance of Amoryus and Cleopes* ... ed. H. Craig (EETS, OS 132; 1916) pp. 124–6; compare also *BA*, i, pp. 182–3, and J.A. Burrow *Gestures and looks in medieval narrative* (Cambridge, 2002) p. 86.

18. M. Pastoureau, 'Le Bestiaire des cinq sens (xii^c–xvie siècle)' *Micrologus* 10 (2002) pp. 133–45, at 138–9; *BA*, i, p. 602; E.C. Rouse and A. Baker, 'The wall-paintings at Longthorpe Tower near Peterborough, Northants' *Archaeologia* 96 (1955) pp. 1–57, at 45; G. Casagrande and C. Kleinhenz, 'Literary and philosophical perspectives on the wheel of the five senses in Longthope Tower' *Traditio* 41 (1985) pp. 311–27, at 312–17.

19. E.g. *Select cases from the coroners' rolls* AD *1265–1413* ed. C. Gross (Selden Society, 9; 1896) p. 10.

20. *Etymologies*, i: lib. V.xxxi, 1.

21. 'DH', p. 22.

22. E.g. *BA*, i, p. 614.

23. Walter Map *De nugis curialium. Courtiers' trifles* ed. M.R. James, revised by C.N.L. Brooke and R.A.B. Mynors (Oxford, 1983) pp. 12–13.

24. *PSQ*, pp. xi, 228–9.

25. *BA*, ii, p. 1228.

26. Lewis, *Discarded image*, pp. 111–12.

27. S. Wenzel, 'Robert Grosseteste's treatise on confession, "Deus est"' *Franciscan Studies* 30 (1970) pp. 218–93, at 220, 232–3, 239–40.

28. K.H. Tachau *Vision and certitude in the age of Ockham: optics, epistemology and the foundations of semantics* (Leiden, 1988) pp. 1–150.

29. *BA*, i, pp. 505–16.

30. *BA*, ii, p. 834.

31. B. Nilson *Cathedral shrines of medieval England* (Woodbridge, 1998) pp. 36–8.

32. F. Lachaud, 'Vêtement et pouvoir à la cour d'Angleterre sous Philippa de Hainaut' in *Au cloître et dans le monde: femmes, hommes et sociétés (IX^c– XVe siècle). Mélanges en l'honneur de Paulette l'Hermite-Leclercq* ed. P. Henriet and A.-M. Legras (Paris, 2000) pp. 217–33, at 219.

33. Hinton, *Gold and gilt*, p. 162.

34. E.g. *La Estoire de Seint Aedward le rei attributed to Matthew Paris* ed. K. Young Wallace (ANTS, 41; 1983) pp. 8, 11.

35. E.g. *Inventory of Henry VIII*, 13179–84, 15351–3.

36. Hinton, *Gold and gilt*, pp. 211–12; D.G. Denery II *Seeing and being seen in the later medieval world: optics, theology and religious life* (Cambridge, 2005) pp. 75–6.

37. Ö. Södergård, 'Un Art d'aimer anglo-normand' *Romania* 77 (1956) pp. 289–330, at 301, 316, 319–20.

38. B.A. Hanawalt *The ties that bound: peasant families in medieval England* (Oxford, 1986) p. 62.

39. Iac 1: 22–3; *EWS*, i, pp. 588–9; v, pp. 55–6; Denery, *Seeing and being seen*, pp. 103–4.

40. *N-town*, i, pp. 157–8.

41. *Mirk's Festial* ed. T. Erbe (EETS, ES 96; 1905) p. 264.
42. *EWS*, iii, pp. 10–12.
43. M. Twycross and S. Carpenter, 'Masks in the medieval English theatre' *Medieval English Theatre* 3 (1981) pp. 7–44, 69–113.
44. *Lollard sermons* ed. G. Cigman (EETS, OS 294; 1989) p. 240.
45. *CS 1205–1313*, i, p. 269, 'semper lumine precedente, cum sit candor lucis eterne'; more generally, pp. 28–9, 126, 170–1, 211, etc.
46. *CS 1205–1313*, i, pp. 404, 592–3.
47. E.g. *MCL*, pp. 42–9.
48. *LEJ*, p. 11.
49. *CS 1205–1313*, i, p. 150.
50. *Jacob's Well*, pp. 63–4.
51. *Swithun*, pp. 684–6.
52. *CS 1205–1313*, i, p. 512.
53. Salisbury Cathedral MS 103, ff. 163vb–164ra.
54. *The book of St Gilbert* ed. R. Foreville and G. Keir (Oxford, 1987) pp. 305–6.
55. E.g. *MCL*, pp. 184–5.
56. BL Add. MS 21480, f. 20r.
57. E.g. *Erkenwald*, pp. 94–6, 212 n. 37; BAV MS Lat. Vat. 4015, ff. 213r–213v.
58. *Osmund*, p. 40.
59. *CS 1205–1313*, i, p. 174.
60. M.R. James, 'Twelve medieval ghost-stories' *EHR* 37 (1922) pp. 413–22, at 415.
61. BAV MS Lat. Vat. 4015, f. 158r.
62. *EWS*, ii, pp. 140–1.
63. *Mirk's Festial*, pp. 24–5.
64. *Becket*, i, p. 237.
65. *CS 1205–1313*, i, p. 174.
66. *CS 1205–1313*, i, p. 273.
67. *LEJ*, pp. 335–6.
68. PRO E101/375/9, f. 32r; E36/205, f. 9r; E101/392/12, f. 44r; *GH*, p. 188.
69. BL MS Cotton Galba E III, f. 178v.
70. A.R. Myers, 'The captivity of a royal witch: the household accounts of Queen Joan of Navarre, 1419–21' in A.R. Myers *Crown, household and Parliament in fifteenth century England* (1985) pp. 93–133, at 119.
71. Household of Henry VII, PRO E36/214, f. 61v.
72. BAV MS Lat. Vat. 4015, ff. 188r–203v.
73. *HO*, p. 91.
74. *HO*, p. 90.
75. *BA*, i, pp. 330–1.
76. *HBQI*, p. 46.
77. *RWH*, 1285–6, pp. 206–41.
78. *Barnwell*, pp. 164–8.
79. *Barnwell*, p. 154.
80. *Barnwell*, p. 204.
81. *PSQ*, p. 236.
82. *CS 1205–1313*, i, p. 171; ii, p. 1006.
83. *LEJ*, p. 112.
84. JRULM Latin MS 234, f. 8v.
85. Hanawalt, *Ties that bound*, pp. 29–30, 51; *Inquests and indictments from late fourteenth century Buckinghamshire: the superior eyre of Michaelmas 1389 at High Wycombe* ed. L. Boatwright (Buckinghamshire Record Society, 29; 1994) p. 52; *Calendar of coroners' rolls of the city of London AD 1300–1378* ed. R.R. Sharpe (1913) pp. 170–1, 183.

86. A.G. Vince, 'The Saxon and medieval pottery of London: a review' *Medieval Archaeology* 29 (1985) pp. 25–93, at 32.

87. M. Pastoureau, 'L'Église et la couleur des origines à la Réforme' *Bibliothèque de l'École des Chartes* 147 (1989) pp. 203–30, at 205–10; M. Pastoureau *Blue: the history of a color* (Princeton, 2001) pp. 42–5.

88. *Ailred,* p. 69.

89. P. Dronke, 'Tradition and innovation in medieval western colour imagery' in *The realms of colour* ed. A. Portmann and R. Ritsema (Eranos 1972 Yearbook, 41: Leiden, 1974) pp. 51–107, at 66–9, 72–83; R. Marks *Stained glass in England during the Middle Ages* (1993) p. 59.

90. For a parallel view of Antiquity, C. Rowe, 'Conceptions of colour and colour symbolism in the Ancient World' in *The realms of colour*, pp. 327–64, at 327–9.

91. S.Y. Edgerton, Jr., 'Alberti's colour theory: a medieval bottle without Renaissance wine' *JWCI* 32 (1969) pp. 109–34, at 111, 116–18.

92. *BA*, ii, pp. 1276–7.

93. *BA*, ii, pp. 1268–9.

94. *BA*, i, p. 136.

95. *BA*, ii, pp. 1289–90.

96. *BA*, ii, pp. 1279–81.

97. *OED* green *a.*[3]

98. William Langland *Piers Plowman: a facsimile of Bodleian Library, Oxford, MS Douce 104* ed. D. Pearsall and K. Scott (Cambridge, 1992) pp. xlv–xlvii.

99. *OED* livid *a.c.*

100. *PSQ*, pp. 43, 134, 246.

101. T.D. Hill, 'Enide's colored horse and Salernitan color theory: *Erec et Enide*, lines 5268–81' *Romania* 108 (1987) pp. 523–7.

102. J. Gage *Colour and culture: practice and meaning from Antiquity to abstraction* (1993) p. 64; Pastoureau, *Blue*, p. 75.

103. N.F. Barley, 'Old English colour classification: where do matters stand?' *Anglo-Saxon England* 3 (1974) pp. 15–28.

104. P. Toynbee, 'The colour *perse* in Dante and other medieval writers' in P. Toynbee *Dante studies and researches* (1902) pp. 307–14.

105. S. Crane, 'Social aspects of bilingualism in the thirteenth century' in *Thirteenth Century England VI* ed. M. Prestwich, R.H. Britnell and R. Frame (Woodbridge, 1997) p. 110; Walter de Bibbesworth *Le Tretiz* ed. W. Rothwell (ANTS, Plain Texts Series, 6; 1990) pp. 1, 10.

106. S. Skard, 'The use of color in literature: a survey of research' *Proceedings of the American Philosophical Society* 90 (1940) pp. 163–249, at 188–9.

107. F. Curta, 'Colour perception, dyestuffs, and colour terms in twelfth-century French literature' *Medium Aevum* 73 (2004) pp. 43–65, at 43.

108. B. Berlin and P. Kay *Basic color terms: their universality and evolution* (2nd edition, Berkeley, 1991) pp. 1–37.

109. This section is based on *OED*. For orange, Longleat MS Misc. XIII, f. 64r; *Inventory of Henry VIII*, 10669; carnation, *Inventory of Henry VIII*, 10041. Baltic squirrel: E.M. Veale *The English fur trade in the later Middle Ages* (2nd edition, London Record Society, 38; 2003) pp. 223–9.

110. *Etymologies*, ii: lib. XII.i, 45–55; lib. XVIII.xli, 1–3; *BA*, ii, pp. 1186–7.

111. *The wardrobe book of William de Norwell 12 July 1338 to 27 May 1340* ed. M. Lyon, B. Lyon and H.S. Lucas (Brussels, 1983) p. 309.

112. BAV MS Lat. Vat. 4015, f. 158r.

113. J. Bell, 'Aristotle as a source for Leonardo's theory of colour perspective after 1500' *JWCI* 56 (1993) pp. 100–18, at 108.

114. Pastoureau, *Blue*, pp. 72, 74; but see Theophilus *De diversis artibus* ed. C.R. Dodwell (Oxford, 1986) pp. 5–13.

115. *Osmund*, p. 78.
116. 'DH', pp. 60–1.
117. L. Gerschel, 'Couleur et teinture chez divers peuples indo-européens' *Annales ESC* 21 (1966) pp. 608–31; M. Pastoureau, 'Les Teinturiers médiévaux: histoire sociale d'un métier réprouvé' in M. Pastoureau *Une Histoire symbolique du Moyen Âge* (Paris, 2004) pp. 173–95.
118. D.V. Thompson *The materials and techniques of medieval painting* (New York, 1956); H. Howard *Pigments of English medieval wall painting* (2003).
119. D.V. Thompson, Jr., '*Liber de coloribus illuminatorum siue pictorum* from Sloane MS. No. 1754' *Speculum* 1 (1926) pp. 280–307, 448–50, at 294.
120. Howard, *Pigments*, pp. 68–9.
121. Tony Hunt, 'Early Anglo-Norman receipts for colours' *JWCI* 58 (1995) pp. 203–9.
122. Dronke, 'Tradition and innovation', p. 52; M. Pastoureau, 'Du Bleu au noir. Éthiques et pratiques de la couleur à la fin du Moyen Âge' *Médiévales* 14 (1988) pp. 9–21; E.S. Bolman, '*De coloribus*: the meanings of color in Beatus manuscripts' *Gesta* 38 (1999) pp. 22–34; Pastoureau, *Blue*, pp. 49–52, 83, 86.
123. Rowe, 'Conceptions of colour', pp. 355–6.
124. Pastoureau, *Blue*, pp. 18–19.
125. D.H. Verkerk, 'Black servant, black demon: color ideology in the Asburnham Pentateuch' *Journal of Medieval and Early Modern Studies* 31 (2001) pp. 57–77, at 59–60, 63–4.
126. *Vitae patrum. PL* 73–4, i, col. 765; *Jacob's Well*, p. 237.
127. *Ailred*, p. 62 and n. 4.
128. *Hugh*, ii, pp. 206, 218–19.
129. *The historians of the church of York and its archbishops* ed. J. Raine (3 vols, RS, 1879–94) ii, p. 281.
130. E.g. *Becket*, i, pp. 148, 158, 198–9, 232; A.T. Baker, 'Vie anglo-normande de Sainte Foy par Simon de Walsingham' *Romania* 68 (1944–5) pp. 49–84, at 72.
131. *La Vie de Seint Auban: an Anglo-Norman poem of the thirteenth century* ed. A.R. Harden (ANTS, 19; 1968) p. 8.
132. Mc 16: 5; Mt 27: 3; *EWS*, i, p. 431; iv, pp. 302–3.
133. E.g. *Lollard sermons*, pp. 237–8.
134. 'Eynsham', p. 360.
135. *Thurkill*, pp. 10–15.
136. J.-C. Schmitt *Les Revenants: les vivants et les morts dans la société médiévale* (Paris, 1994) p. 224.
137. *Ailred*, p. 10.
138. Ex 25: 4.
139. A.T. Baker, 'La Vie de Saint Edmond archevêque de Cantobéry' *Romania* 55 (1929) pp. 332–81, at 372 and n.
140. 'Eynsham', pp. 299–300.
141. *Becket*, ii, p. 129.
142. E.g. *The life of St Catherine by Clemence of Barking* ed. W. MacBain (ANTS, 18; 1964) pp. 8, 38.
143. Ct 5: 10.
144. *Becket*, ii, p. 27: 'ecce dilectus noster, candidus et rubicundus.'
145. *Hugh*, ii, p. 229.
146. *La Vie Seint Richard evesque de Cycestre by Pierre d'Abernon of Fetcham* ed. D.W. Russell (ANTS, 51; 1995) pp. 81, 85, 138.
147. *VV*, pp. 180–1.
148. *Becket*, i, p. 469.
149. *MK*, pp. 9, 32–5, 211, 273.
150. *Heresy trials in the diocese of Norwich, 1428–31* ed. N.P. Tanner (Camden, 4th series, 20; 1977) p. 75, n. 100; A. Hudson *The premature reformation: Wycliffite texts and Lollard history* (Oxford, 1988) pp. 144–7.

151. *EWS*, iii, p. 84.
152. *Two Wycliffite texts: the sermon of William Taylor 1406; the testimony of William Thorpe 1407* ed. A. Hudson (EETS, OS 301; 1993) p. 73.
153. *Twenty-six political and other poems (including 'Petty Job') from the Oxford MSS Digby 102 and Douce 322* ed. J. Kail (EETS, OS 124; 1904) p. 6.
154. Walter Map, *De nugis curialium*, p. 84.
155. *CS 871–1204*, ii, p. 1070.
156. *CS 871–1204*, ii, p. 676; *CS 1205–1313*, ii, p. 1012.
157. *CS 1205–1313*, i, pp. 63, 151, 306, 565, 602, 646, 710; ii, pp. 754, 1012.
158. *Lollard sermons*, p. 23.
159. T.M. Izbicki, 'Forbidden colors in the regulation of clerical dress from the Fourth Lateran Council (1215) to the time of Nicholas de Cusa (d. 1464)' *Medieval Clothing and Textiles* 1 (2005) pp. 105–14; M. Pastoureau *The Devil's cloth: a history of stripes and striped fabric* (New York, 2001) p. 13; C. Rawcliffe *Medicine & society in later medieval England* (Far Thrupp, 1995) p. 207.
160. T.F. Kirby *Annals of Winchester College from its foundation in the year 1382 to the present time* (1892) pp. 499–500.
161. In general, Pastoureau, *Devil's cloth*, pp. 1–32, 101–8.
162. M. Carlin *Medieval Southwark* (1996) p. 215.
163. *Statutes of the realm* ed. A. Luders, T.E. Tomlins, J. Raithby *et al.* (11 vols, 1810–28) i, pp. 380–2; ii, pp. 399–402, 468–70; iii, pp. 8–9.
164. Hanawalt, *Ties that bound*, p. 62; B.A. Hanawalt *Growing up in medieval London* (Oxford, 1993) p. 37.
165. *Bedfordshire coroners' rolls* ed. R.F. Hunnisett (Bedfordshire Historical Record Society, 41; 1961) p. 57.
166. BAV MS Lat. Vat. 4015, f. 127v.
167. *LEJ*, p. 432.
168. *Towneley*, i, p. 34; ii, p. 450.
169. Gage, *Colour and culture*, pp. 26, 52–3, 56, 58–61.
170. Theophilus, *De diversis artibus*, pp. 15–16.
171. D. Pearsall and E. Salter *Landscapes and seasons of the medieval world* (1973) pp. 25–6, 192.
172. *Three eleventh-century Anglo-Latin saints' lives: Vita S. Birini, Vita et miracula S. Kenelmi and Vita S. Rumwoldi* ed. and trans. R.C. Love (Oxford, 1996) pp. 102–3.
173. *N-town*, i, p. 59.
174. *Etymologies*, ii: lib. XI.vii, 36; *BA*, ii, p. 944.
175. *HAME*, ii, pp. 634–5; F. Piponnier *Costume et vie sociale: la cour d'Anjou xive–xve siècle* (Paris, 1970) pp. 163–94, especially 188–9; F. Piponnier and P. Mane *Dress in the Middle Ages* trans. C. Beamish (New Haven, 1997) pp. 70–4; Pastoureau, *Blue*, p. 99.
176. Gage, *Colour and culture*, pp. 16, 25–6.
177. *Becket*, i, p. 298.
178. *Towneley*, i, p. 100; ii, p. 478.
179. *Book of St Gilbert*, pp. 282–3.
180. *MK*, p. 8.
181. *Hugh*, ii, p. 173.
182. E.g *Inventory of Henry VIII*, 1, 6 (crowns); 9043–4 (chairs); 9218 (celure).
183. *HO*, p. 119.
184. Ex 37: 35; *Becket*, i, p. 366.
185. *Hugh*, i, p. 16.
186. The gown of glory: Sir 15: 5; G. Constable, 'The vision of Gunthelm and other visions attributed to Peter the Venerable' *Revue Bénédictine* 66 (1956) pp. 92–114, at 99, 108–9; *Thurkill*, pp. 34–6. Thurkill's vision was more widely known, e.g. from Matthew Paris *Chronica majora* ed. H.R. Luard (7 vols, RS, 1872–83) ii, pp. 497–511, at 509–10. See also

C. Roussel, 'Le Jeu des formes et des couleurs: observations sur "la beste glatissant"' *Romania* 104 (1983) pp. 49–82.

187. 'DH', p. 61.

188. J. Leclercq *The love of learning and the desire for God: a study of monastic culture* trans. Catharine Misrahi (New York, 1961) pp. 97–8.

189. Pastoureau, 'L'Église et la couleur', pp. 217–22, although in part contradicted by Innocent's text.

190. Innocent III *De sacro altaris mysterio. PL*, 217, cols 799–802.

191. J. Wickham Legg, 'On an inventory of the vestry in Westminster Abbey, taken in 1388' *Archaeologia* 52 part 1 (1890) pp. 195–286, at 205; J. Wickham Legg *Notes on the history of liturgical colours* (1882); P. Johnstone *High fashion in the Church: the place of church vestments in the history of art from the ninth to the nineteenth century* (Leeds, 2002) pp. 21–5.

192. *Inventory of Henry VIII*, 8911–17.

193. Salisbury Cathedral MS 103, ff. 121va–121vb.

194. F. Lachaud, 'Dress and social status in England before the sumptuary laws' in *Heraldry, pageantry and social display in medieval England* ed. P. Coss and M. Keen (Woodbridge, 2002) pp. 105–23.

195. M.S. Giuseppi, 'The wardrobe and household accounts of Bogo de Clare, AD 1284–6' *Archaeologia* 70 (1918–20) pp. 1–56, at 50–1.

196. *The floure and the leafe; the assembly of ladies; the isle of ladies* ed. D. Pearsall (Kalamazoo, 1990) pp. 29, 35, 44.

197. *Macro*, pp. xxx, xxxiv, 114–15, 119; M. Pastoureau, 'L'Homme roux: iconographie médiévale de Judas' in M. Pastoureau *Une Histoire symbolique du Moyen Âge occidental* (Paris, 2004) pp. 197–209.

198. *Macro*, pp. 1, 30, 98.

199. *The late medieval religious plays of Bodleian MSS Digby 133 and E Museo 160* ed. D.C. Baker, J.L. Murphy and L.B. Hall Jr. (EETS, OS 283; 1982) pp. xxx, xxxii, 41.

200. Apc 21: 9–27; Dronke, 'Tradition and innovation', pp. 77–82.

201. G. Kipling *Enter the King: theatre, liturgy, and ritual in the medieval civic triumph* (Oxford, 1998) pp. 36, 144.

202. *EWS*, i, pp. 602–3; v, p. 63.

203. Salisbury Cathedral MS 103, f. 182vb.

204. *Becket*, i, p. 465; *Wulfstan*, p. 160; BAV MS Lat. Vat. 4015, ff. 53r, 54r.

205. *Becket*, ii, p. 54.

206. *Book of St Gilbert*, pp. 286–7.

207. *La Vie Seint Richard*, p. 89.

208. *Wulfstan*, pp. 152–3.

209. *Book of St Gilbert*, p. 289; *Becket*, ii, p. 128; *Wulfstan*, pp. 124–5, 127–8.

210. Dronke, 'Tradition and innovation', p. 69; J. Gage *Colour and meaning: art, science and symbolism* (1999) pp. 24–6; Pastoureau, *Blue*, p. 78; Theophilus, *De diversis artibus*, pp. 14–15.

211. *BA*, i, pp. 580–2.

212. *PSQ*, pp. 235–6.

213. *EWS*, v, p. 64.

214. *The York plays* ed. R. Beadle (1982) p. 285.

215. *VV*, p. 60.

216. *Jacob's Well*, p. 151.

217. *BA*, ii, pp. 1159–61.

218. *PSQ*, pp. 128, 263–4, 330.

219. Rawcliffe, *Medicine & society*, pp. 29–33.

220. *Cyrurgie of Guy de Chauliac*, pp. 430–1.

221. Rawcliffe, *Medicine & society*, pp. 47–8.

222. *BA*, i, pp. 157–9, 195–6, 198–9, 285–7; *Works of John Metham*, pp. 143–5.

223. E.g. *The life of St Catherine* p. 63; K. Reichl, 'An Anglo-Norman legend of Saint Margaret (MS BM. Add. 38664)' *Romania* 96 (1975) pp. 53–66, at 59.

224. 'The squire of low degree' in *Middle English metrical romances* ed. W.H. French and C.B. Hale (2 vols, New York, 1964) ii, pp. 721–55, at 743. 'White as whale's bone' is a not uncommon simile, e.g. *The Harley Lyrics: the Middle English lyrics of Ms. Harley 2253* ed. G.L. Brook (4th edition, Manchester, 1968) p. 40.

225. *N-town*, i, p. 64.

226. *N-town*, i, p. 104.

227. G.E. Trease, 'The spicers and apothecaries of the royal household in the reigns of Henry III, Edward I and Edward II' *Nottingham Medieval Studies* 3 (1959) pp. 19–52, at 22–38.

228. P. Ruelle *L'Ornement des dames (ornatus mulierum): texte anglo-normand du xiiie siècle* (Brussels: Université Libre de Bruxelles, Travaux de la Faculté de Philosophie et Lettres, 36; 1967) p. 56.

229. J.R. Gilleland, 'Eight Anglo-Norman cosmetic recipes: MS. Cambridge, Trinity College 1044' *Romania* 109 (1988) pp. 50–69, at 50; Ruelle, *L'Ornement des dames*, p. 56.

230. Gilleland, 'Eight Anglo-Norman cosmetic recipes', p. 59.

231. Ibid., pp. 56–7. Other sources of rouge: Tony Hunt *Popular medicine in thirteenth-century England: introduction and texts* (Cambridge, 1990) p. 272, number 40.

232. Ruelle, *L'Ornement des dames*, pp. 34–6, 40–2, 46–8, 58–60, 66–8; *Macer*, pp. 141–2.

233. Gilleland, 'Eight Anglo-Norman cosmetic recipes', p. 59; Ruelle, *L'Ornement des dames*, pp. 42–6; Hunt, *Popular medicine*, p. 202; *Lanfrank's 'Science of cirurgie'* ed. R. von Fleischhacker (EETS, OS 102; 1894) pp. 178–80.

234. *The local port book of Southampton for 1439–40* ed. H.S. Cobb (Southampton Records Series, 5; 1961) pp. 76, 81.

235. *HAME*, ii, p. 677; compare also: Myers, 'Captivity of a royal witch', p. 125. Late fourteenth-century evidence for them is more ambiguous: Joan Holland, Duchess of Britanny, on 2 October 1378, at Castle Rising, bought two boxes of *pinonade* and one of *char de coins* for 7s. 10d.; and she bought various other electuaries and confections at the same time: Archives Départementales, Loire-Atlantique, E206/3, m. 2r. One cannot rule out entirely the possibility that they were foodstuffs or for perfuming cloth. For citrinade as a cosmetic, *OED, sub* citrinade.

236. J.L. Flandrin, 'Soins de beauté et recueils de secrets' in *Les Soins de beauté* (Nice, 1987) pp. 13–29; C. Martineau-Genieys, 'Modèles, maquillage et misogynie, à travers les textes littéraires français du Moyen Âge' in *Les Soins de beauté*, pp. 31–50; Paulino Iradiel, 'Cuidar el cuerpo, cuidar la imagen: los paradigmas de la belleza femenina en la Valencia bajomedieval' in *Les Soins de beauté*, pp. 61–86.

237. *PI*, p. 163.

238. M.-T. Lorcin, 'Rides et cheveux gris dans les ouvrages de Roger Bacon' in *Les Soins de beauté*, pp. 253–9; M.-A. Polo de Beaulieu, 'La Condamnation des soins de beauté par les prédicateurs du Moyen Âge (xiiième–xvème siècles)' in *Les Soins de beauté*, pp. 297–309.

239. *VV*, pp. 179–80.

240. *The book of the knight of La Tour-Landry* . . . ed. T. Wright (EETS, OS 33; 1868) pp. 69–70.

241. J. Goering and P.J. Payer, 'The "Summa penitentie fratrum predicatorum": a thirteenth-century confessional formula' *Mediaeval Studies* 55 (1993) pp. 1–50, at 37.

242. L.F. Sandler *Omne bonum: a fourteenth-century encyclopedia of universal knowledge, British Library MSS Royal 6 E VI– 6 E VII* (2 vols, 1996) ii, p. 142.

243. J. Goering and F.A.C. Mantello, 'The "Perambulauit Iudas . . ." (Speculum Confessionis) attributed to Robert Grosseteste' *Revue Bénédictine* 96 (1986) pp. 125–68, at 136–7, 149–50. As the text mentions a prior rather than an abbot, it is more likely to have been prepared for a Cluniac rather than a Benedictine house: I am grateful to Barbara Harvey for this point.

244. *EWS*, ii, pp. 117–19.

245. *Seyntz medicines*, pp. 12, 72.

246. Salisbury Cathedral MS 103, ff. 131ra-131rb. For twinkling and winking, Burrow, *Gestures and looks*, pp. 103–13.
247. *RB*, pp. 200–1.
248. *Barnwell*, pp. 86, 164.
249. E. Mikkers, 'Un "Speculum Novitii" inédit d'Étienne de Salley' *Collectanea Cisterciensis Ordinis Reformatorum* 8 (1946) pp. 17–68, at 49.
250. II Cor 4: 18.
251. Aelred of Rievaulx *Speculum caritatis*, in *Aelredi Rievallensis Opera omnia, 1: Opera ascetica* ed. A. Hoste and C.H. Talbot (CCCM 1; 1971) pp. 99–100.
252. *Lollard sermons*, p. 119.
253. *WN*, p. 54.
254. Burke, *Vision . . . Celestina*, pp. 22, 24, 25; Frank, *Memory of the eyes*, pp. 150–4; Burrow, *Gestures and looks*, pp. 91, 130; Hahn, 'Visio Dei', pp. 169–96.
255. *Eulogium historiarum sive temporis* ed. F.S. Haydon (3 vols, RS, 1858–63) iii, p. 378; N. Saul *Richard II* (New Haven, 1997) pp. 338–45, 391.
256. *The Westminster Chronicle 1381–1394* ed. L.C. Hector and B.F. Harvey (Oxford, 1982) pp. 450–1.
257. *English mediaeval lapidaries* ed. J. Evans and M.S. Serjeantson (EETS, OS 190; 1933) p. 20.
258. *Early English versions of the Gesta Romanorum*, pp. 1–2.
259. H.H. Lauer, 'Böser Blick' *Lexikon des Mittelalters* (10 vols, Munich, 1981–3) ii, cols 470–2.
260. Sap 4: 12.
261. Frank, *Memory of the eyes*, pp. 129–31; Burke, *Vision . . . Celestina*, pp. 63–5, 73, 75.
262. Hinton, *Gold and gilt*, pp. 159–60, 327 n. 118.
263. *Barnwell*, p. 196.
264. *Barnwell*, p. 204.
265. *MCL*, p. 166.
266. 'Liber de sinthomatibus mulierum' in *Anglo-Norman medicine* ed. Tony Hunt (2 vols, Cambridge, 1994–7) i, pp. 116–28, at 127; *The Trotula: a medieval compendium of women's medicine* ed. M.H. Green (Philadelphia, 2001) pp. 101–2, 236, 250.
267. Denery, *Seeing and being seen*, pp. 53–4.
268. W. St J. Hope, 'The Sarum consuetudinary and its relation to the cathedral church of Old Sarum' *Archaeologia* 68 (1916–17) pp. 111–26, at 117–18.
269. *MCL*, pp. 28–41.
270. C.J. Holdsworth, 'Eleven visions connected with the Cistercian monastery of Stratford Langthorne' *Cîteaux* 13 (1962) pp. 185–204, at 196–7.
271. *Bedfordshire coroners' rolls*, pp. x–xi.
272. *LEJ*, p. 88.
273. *Henry VI*, pp. 98–100, 188; for pilgrim dress, B.W. Spencer, 'Medieval pilgrim badges: some general observations illustrated mainly from English sources' in *Rotterdam papers: a contribution to medieval archaeology* ed. J.G.N. Renaud (Rotterdam, 1968) pp. 137–53, at 144.
274. M. Prestwich *Armies and warfare in the Middle Ages: the English experience* (New Haven, 1996) pp. 18, 220–4.
275. G.J. Brault *Early blazon: heraldic terminology in the twelfth and thirteenth centuries with special reference to Arthurian literature* (Oxford, 1972) pp. 5–8, 24.
276. A.R. Wagner *Heralds and heraldry in the Middle Ages: an inquiry into the growth of the armorial function of heralds* (1939) p. 18.
277. Ibid., pp. 20–2; *The roll of arms of the princes, barons and knights who attended King Edward I to the siege of Caerlaverock, in 1300 . . .* ed. T. Wright (London, 1864) pp. 15–16; Brault, *Early blazon*, p. 6.
278. *HKW*, i, pp. 130, 247–8.
279. *Inventory of Henry VIII*, 8996.

280. *Trotula*, pp. 106–9, 237; *BA*, i, pp. 298–9.
281. *PSQ*, p. 165.
282. E. Rosen, 'The invention of eyeglasses' *Journal of the History of Medicine and Allied Sciences* 9 (1956) pp. 13–46, 183–218.
283. *MK*, p. 5.
284. *Inventory of Henry VIII*, e.g. 2088, 2285–9, 2291, 2295, 2312–13, 2391, 2530; 2540, 2546 (with books); 2552, 2598; 2612 (backgammon), etc.
285. BAV MS Lat. Vat. 4015, ff. 71v–72v.
286. *MK*, pp. 28–9, 271.
287. Tb 11: 1–17. The father is also called Tobias.
288. *Erkenwald*, pp. 135–9.
289. *Osmund*, pp. 59–60.
290. *Swithun*, pp. 692–4.
291. *Becket*, ii, p. 115.
292. *Becket*, ii, pp. 116, 150.
293. BAV MS Lat. Vat. 4015, ff. 227v–230v.
294. Edwards, 'Translation of St Milburga', pp. 148–50.
295. *Erkenwald*, pp. 140–1.
296. *Becket*, i, p. 243; ii, p. 126.
297. BAV MS Lat. Vat. 4015, ff. 27r–28r.
298. BAV MS Lat. Vat. 4015, ff. 158v-159v, 163r.
299. *HO*, pp. 129–30; H. Farmer, 'The canonization of St Hugh of Lincoln' *Lincolnshire Architectural and Archaeological Society Reports and Papers* 6 (1956) p. 99.
300. Leclercq, *Love of learning*, pp. 93–4; M. Cassidy-Welch *Monastic spaces and their meanings: thirteenth-century English Cistercian monasteries* (Turnhout, 2001) pp. 23–71, 245–9.
301. M.J. Carruthers *The book of memory: a study of memory in medieval culture* (Cambridge, 1990) pp. 12–13.
302. Ibid., p. 150; M. Carruthers *The craft of thought: meditation, rhetoric, and the making of images, 400–1200* (Cambridge, 1998) pp. 7–24, 237–76.
303. Carruthers, *Book of memory*, pp. 130–7, 281–8; B. Rowland, 'Bishop Bradwardine on the artificial memory' *JWCI* 41 (1978) pp. 307–12.
304. Carruthers, *Book of memory*, pp. 8, 94; M. Camille *Mirror in parchment: the Luttrell Psalter and the making of medieval England* (1998) pp. 160–77.
305. D. Renevey, 'Household chores in *The doctrine of the hert*: affective spirituality and subjectivity' in *The medieval household in Christian Europe, c.850–c.1550: managing power, wealth, and the body* ed. C. Beattie, A. Maslakovic and S. Rees Jones (Turnhout, 2003) pp. 167–85.
306. A.I. Doyle, '"Lectulus noster floridus": an allegory of the penitent soul' in *Literature and religion in the later Middle Ages: philological studies in honor of Siegfried Wenzel* ed. R.G. Newhauser and J.A. Alford (Binghamton, 1995) pp. 179–90.
307. J.M. Hamburger, 'The visual and the visionary: the image in late medieval monastic devotions' *Viator* 20 (1989) pp. 161–83, at 180–2.
308. *Medieval English prose for women: selections from the Katherine Group and Ancrene Wisse* ed. B. Millett and J. Wogan-Browne (Oxford, 1990) pp. xxvi-xxvii, 86–109; C. Whitehead *Castles of the mind: a study of medieval architectural allegory* (Cardiff, 2003) pp. 119–20. The work is based on a Latin dialogue of about a century earlier, 'De custodia interioris hominis', *Memorials of St Anselm* ed. R.W. Southern and F.S. Schmitt (ABMA, 1; 1969) pp. 354–60.
309. *Seyntz medicines*, p. 115.

CHAPTER 9: SENSORY ENVIRONMENTS AND EPISCOPAL HOUSEHOLDS OF THE THIRTEENTH AND EARLY FOURTEENTH CENTURIES

1. *Lollard sermons* ed. G. Cigman (EETS, OS 294; 1989) pp. 1–2, 23, 137–8; Owst, *Literature*, pp. 282–4, 293–4.
2. *Mirk's Festial* ed. T. Erbe (EETS, ES 96; 1905) pp. 39–40, 197.
3. Corpus Christi College, Oxford, MS 154, pp. 375–85, printed in *Edmund*, pp. 187–202; *Oxford Dictionary of National Biography*, ed. H.C.G. Matthew and B. Harrison (61 vols, Oxford, 2004), *sub* Edmund of Abingdon.
4. *Edmund*, pp. 187, 189–91, 195.
5. *Edmund*, pp. 248–53.
6. *Edmund*, p. 187; for the meaning of *pulmentum* at this date in monastic circles, see B.F. Harvey, 'Monastic pittances in the Middle Ages' in *Food in medieval England: diet and nutrition* ed. C.M. Woolgar, D. Serjeantson and T. Waldron (Oxford, 2006) pp. 215–27, at 215–17.
7. *Edmund*, p. 192.
8. *Edmund*, p. 187.
9. *Edmund*, p. 192.
10. *Edmund*, p. 187.
11. *Edmund*, pp. 190, 249.
12. *Edmund*, p. 192.
13. *Edmund*, p. 193.
14. *Edmund*, pp. 192, 249.
15. *Edmund*, pp. 187–8, 190.
16. *Edmund*, pp. 191–2.
17. *Edmund*, pp. 248–9.
18. *Edmund*, p. 250.
19. *Edmund*, pp. 187–8, 190, 192–3, 197. Matins were the equivalent of the modern lauds: see, for example, *MCL*, pp. xx–xxiii.
20. *Edmund*, pp. 188, 190.
21. *Edmund*, pp. 188, 190–1.
22. *Edmund*, p. 191.
23. *Edmund*, p. 252.
24. *Edmund*, pp. 191–2.
25. *Edmund*, p. 192.
26. *Edmund*, p. 198.
27. *Edmund*, p. 199.
28. *Edmund*, p. 201.
29. *Edmund*, p. 188.
30. *Edmund*, p. 249.
31. *Edmund*, p. 188.
32. *Edmund*, p. 189.
33. *Walter of Henley and other treatises on estate management and accounting* ed. D. Oschinsky (Oxford, 1971) pp. 408–9; J. Goering and F.A.C. Mantello, '"Notus in Iudea Deus": Robert Grosseteste's confessional formula in Lambeth Palace MS 499' *Viator* 18 (1987) pp. 253–73.
34. Goering and Mantello, 'Notus in Iudea Deus', pp. 266–7, 271–2.
35. N.D.S. Martin, 'The life of St Thomas of Hereford' in *St Thomas Cantilupe Bishop of Hereford: essays in his honour* ed. M. Jancey (Hereford, 1982) pp. 15–19; D. Carpenter, 'St Thomas Cantilupe: his political career' in *St Thomas Cantilupe*, pp. 57–72; *Oxford Dictionary of National Biography*; for relative levels of income, B.F. Harvey, 'The aristocratic consumer in England in the long thirteenth century' in *Thirteenth-Century England VI* ed. M. Prestwich, R.H. Britnell and R. Frame (Woodbridge, 1997) pp. 17–37.

36. BAV MS Lat. Vat. 4015, ff. 18r, 18v, 32v, 56v, 93r.
37. R.C. Finucane, 'The Cantilupe–Pecham controversy' in *St Thomas Cantilupe*, pp. 103–23, at 121–2.
38. BAV MS Lat. Vat. 4015, f. 56v.
39. BAV MS Lat. Vat. 4015, ff. 18v, 19v.
40. BAV MS Lat. Vat. 4015, ff. 24v, 45r, 58r, 60r.
41. *Matutine* in the text: see note 19, above. BAV MS Lat. Vat. 4015, ff. 34r–v.
42. BAV MS Lat. Vat. 4015, f. 79r.
43. BAV MS Lat. Vat. 4015, f. 6r.
44. BAV MS Lat. Vat. 4015, ff. 35v, 59r–v.
45. BAV MS Lat. Vat. 4015, f. 58v.
46. 1 Tim 3: 1–7; BAV MS Lat. Vat. 4015, f. 78v.
47. BAV MS Lat. Vat. 4015, f. 37r.
48. BAV MS Lat. Vat. 4015, f. 60r.
49. BAV MS Lat. Vat. 4015, ff. 24r–v.
50. BAV MS Lat. Vat. 4015, f. 45r.
51. BAV MS Lat. Vat. 4015, f. 60r.
52. BAV MS Lat. Vat. 4015, f. 109v.
53. BAV MS Lat. Vat. 4015, f. 32v.
54. BAV MS Lat. Vat. 4015, f. 40v.
55. BAV MS Lat. Vat. 4015, f. 61r.
56. BAV MS Lat. Vat. 4015, f. 83r.
57. BAV MS Lat. Vat. 4015, f. 6r.
58. BAV MS Lat. Vat. 4015, ff. 15r–v.
59. BAV MS Lat. Vat. 4015, f. 31r.
60. BAV MS Lat. Vat. 4015, ff. 20r–v.
61. BAV MS Lat. Vat. 4015, ff. 20v–21r.
62. BAV MS Lat. Vat. 4015, ff. 34v–35r.
63. BAV MS Lat. Vat. 4015, ff. 49v–50r.
64. BAV MS Lat. Vat. 4015, f. 57v.
65. BAV MS Lat. Vat. 4015, ff. 95v–96r.
66. BAV MS Lat. Vat. 4015, f. 107v.
67. BAV MS Lat. Vat. 4015, f. 108r.
68. BAV MS Lat. Vat. 4015, f. 21r.
69. BAV MS Lat. Vat. 4015, f. 23v.
70. BAV MS Lat. Vat. 4015, f. 36v.
71. BAV MS Lat. Vat. 4015, ff. 88r–v.
72. BAV MS Lat. Vat. 4015, f. 98v.
73. BAV MS Lat. Vat. 4015, ff. 24v–25r.
74. BAV MS Lat. Vat. 4015, f. 60v.
75. BAV MS Lat. Vat. 4015, f. 20r.
76. BAV MS Lat. Vat. 4015, ff. 57v–58r.
77. BAV MS Lat. Vat. 4015, ff. 103r–v.
78. BAV MS Lat. Vat. 4015, f. 108r.
79. BAV MS Lat. Vat. 4015, ff. 34v, 108r.
80. BAV MS Lat. Vat. 4015, ff. 96r–v.
81. BAV MS Lat. Vat. 4015, ff. 5v, 15r, 30v.
82. BAV MS Lat. Vat. 4015, f. 18v.
83. BAV MS Lat. Vat. 4015, f. 6r.
84. BAV MS Lat. Vat. 4015, ff. 19r, 30v.
85. BAV MS Lat. Vat. 4015, f. 19v.
86. BAV MS Lat. Vat. 4015, f. 31v.
87. BAV MS Lat. Vat. 4015, f. 82r.

88. BAV MS Lat. Vat. 4015, f. 15v.
89. C.M. Woolgar, 'Group diets in late medieval England' in *Food in medieval England*, pp. 191–200, at 191–6.
90. BAV MS Lat. Vat. 4015, ff. 5v–6r, 16r, 34r, 56v–57r.
91. BAV MS Lat. Vat. 4015, f. 94r.
92. BAV MS Lat. Vat. 4015, f. 15v.
93. BAV MS Lat. Vat. 4015, ff. 23v–24r, 30v–31r.
94. BAV MS Lat. Vat. 4015, ff. 49v, 108r.
95. BAV MS Lat. Vat. 4015, f. 59v.
96. BAV MS Lat. Vat. 4015, f. 103v.
97. BAV MS Lat. Vat. 4015, f. 24r.
98. BAV MS Lat. Vat. 4015, f. 57r.
99. *GH*, pp. 90–2.
100. BAV MS Lat. Vat. 4015, ff. 19v, 34r, 57r, 103v, 108r.
101. BAV MS Lat. Vat. 4015, ff. 114r–v.
102. See p. 163, above. BAV MS Lat. Vat. 4015, f. 49v.
103. BAV MS Lat. Vat. 4015, f. 23r.
104. BAV MS Lat. Vat. 4015, f. 50r.
105. BAV MS Lat. Vat. 4015, f. 114r.
106. BAV MS Lat. Vat. 4015, f. 44v.
107. BAV MS Lat. Vat. 4015, f. 59r.
108. BAV MS Lat. Vat. 4015, ff. 37v, 61r, 97v, 109r.
109. BAV MS Lat. Vat. 4015, f. 15r.
110. BAV MS Lat. Vat. 4015, f. 23v.
111. BAV MS Lat. Vat. 4015, f. 59v.
112. BAV MS Lat. Vat. 4015, f. 24r.
113. BAV MS Lat. Vat. 4015, f. 43v.
114. BAV MS Lat. Vat. 4015, f. 50r.
115. BAV MS Lat. Vat. 4015, ff. 96v, 108v–109r.
116. BAV MS Lat. Vat. 4015, ff. 31v–32r, 59v–60r.
117. BAV MS Lat. Vat. 4015, f. 25v.
118. BAV MS Lat. Vat. 4015, f. 88r.
119. BAV MS Lat. Vat. 4015, ff. 104v–105r.
120. BAV MS Lat. Vat. 4015, f. 6v.
121. BAV MS Lat. Vat. 4015, ff. 36v, 59r.
122. BAV MS Lat. Vat. 4015, f. 25v.
123. BAV MS Lat. Vat. 4015, ff. 78v–79r.
124. BAV MS Lat. Vat. 4015, ff. 26r–v.
125. BAV MS Lat. Vat. 4015, f. 36v.
126. BAV MS Lat. Vat. 4015, f. 25v.
127. BAV MS Lat. Vat. 4015, f. 36v.
128. BAV MS Lat. Vat. 4015, f. 18r.
129. BAV MS Lat. Vat. 4015, f. 33r.
130. BAV MS Lat. Vat. 4015, f. 56r; *A roll of the household expenses of Richard de Swinfield, Bishop of Hereford, during part of the years 1289 and 1290* ed. J. Webb (Camden Society, OS 59, 62; 1854–5) i, pp. 166, 194; ii, p. xxxi.
131. BAV MS Lat. Vat. 4015, f. 99r; *Household expenses of . . . Swinfield*, ii, pp. lxxxiii–lxxxvi.
132. BAV MS Lat. Vat. 4015, ff. 107v, 110r–110v; *Household expenses of . . . Swinfield*, i, pp. 168, 194.
133. BAV MS Lat. Vat. 4015, ff. 102r–v; *Household expenses of . . . Swinfield*, i, pp. 146–7.
134. *Household expenses of . . . Swinfield*, ii, p. xxxi.
135. C.E. Woodruff, 'The will of Peter de Aqua Blanca Bishop of Hereford (1268)' in *Camden Miscellany XIV* (Camden, 3rd series, 37; 1926) pp. v–vi, 4–8.

136. Finucane, 'Cantilupe–Pecham controversy', pp. 118–19.
137. BAV MS Lat. Vat. 4015, ff. 19r, 37v, 113r.
138. BAV MS Lat. Vat. 4015, f. 113r.
139. *Registrum Thome de Cantilupo episcopi Herefordensis, AD MCCLXXV–MCCLXXXII* ed. R.G. Griffiths and W.W. Capes (Canterbury and York Society, 2; 1907) pp. xli–xlii.
140. *Registrum Ricardi de Swinfield, episcopi Herefordensis, AD MCCLXXXIII–MCCCXVII* ed. W.W. Capes (Canterbury and York Society, 6; 1909) p. iv.
141. J. Blair, 'The 12th-century Bishop's Palace at Hereford' *Medieval Archaeology* 31 (1987) pp. 59–72; for its date and evidence for the other residences, A. Emery *Greater medieval houses of England and Wales 1300–1500* (3 vols, Cambridge, 1996–2006) ii, pp. 512–15.
142. Aelred of Rievaulx *Speculum caritatis*, in *Aelredi Rievallensis omnia opera, 1: Opera ascetica*, ed. A. Hoste and C.H. Talbot (CCCM 1; 1971) pp. 99–100.
143. Hugh of Fouilloy *De claustro anime. PL*, 176, cols 1019–20. On this work generally, C. Whitehead *Castles of the mind: a study of medieval architectural allegory* (Cardiff, 2003) pp. 62–7, 278.
144. M. Thompson *Medieval bishops' houses in England and Wales* (Aldershot, 1998) pp. 14–103, 125–62; also sections on individual sees in Emery, *Greater medieval houses*, i, pp. 51–4, 298–303; ii, 89–90, 131–4, 269–72, 463–6, 512–15, 641–9; J. Schofield *Medieval London houses* (New Haven, 1994) pp. 34–51.
145. Henry of Huntingdon *Historia Anglorum* ed. T. Arnold (RS, 1879) p. 265.
146. *HAME*, i, pp. 261–430; N.D. McGlashan and R.E. Sandell, 'The Bishop of Salisbury's house at his manor of Potterne' *Wiltshire Archaeological and Natural History Magazine* 69 (1974) pp. 85–96.
147. H. Brakspear, 'The Bishop's Palace, Sonning' *Berks., Bucks., and Oxon. Archaeological Journal* 22 (1916) pp. 7–21; Thompson, *Medieval bishops' houses*, pp. 129–30.
148. M.M. Reeve, 'The former painted cycle of the life of Edward I at the Bishop's Palace, Lichfield' *Nottingham Medieval Studies* 46 (2002) pp. 70–83.
149. J.H. Winstone 'The Bishop's Palace at Wookey' *Somerset Archaeology and Natural History* 141 (1998) pp. 91–101.
150. *English episcopal acta: X. Bath and Wells 1061–1205* ed. F.M.R. Ramsey (Oxford, 1995) pp. xxiv–xxv, xxvii, 136–7; *Calendar of Close Rolls 1227–1231*, p. 23.
151. Lambeth Palace Library ED 363, m. 2r.
152. Lambeth Palace Library ED 364, mm. 1r–2r, 2d.
153. J.A. Robinson, 'Household roll of Bishop Ralph of Shrewsbury 1337–8' in *Collectanea I* ed. T.F. Palmer (Somerset Record Society, 39; 1924) pp. 72–174, at 155, 157; PRO E101/396/2, f. 31r.
154. *The receyt of the Ladie Kateryne* ed. G. Kipling (EETS, OS 296; 1990) pp. 7–8, 115.
155. Hants RO 5M53/767, f. 11v.
156. G. Williams *Henry de Gower, bishop and builder* (St David's, n.d. *c.* 1980) p. 11; J.W. Evans and R. Turner *St David's Bishop's Palace; St Non's Chapel* (revised edition, Cardiff, 1999) pp. 14–17.
157. R. Turner *Lamphey Bishop's Palace; Llawhaden Castle; Carswell medieval house; Carew Cross* (2nd edition, Cardiff, 2000) pp. 12–23.
158. *GH*, p. 61.
159. *Oxford Dictionary of National Biography*; BL Add. MS 57334, mm. 1r, 2r; 7, rot. 1d–2d.
160. *Executors*, pp. ii–iii, 1, 45; *Oxford Dictionary of National Biography, sub* William Button.
161. *Executors*, pp. viii, 47; *Oxford Dictionary of National Biography*.
162. *Executors*, pp. 47–99, 138.
163. *Executors*, drawing opposite p. i; p. 1.
164. *Executors*, p. 55; Hinton, *Gold and gilt*, pp. 214–15.
165. BL Add. MS 57334, m. 7, rot 2r: 'cum safiro bono ut creditur virtuoso'.
166. *Executors*, p. 2.
167. BL Add. MS 57334, m. 7, rot. 2r.

168. *Executors*, p. 2.
169. *Executors*, p. 55.
170. BL Add. MS 57334, m. 7, rot. 2r.
171. *Executors*, pp. 2, 30–1.
172. *Executors*, p. 3.
173. *Executors*, p. 53.
174. BL Add. MS 57334, m. 7, rot. 2r.
175. *Executors*, pp. 8–9.
176. BL Add. MS 57334, m. 6d.
177. *Executors*, p. 7.
178. *Executors*, p. 43.
179. *Executors*, p. 11.
180. BL Add. MS 57334, m. 7, rot. 1d.
181. E.g. *Executors*, p. 2.
182. *Executors*, p. 55.
183. *Executors*, pp. 5–6.
184. *Executors*, p. 7.
185. *Executors*, pp. 55–6.
186. BL Add. MS 57334, mm. 6d; 7, rot. 1r–1d.
187. That he had a nephew called Robert de Gosinton is suggestive of this link: *Executors*, pp. 31, 43.
188. *Executors*, pp. 56–7.
189. *Executors*, p. 6.
190. *Executors*, p. 57.
191. BL Add. MS 57334, mm. 6d; 7, rot. 1r.
192. *Executors*, p. 7.
193. *Executors*, pp. 50–2.
194. BL Add. MS 57334, mm. 6d; 7, rot. 1d.
195. *Executors*, p. 6.
196. *Executors*, p. 6.
197. *Executors*, pp. 57–8.
198. BL Add. MS 57334, mm. 6d; 7, rot. 2r.
199. *Executors*, p. 49.
200. BL Add. MS 57334, m. 7, rot. 1r.
201. BL Add. MS 57334, mm. 6d; 7, rot. 1r. For the different types of vestment and their functions, P. Johnstone *High fashion in the Church: the place of church vestments in the history of art from the ninth to the nineteenth century* (Leeds, 2002) pp. 5–20.
202. BL Add. MS 57334, m. 7, rot. 1r.
203. BL Add. MS 57334, mm. 6d; 7, rot. 1r.
204. *Executors*, pp. 4–5.
205. *Executors*, pp. 47–9.
206. *Executors*, pp. 11, 58; BL Add. MS 57334, m. 7, rot. 1d.
207. BL Add. MS 57334, m. 7, rot. 1d, 2r.
208. *Executors*, pp. 12, 43, 59.
209. *The registers of John de Sandale and Rigaud de Asserio, Bishops of Winchester (AD 1316–1323)* ed. F.J. Baigent (Hampshire Record Society, 8; 1897) pp. 624–6.
210. *Registers of John de Sandale . . .*, pp. 622–3.
211. *Executors*, pp. 32–5.
212. *Executors*, pp. 109, 114.

CHAPTER 10: HOUSEHOLDS OF LATE MEDIEVAL QUEENS OF ENGLAND

1. Phil 3: 20.
2. M.D. Legge, 'John Pecham's *Jerarchie*' *Medium Aevum* 11 (1942) pp. 77–84.
3. J.C. Parsons, 'Ritual and symbol in the English medieval queenship to 1500' in *Women and sovereignty* ed. L.O. Fradenburg (Edinburgh, 1992) pp. 60–77, at 66–7, citing T. Borenius, 'The cycle of images in the palaces and castles of Henry III' *JWCI* 6 (1943) pp. 40–50, at 49.
4. E.W. Safford, 'An account of the expenses of Eleanor, sister of Edward III, on the occasion of her marriage to Reynald, Count of Guelders' *Archaeologia* 77 (1927) pp. 111–40; for Joan, N.H. Nicolas, 'Observations on the institution of the most noble Order of the Garter' *Archaeologia* 31 (1846) pp. 1–163; W. Paley Baildon, 'The trousseaux of Princess Philippa, wife of Eric, King of Denmark, Norway, and Sweden' *Archaeologia* 67 (1916) pp. 163–88.
5. *EC*, p. 12; Matthew Paris *Chronica majora* ed. H.R. Luard (7 vols, RS, 1872–83) v, pp. 513–14.
6. One can see how far this change had gone by the 1320s from the privy wardrobe account of John de Flete, listing materials he held in the Tower: BL Add. MS 60584, ff. 4r–5r.
7. H. Johnstone, 'The Queen's household' in T.F. Tout *Chapters in the administrative history of mediaeval England: the wardrobe, the chamber and the small seals* (6 vols, Manchester, 1920–33) v, pp. 231–89, at 264–84.
8. PRO E101/375/9, f. 2r.
9. PRO E101/393/5, ff. 9r, 11r, 13r.
10. Society of Antiquaries MS 216, f. 17v.
11. A.R. Myers, 'The household of Queen Margaret of Anjou, 1452–3' in A.R. Myers *Crown, household and Parliament in fifteenth century England* (1985) pp. 135–209, at 137–42, 181.
12. A.R. Myers, 'The household of Queen Elizabeth Woodville, 1466–7' in Myers, *Crown, household and Parliament*, pp. 251–318, at 252–5.
13. *HKW*, ii, pp. 671–2.
14. T.H. Turner, 'Liberationes factae per executores dominae Alianorae consortis Edwardi regis Angliae primi' in *Manners and household expenses of England in the thirteenth and fifteenth centuries* ed. T.H. Turner (Roxburghe Club, 1841) pp. 93–145, at 106, 117.
15. Ibid., pp. 97, 102.
16. *RWH 1286–9*, p. 372, no. 3154.
17. *HKW*, ii, pp. 695–9; *EC*, p. 70; Turner, 'Liberationes facte per executores', p. 102.
18. E.g. PRO E36/205, ff. 2r, 6r, 9r.
19. *EC*, p. 104; *HKW*, ii, pp. 970–7.
20. *HKW*, ii, pp. 730–1.
21. H. Johnstone, 'The wardrobe and household of Henry, son of Edward I' *Bulletin of the John Rylands Library* 7 (1922–3) pp. 384–420, at 406–7.
22. BL Add. MS 37656, f. 1r.
23. BL Add. MS 32050, ff. 5r–6r.
24. *HKW*, ii, p. 731; P. Ellis *et al.*, *Ludgershall Castle Wiltshire: a report on the excavations by Peter Addyman, 1964–1972* (Wiltshire Archaeological and Natural History Society Monograph Series, 2; 2000) pp. 17–18, 152–3, 236; microfiche 1:D2, 1:E11.
25. *HKW*, ii, p. 767.
26. *HKW*, i, p. 76; ii, pp. 766–8; PRO E101/478/26, mm. 1r–4r; *EC*, p. 79.
27. *HKW*, ii, pp. 677–81; PRO E101/393/5, f. 2r; BL MS Cotton Galba E XIV, ff. 6r, 8r, 13r, 15r, 16r, 24r; E.A. Bond, 'Notices of the last days of Isabella, Queen of Edward the Second, drawn from an account of her household' *Archaeologia* 35 (1853–4) pp. 453–69.
28. PRO E101/393/5, ff. 2v, 5r; B. Morley and D. Gurney *Castle Rising Castle, Norfolk* (East Anglian Archaeology Report, 81; Dereham, 1997) pp. 56–82, 133–8.
29. A. Richardson, 'Gender and space in English royal palaces *c.* 1160–*c.* 1547: a study in access analysis and imagery' *Medieval Archaeology* 47 (2003) pp. 131–65.
30. *HKW*, ii, p. 1041.

31. PRO E101/92/23; also J.C. Ward *English noblewomen in the later Middle Ages* (Harlow, 1992) p. 52; *HAME*, ii, pp. 651–5.

32. *GH*, pp. 34–6.

33. M.A. Michael, 'A manuscript wedding gift from Philippa of Hainault to Edward III' *Burlington Magazine* 127 (1985) pp. 582–9, at 582.

34. Johnstone, 'The Queen's household', pp. 231–89.

35. P. Strohm *Hochon's arrow: the social imagination of fourteenth-century texts* (Princeton, 1992) pp. 95–7; G. Kipling *Enter the King: theatre, liturgy, and ritual in the medieval civic triumph* (Oxford, 1998) pp. 183, 191–201, 289–316, 342–5; Parsons, 'Ritual and symbol in the English medieval queenship', p. 60.

36. JRULM Latin MS 234, ff. 13v–18r.

37. JRULM Latin MS 234, ff. 29v–31v.

38. JRULM Latin MS 234, ff. 2v–3r.

39. F. Lachaud, 'Vêtement et pouvoir à la cour d'Angleterre sous Philippa de Hainaut' in *Au Cloître et dans le monde: femmes, hommes et sociétés (IXᵉ–XVᵉ siècle). Mélanges en l'honneur de Paulette L'Hermite-Leclercq* ed. P. Henriet and A.M. Legras (Paris, 2000) pp. 217–33, at 231–2.

40. Myers, 'Household of Margaret of Anjou', pp. 181–9.

41. G.L. Harriss, 'The court of the Lancastrian kings' in *The Lancastrian court* ed. J. Stratford (Harlaxton Medieval Studies, 13; 2003) pp. 1–18, at 8.

42. Myers, 'Household of Margaret of Anjou', p. 197; A.R. Myers, 'The jewels of Queen Margaret of Anjou' in Myers, *Crown, household and Parliament*, pp. 211–29, at 228.

43. Society of Antiquaries MS 208, f. 7v; JRULM Latin MS 236, f. 6v.

44. *The coronation of Richard III: the extant documents* ed. A.F. Sutton and P.W. Hammond (Gloucester, 1983) pp. 101, 160–2, 164.

45. S.M. Newton *Fashion in the age of the Black Prince. A study of the years 1340–1365* (Woodbridge, 1980) pp. 14–18, dating the changes to between 1333 and 1338, running into the 1340s, based on the accounts of the King's wardrobe; J. Scattergood, 'Fashion and morality in the late Middle Ages' in *England in the fifteenth century: proceedings of the 1986 Harlaxton symposium* ed. D. Williams (Woodbridge, 1987) pp. 255–72, at 258–9.

46. Lachaud, 'Vêtement et pouvoir', pp. 221–2.

47. Ibid., p. 222.

48. E.g. JRULM Latin MS 234, f. 6r; see also *RWH 1286–9*, p. 393, no. 3238.

49. BL MS Cotton Galba E III, ff. 176v–177r; JRULM Latin MS 234, ff. 7r–8r.

50. JRULM Latin MS 234, ff. 7r–8r, 19v.

51. JRULM Latin MS 234, ff.2r-3v. Compare also the purchases of Queen Isabella: PRO E101/375/9, ff. 22v, 24r.

52. JRULM Latin MS 234, ff. 1r, 13r, 19r–19v.

53. JRULM Latin MS 234, ff. 2r, 13r.

54. JRULM Latin MS 234, ff. 1r, 13v.

55. JRULM Latin MS 234, ff. 1r, 14r–14v.

56. JRULM Latin MS 234, ff. 19v, 32r–32v.

57. JRULM Latin MS 234, f. 8r; BL MS Cotton Galba E III, f. 177r.

58. JRULM Latin MS 234, ff. 5v, 33r; BL MS Cotton Galba E III, f. 175r.

59. PRO E101/376/7, f. 11r.

60. JRULM Latin MS 234, f. 2r.

61. Lachaud, 'Vêtement et pouvoir', pp. 217–33, especially p. 223. Seamstress of the chamber: Society of Antiquaries MS 208, f. 9r.

62. JRULM Latin MS 235, ff. 6r, 12v. Compare *HBQI*, p. 148.

63. JRULM Latin MS 235, f. 18r.

64. Paley Baildon, 'Trousseaux of Princess Philippa', p. 183.

65. Similar displays, probably with mumming, took place within the King's household: e.g. K. Staniland, 'Clothing and textiles at the court of Edward III 1342–1352' in *Collectanea Londiniensia. Studies in London archaeology and history presented to Ralph Merrifield* ed.

J. Bird, H. Chapman and J. Clark (London and Middlesex Archaeological Society, special paper, 2; 1978) pp. 223–34, at 228–9. For the Queen's household: JRULM Latin MS 234, ff. 8v–9r. Compare the costs of a disguising for Margaret of Anjou and Henry VI: Myers, 'Household of Margaret of Anjou', p. 200.

66. PRO E101/375/9, f. 19r.
67. F. Lachaud, 'Embroidery for the court of Edward I (1272–1307)' *Nottingham Medieval Studies* 37 (1993) pp. 33–52, at 36–7, 41–33, 46.
68. Paley Baildon, 'Trousseaux of Princess Philippa', pp. 171, 173.
69. Ibid., p. 174.
70. A.R. Myers, 'The captivity of a royal witch: the household accounts of Queen Joan of Navarre, 1419–21' in Myers, *Crown, household and Parliament*, pp. 93–133, at 120–1, 131; Society of Antiquaries MS 216, ff. 51r–v.
71. PRO E101/393/4, ff. 10v–11r.
72. PRO E101/393/4, f. 4v.
73. *RWH 1286–9*, p. 384, no. 3225; p. 390, no. 3233.
74. BL MS Cotton Galba E III, f. 176r.
75. JRULM Latin MS 235, f. 31r.
76. PRO E101/375/9, f. 9r.
77. BL Add. MS 60584, f. 24r.
78. PRO E101/393/4, ff. 11r–11v.
79. Paley Baildon, 'Trousseaux of Princess Philippa', p. 183.
80. PRO E101/393/4, f. 11v.
81. JRULM Latin MS 234, f. 6v.
82. JRULM Latin MS 234, f. 8v.
83. BL MS Cotton Galba E III, ff. 177r–177v.
84. BL MS Cotton Galba E III, ff. 182v, 190v.
85. *EC*, p. 125.
86. PRO E101/393/4, f. 12r.
87. BL Add. MS 60584, ff. 13r, 14r.
88. BL Add. MS 60584, f. 5r.
89. Nicolas, 'Observations on the . . . Order of the Garter', p. 57.
90. *RWH 1286–9*, p. 318, no. 3219; p. 383, no. 3223; *EC*, pp. 79, 107.
91. BL Add. MS 60584, f. 16v.
92. Nicolas, 'Observations on the . . . Order of the Garter', p. 76; Paley Baildon, 'Trousseaux of Princess Philippa', p. 176.
93. Nicolas, 'Observations on the . . . Order of the Garter', pp. 75–6; *RWH 1286–9*, p. 381, no. 3219; p. 384, no. 3225.
94. PRO E101/393/4, ff. 8r–10r.
95. BL Add. MS 60584, ff. 20r, 21r, 22r. For the collections of royal relics, W.M. Ormrod, 'The personal religion of Edward III' *Speculum* 64 (1989) pp. 849–77, at 855–7.
96. BL Add. MS 60584, ff. 28r, 61v.
97. JRULM Latin MS 234, ff. 23r–24r. Isabella's establishment is calculated from her purchases of cloth for winter livery, BL MS Cotton Galba E XIV, f. 45r: 5 cloths, 10 ells for clerks; and 1½ cloths for sub-clerks; allowing an average of 10 ells per person (clerks in the household of Queen Philippa in 1330–1 received between 7½ and 12 ells per person: JRULM Latin MS 234, ff. 23r–24v) and 24 ells per cloth (short cloths and unspecified cloths generally of this length, long cloths of 27 ells: BL MS Cotton Galba E III, f. 174r).
98. JRULM Latin MS 236, f. 2v.
99. PRO E101/393/4, f. 5r.
100. *EC*, p. 80 and n. 99.
101. PRO E101/393/4, ff. 5r–6v.
102. *HBQI*, pp. 234, 236.
103. PRO E36/205, f. 6r.

104. PRO E101/393/4, f. 7r.
105. Ormrod, 'Personal religion of Edward III', p. 864, n. 91.
106. Parsons, 'Ritual and symbol in the English medieval queenship', pp. 63–6.
107. JRULM Latin MS 235, ff. 25r, 30v.
108. JRULM Latin MS 235, ff. 14r.
109. BL MS Cotton Galba E XIV, f. 48r; Bond, 'Notices of the last days of Isabella', p. 466: £105 is a typographical slip.
110. BL Add. MS 60584, f. 60v.
111. PRO E101/375/9, f. 35r.
112. JRULM Latin MS 235, f. 23r.
113. Myers, 'Captivity of a royal witch', pp. 129–30.
114. PRO E101/393/4, f. 4r.
115. Myers, 'Jewels of Margaret of Anjou', p. 217.
116. BL Add. MS 60584, f. 20r.
117. Nicolas, 'Observations on the . . . Order of the Garter', p. 79.
118. PRO E101/393/4, f. 1r.
119. PRO E101/393/4, ff. 2v–3r.
120. JRULM Latin MS 235, ff. 24r–27r.
121. Nicolas, 'Observations on the . . . Order of the Garter', pp. 79, 81.
122. Paley Baildon, 'Trousseaux of Princess Philippa', pp. 187–8.
123. JRULM Latin MS 235, ff. 18v–19r.
124. Society of Antiquaries MS 208, ff. 4v, 5r.
125. JRULM Latin MS 236, f. 3v.
126. BL MS Cotton Galba E XIV, ff. 52r–52v.
127. Myers, 'Captivity of a royal witch', p. 125.
128. PRO E101/393/4, f. 13r.
129. Bond, 'Notices of the last days of Isabella', p. 465.
130. BL Add. MS 60584, f. 8r.
131. JRULM Latin MS 234, ff. 3r, 33r.
132. Paley Baildon, 'Trousseaux of Princess Philippa', p. 181.
133. Ibid., p. 181.
134. PRO E101/375/9, f. 28r.
135. BL Add. MS 60584, ff. 16v, 24r.
136. PRO E101/375/9, f. 24r.
137. BL Add. MS 38006, f. 5v.
138. *RWH 1286–9*, p. 393, no. 3238; *EC*, p. 112.
139. Myers, 'Captivity of a royal witch', p. 125.
140. Bond, 'Notices of the last days of Isabella', p. 465.
141. *EC*, p. 63.
142. L.W. Stone, 'Jean de Howden, poète anglo-normand du xiiie siècle' *Romania* 69 (1946–7) pp. 496–519.
143. Ibid., pp. 513–14.
144. BL Add. MS 60584, f. 20r.
145. R. Averkorn, 'Les Nobles, sont-ils toujours beaux? Quelques remarques sur les descriptions de personnages dans les chroniques médiévales de la Péninsule Ibérique' in *Le Beau et le laid au Moyen Âge* (Aix-en-Provence, 2000) pp. 27–44.
146. *HBQI*, p. 114.
147. BL Add. MS 60584, ff. 16v (receipt), 24r (issue).
148. BL MS Cotton Galba E III, f. 191r.
149. PRO E36/205, f.12r.
150. Myers, 'Captivity of a royal witch', p. 122.
151. PRO E101/393/4, f. 10r; compare as well the strong personal connections with vessels used for washing parts of the body: *GH*, p. 167.

152. PRO E101/375/9, f. 21v.
153. *EC*, p. 85; also Hinton, *Gold and gilt*, p. 351, n. 30.
154. BL Add. MS 60584, f. 21r.
155. PRO E101/393/4, f. 9r.
156. JRULM Latin MS 234, f. 28v; Latin MS 235, ff. 14r, 35r; PRO E E36/205, f. 15r; Society of Antiquaries MS 208, f. 9r; JRULM Latin MS 236, f. 8r.
157. PRO E101/375/9, ff. 22r–v, 24r–v.
158. PRO E101/376/7, f. 13v.
159. *HBQI*, pp. 70, 108.
160. Myers, 'Captivity of a royal witch', pp. 125, 130.
161. PRO E101/375/9, f. 25r.
162. JRULM Latin MS 234, f. 27v; Latin MS 235, f. 12v; PRO E36/205, f. 15r; Society of Antiquaries MS 208, f. 9r; JRULM Latin MS 236, f. 8r.
163. JRULM Latin MS 235, f. 11r; PRO E101/375/9, f. 22r.
164. *EC*, p. 133.
165. Turner, 'Liberationes factae per executores', pp. 93–145.
166. M. Prestwich *Edward I* (new edition, New Haven, 1997) pp. 507–8.
167. *HO*, pp. 109–28, at 125.
168. *HO*, pp. 123–5.

CHAPTER 11: THE GREAT HOUSEHOLD AT THE END OF THE MIDDLE AGES

1. Before this date, the functions the gentlemen ushers were to undertake in more elaborate form were typically in the charge of the marshal of the hall: *A fifteenth-century courtesy book* ed. R.W. Chambers (EETS, OS 148; 1914) pp. 1–22.
2. *GH*, pp. 197–204.
3. BL MS Harley 6815, ff. 25r–56v, 16r; copied twice in two sixteenth-century hands: this discussion uses the first copy, ff. 26r–41v and 16r, a detached leaf. K.M. Phillips, 'The invisible man: body and ritual in a fifteenth-century noble household' *Journal of Medieval History* 31 (2005) pp. 143–62, discusses the text and its dating.
4. Bodl. MS Eng. hist. b. 208.
5. *NHB*.
6. By someone who wrote part of the First Household Book, including material dated 1524: D.M. Barratt, 'A second Northumberland Household Book' *Bodleian Library Record* 8 (1967–72) pp. 93–8, at 94.
7. I. Lancashire, 'Orders for Twelfth Day and Night *circa* 1515 in the Second Northumberland Household Book', *English Literary Renaissance* 10 (1980) pp. 7–45, at 11. Suggestions for linking its contents to the household of the Percies argue that they derive from the period 1500 to 1519. This link, however, may not be sufficiently close to allow the text to be dated in this way: the events outside the daily course that the book describes also feature in other, contemporary discussions of household organisation, such as the so-called household ordinance of Henry VII of 1493, and therefore may have been standard components of this genre of document: Lancashire, 'Orders for Twelfth Day', pp. 8–10; *HO*, pp. 109–28.
8. My interpretation of the Second Household Book differs from both Barratt and Lancashire. A key point is the structure of the volume: it is no longer in its original order. The first item was almost certainly number twenty in Molly Barratt's list, ff. 62–78, 'This is the book of thordour of the gentillmen uschers, how they shall ordour theme to their lord and maister at certein tymes concernyng certain causis, when they shall exercise it at the times and seasounes when they shall doo it more plainly hereaftir followeth, how they shall order there selffes and observe the said causis'. This title is echoed on the label of the original binding: 'The booke of all maner of orders concernynge an erles hous perteignynge to his

gentyllmen usshers and all manner his officers of householde'. Confirmation of this arrangement comes from some of the many notes and pen trials that were made in the volume in
the late sixteenth century. Not all the hands that made these additions can be found
throughout the text. One hand, mainly writing medical notes and prescriptions, is found in
only two places, on the gatherings that form ff. 62–78 and ff. 1–6, on 'The ordouring and
apariling of the chapell within an erle's house against the lady in the estate of a countesse
schall taik hir rightes befoir sche shall taike hir chaumbre . . .' It is reasonable to assume that
these two sections were once placed together. It follows from this that the numbering of the
orders in the manuscript could not have been carried out before the manuscript was
rearranged and, as it does not reflect the original arrangement, it cannot be earlier than the
late sixteenth century. While one might set aside any preconception about the order which
the sections were intended to follow, transposing the twentieth section to the beginning of
the volume makes a coherent text.

9. BL MS Harley 6815, f. 26r.
10. Bodl. MS Eng. hist. b. 208, f. 76v.
11. *Fifteenth-century courtesy book*, p. 15.
12. Bodl. MS Eng. hist. b. 208, ff. 62r, 64v.
13. Bodl. MS Eng. hist. b. 208, f. 62r.
14. Bodl. MS Eng. hist. b. 208, ff. 62v–64r.
15. Bodl. MS Eng. hist. b. 208, f. 64r.
16. Bodl. MS Eng. hist. b. 208, f. 64v.
17. Bodl. MS Eng. hist. b. 208, f. 64v.
18. BL MS Harley 6815, f. 26r.
19. Bodl. MS Eng. hist. b. 208, ff. 66r–v; *NHB*, pp. 297–301.
20. Bodl. MS Eng. hist. b. 208, ff. 66v–68r.
21. BL MS Harley 6815, f. 26r; Bodl. MS Eng. hist. b. 208, ff. 8v, 9v.
22. Bodl. MS Eng. hist. b. 208, ff. 66v–68v.
23. BL MS Harley 6815, ff. 26v–27r.
24. BL MS Harley 6815, ff. 27r–28r.
25. Bodl. MS Eng. hist. b. 208, f. 72r.
26. Bodl. MS Eng. hist. b. 208, f. 72v.
27. Bodl. MS Eng. hist. b. 208, f. 74v.
28. Bodl. MS Eng. hist. b. 208, f. 78v.
29. Bodl. MS Eng. hist. b. 208, ff. 13r–v, 31v.
30. Bodl. MS Eng. hist. b. 208, f. 13v.
31. Bodl. MS Eng. hist. b. 208, ff. 14r–v.
32. Bodl. MS Eng. hist. b. 208, f. 10r.
33. Bodl. MS Eng. hist. b. 208, ff. 12r–v.
34. E.g. Bodl. MS Eng. hist. b. 208, ff. 74v–76v.
35. BL MS Harley 6815, ff. 28v–29r.
36. BL MS Harley 6815, f. 34r.
37. BL MS Harley 6815, f. 32r.
38. BL MS Harley 6815, f. 41v.
39. BL MS Harley 6815, f. 37v.
40. BL MS Harley 6815, f. 38r.
41. Bodl. MS Eng. hist. b. 208, ff. 78r–v.
42. Bodl. MS Eng. hist. b. 208, ff. 45r–46r.
43. Bodl. MS Eng. hist. b. 208, f. 46r.
44. Bodl. MS Eng. hist. b. 208, f. 46v.
45. Bodl. MS Eng. hist. b. 208, f. 47r.
46. Bodl. MS Eng. hist. b. 208, f. 47r.
47. Bodl. MS Eng. hist. b. 208, f. 47v. Henry VII: PRO E101/414/16, f. 60v; E36/214, ff. 27r, 73r, 128r, 165v.

48. Bodl. MS Eng. hist. b. 208, f. 47v.
49. Io 13: 4.
50. Bodl. MS Eng. hist. b. 208, f. 48r.
51. Bodl. MS Eng. hist. b. 208, ff. 48v–49r.
52. Bodl. MS Eng. hist. b. 208, ff. 49r–v.
53. Bodl. MS Eng. hist. b. 208, f. 49v.
54. Bodl. MS Eng. hist. b. 208, f. 74r.
55. Bodl. MS Eng. hist. b. 208, ff. 34r–40r.
56. Bodl. MS Eng. hist. b. 208, f. 43r.
57. *HAME*, ii, pp. 644–5, 651–2; *GH*, p. 177.
58. *NHB*, pp. 354–63; Bodl. MS Eng. hist. b. 208, ff. 50r–v.
59. Bodl. MS Eng. hist. b. 208, f. 8r.
60. The other duchies were those of Norfolk and Suffolk. For a general discussion of Buckingham's household and finances, see C. Rawcliffe *The Staffords, Earls of Stafford and Dukes of Buckingham 1394–1521* (Cambridge, 1978) pp. 129–43.
61. PRO E35/150, p. 19. The Thornbury entry is printed in J. Gage, 'Extracts from the household book of Edward Stafford, Duke of Buckingham' *Archaeologia* 25 (1834) pp. 311–41, at 311–13.
62. PRO E35/150, p. 121.
63. PRO E35/150, p. 38.
64. PRO E35/150, p. 43.
65. S. Thurley *The royal palaces of Tudor England: architecture and court life 1460–1547* (New Haven, 1993) pp. 212–17.
66. PRO E35/150, pp. 24, 55, 65, 68–9, 111, 115, 127.
67. Longleat MS Misc. XIII, ff. 11r, 17r, 24r.
68. Longleat MS Misc. XIII, ff. 18v, 32r.
69. Longleat MS Misc. XIII, f. 38r.
70. Longleat MS Misc. XIII, ff. 6r, 47r.
71. Longleat MS Misc. XIII, ff. 71r, 18r, 31r.
72. Longleat MS Misc. XIII, ff. 30r–v.
73. Longleat MS Misc. XIII, ff. 1v, 2r, 12r, 20r–v, 25r, 34v, 37v.
74. Longleat MS Misc. XIII, f. 3r.
75. Longleat MS Misc. XIII, f. 58r.
76. Longleat MS Misc. XIII, f. 4r.
77. Longleat MS Misc. XIII, f. 77v.
78. Longleat MS Misc. XIII, f. 25r.
79. Longleat MS Misc. XIII, f. 5v.
80. Longleat MS Misc. XIII, ff. 7r, 23r, 49r, 56v.
81. Longleat MS Misc. XIII, f. 3v.
82. Longleat MS Misc. XIII, f. 15v.
83. Longleat MS Misc. XIII, f. 86v.
84. Longleat MS Misc. XIII, f. 10v.
85. Longleat MS Misc. XIII, f. 10r.
86. Longleat MS Misc. XIII, f. 16r.
87. Longleat MS Misc. XII, ff. 115v–116r.
88. Longleat MS Misc. XIII, ff. 26r, 29r, 31v.
89. Longleat MS Misc. XIII, f. 33v.
90. *Inventory of Henry VIII*, censers and containers for incense, 387–426; images, see above, Chapter 7.
91. *Inventory of Henry VIII*, 1929, 1935, 1952, 1954, 1972, 1979, 1982–3, 1986, 1988, 1993, 1996–8, 2445–7, 3052, 3447, 10430, 10477, 10487, 10495, 12355, 12827, 15799, 15839, 15849, 15857, 16570, 16669.
92. *Inventory of Henry VIII*, 1982–3.

93. *Inventory of Henry VIII*, 12827.
94. PRO SP1/228, ff. 142–59. I am grateful to Maria Hayward for this reference.
95. *Inventory of Henry VIII*, 10492.
96. *Inventory of Henry VIII*, 2452, 9686, 15880.
97. *Inventory of Henry VIII*, casting bottles, e.g. 2430–8, 2787, 2906, 2956–9, 2961, 3029–30, 3073–4, 3289, 3372, 3390; pomanders, 2545, 2654, 2778, 2879, 2905, 2998, 3015, 3017, 3097, 3467; bottle 'for swete waters', 3446.
98. *Inventory of Henry VIII*, gloves, 3429–32; walking sticks, 14355–8.
99. G. Manners, 'On an apothecary's account for James I, 1622' *Proceedings of the Society of Antiquaries of London*, 2nd series, 4 (1867–70) pp. 435–7.

CHAPTER 12: CHANGING PERCEPTIONS

1. The division made by C. Walker Bynum, 'Wonder' *American Historical Review* 102 (1997) pp. 1–26, is in many ways a parallel.
2. Hinton, *Gold and gilt*, pp. 203–4.
3. M.R. McVaugh, 'Smells and the medieval surgeon' *Micrologus* 10 (2002) pp. 113–32.
4. *Heresy trials in the diocese of Norwich, 1428–31* ed N.P. Tanner (Camden, 4th series, 20; 1977) pp. 107–12. Compare also K. Thomas *Religion and the decline of magic: studies in popular beliefs in sixteenth- and seventeenth-century England* (Harmondsworth, 1991) pp. 27–89.
5. C. Morris *The discovery of the individual 1050–1200* (1972) pp. 139–57; D.G. Denery II *Seeing and being seen in the later medieval world: optics, theology and religious life* (Cambridge, 2005) pp. 14–15.
6. M. Carruthers *The craft of thought: meditation, rhetoric, and the making of images, 400–1200* (Cambridge, 1998) pp. 257–9, 272–3, 276; C. Waddell, 'The early Cistercian experience of the liturgy' in *Rule and life: an interdisciplinary symposium* ed. M.B. Pennington (Shannon: Cistercian studies series, 12; 1971) pp. 77–116, at 101.
7. D. Pearsall and E. Salter *Landscapes and seasons of the medieval world* (1973) pp. 1, 25, 177, 190–1; Denery, *Seeing and being seen*, pp. 9–10, 14.
8. PRO E36/214, ff. 8r, 42r, 58v, 75v, 85r, 101r, 124v, 127v, 133v, 145r, 155v, 164v.
9. C.B. Hieatt, 'Medieval Britain' in *Regional cuisines of medieval Europe: a book of essays* ed. M. Weiss Adamson (New York, 2002) pp. 24, 37–9.
10. *Food in medieval England: diet and nutrition* ed. C.M. Woolgar, D. Serjeantson and T. Waldron (Oxford, 2006), pp. 267–77.
11. *GH*, pp. 133–4.
12. *The eclogues of Alexander Barclay from the original edition by John Cawood* ed. B. White (EETS, OS 175; 1928) pp. 79–80.
13. *Southampton probate inventories 1447–1575* ed. E. Roberts and K. Parker (2 vols, Southampton Records Series, 34–5; 1992) ii, pp. 290–8.
14. *HAME*, ii, p. 534.
15. *La Vie Seint Richard evesque de Cycestre by Pierre d'Abernon of Fetcham* ed. D.W. Russell (ANTS, 51; 1995) p. 82; D. Howes, 'Sensorial anthropology', in *The varieties of sensory experience: a sourcebook in the anthropology of the senses* ed. D. Howes (Toronto, 1991) pp. 167–91, at 167–8.

BIBLIOGRAPHY

MANUSCRIPTS: UK

Kew: Public Record Office (The National Archives)

E35
E36
E101
SC6
SP1

London: The British Library

MS Cotton Galba E III
MS Cotton Galba E XIV
MS Harley 6815
MS Royal 6 E VI
MS Royal 6 E VII
MS Royal 12 B XIX
MS Royal 15 E II
MS Royal 15 E III
MS Sloane 3550
Add. MS 21480
Add. MS 32050
Add. MS 35115
Add. MS 37656
Add. MS 38006
Add. MS 47682
Add. MS 57334
Add. MS 60584

London: Lambeth Palace Library

ED 363
ED 364

London: Society of Antiquaries

MS 208
MS 216

Longleat House

MS Misc. IX
MS Misc. XII
MS Misc. XIII

Manchester: John Rylands University Library

Latin MS 234
Latin MS 235
Latin MS 236

Oxford: Bodleian Library

MS Eng. hist. b. 208

Oxford: Corpus Christi College

MS 154

Salisbury: Cathedral Library

MS 103

Southampton: University Library

MS 340/3

Truro: Cornwall Record Office

AR12/25

Winchester: Hampshire Record Office

5M53/767
Winchester City Archives W/D1

Winchester: Winchester College

Muniments, 1

MANUSCRIPTS: OVERSEAS

Nantes: Archives Départementales, Loire-Atlantique

E206/3

The Vatican: Biblioteca Apostolica Vaticana

MS Lat. Vat. 4015

PRIMARY SOURCES

The 1542 inventory of Whitehall: the palace and its keeper ed. M. Hayward (2 vols, 2004).

Account of the executors of Richard, Bishop of London, 1303, and of the executors of Thomas, Bishop of Exeter, 1310 ed. W.H. Hale and H.T. Ellacombe (Camden Society, NS 10; 1874).

Adam of Eynsham. *Magna vita sancti Hugonis: the life of St Hugh of Lincoln [by Adam of Eynsham]* ed. D.L. Douie and D.H. Farmer (2nd impression, 2 vols, Oxford, 1985).

Aelred of Rievaulx. *De anima* ed. C.H. Talbot (*Mediaeval and Renaissance Studies*, Supplement 1; 1952).

Aelred of Rievaulx. *De spirituali amicitia*, in *Aelredi Rievallensis opera omnia, 1: Opera ascetica* ed. A. Hoste and C.H. Talbot (CCCM 1; 1971) pp. 287–350.

Aelred of Rievaulx. *Speculum caritatis*, in *Aelredi Rievallensis opera omnia, 1: Opera ascetica* ed. A. Hoste and C.H. Talbot (CCCM 1; 1971) pp. 5–161.

Alan of Lille. *Anticlaudianus* ed. R. Bossuat (Paris, 1955).

An alphabet of tales, an English 15th-century translation of the Alphabetum narrationum of Etienne de Besançon from Additional MS. 25,719 of the British Museum ed. M.M. Banks (2 vols, EETS, OS 126–7; 1904–5).

Anglo-Norman dictionary ed. L.W. Stone, W. Rothwell *et al.* (Publications of the Modern Humanities Research Association, 8; 1977–92).

Anglo-Norman lapidaries ed. P. Studer and J. Evans (Paris, 1924).

Anglo-Norman medicine ed. Tony Hunt (2 vols, Cambridge, 1994–7).

The Anonimalle Chronicle 1333 to 1381, . . . ed. V.H. Galbraith (Manchester, 1927).

Augustine. *De doctrina christiana* ed. I. Martin (CCSL 32; 1962).

Augustine. *Confessiones* ed. L. Verheijen (CCSL 27; 1981).

Bacon, Roger. *Opera quaedam hactenus inedita* ed. J.S. Brewer (RS, 1859).

Baker, A.T. 'La Vie de Saint Edmond archevêque de Cantobéry' *Romania* 55 (1929) pp. 332–81.

Baker, A.T. 'Vie anglo-normande de Sainte Foy par Simon de Walsingham' *Romania* 68 (1944–5) pp. 49–84.

Baldwin of Ford. *Sermones de commendatione fidei* ed. D.N. Bell (CCCM 99; 1991).

Bartholomew the Englishman. *On the properties of things: John Trevisa's translation of Bartholomaeus Anglicus De proprietatibus rerum* ed. M.C. Seymour (3 vols, Oxford, 1975–88).

The Beauchamp cartulary: charters 1100–1268 ed. E. Mason (Pipe Roll Society, NS 43; 1980).

The Bedford inventories: the worldly goods of John, Duke of Bedford, Regent of France (1389–1435) ed. J. Stratford (Reports of the Research Committee of the Society of Antiquaries of London, 49; 1993).

Bedfordshire coroners' rolls ed. R.F. Hunnisett (Bedfordshire Historical Record Society, 41; 1961).

Biblia sacra iuxta vulgatam versionem ed. R. Weber *et al.* (4th edition, Stuttgart, 1994).

Blund, John. *Tractatus de anima* ed. D.A. Callus and R.W. Hunt (ABMA, 2; 1970).

The book of the knight of La Tour-Landry . . . ed. T. Wright (EETS, OS 33; 1868).

The book of Margery Kempe: the text from the unique MS. owned by Colonel W. Butler Bowdon ed. S.B. Meech and H.E. Allen (EETS, OS 212; 1940).

The book of quinte essence or the fifth being; that is to say, man's heaven . . . ed. F.J. Furnivall (EETS, OS 16; 1866).

The book of St Gilbert ed. R. Foreville and G. Keir (Oxford, 1987).

The book of vices and virtues: a fourteenth century English translation of the Somme le Roi of Lorens d'Orléans ed. W. Nelson Francis (EETS, OS 217; 1942).

Braekman, W.L. 'Fortune-telling by the casting of dice: a Middle English poem and its background' *Studia Neophilologica* 52 (1980) pp. 3–29.

Braunschweig, Hieronymus. *Liber de arte distillandi, de simplicibus* (Strasbourg, 1500).

Braunschweig, Hieronymus. *Liber de arte distillandi, de compositis* (Strasbourg, 1512).

Braunschweig, Hieronymus. *The vertuouse boke of distyllacyon* (1527).

Burnett, C. 'The earliest chiromancy in the West' *JWCI* 50 (1987) pp. 189–95.

Burnett, C. 'The superiority of taste' *JWCI* 54 (1991) pp. 230–8.

Calendar of coroners' rolls of the city of London AD 1300–1378 ed. R.R. Sharpe (1913).

Calendar of Nottinghamshire coroners' inquests 1485–1558 ed. R.F. Hunnisett (Thoroton Society, Record Series, 25; 1969).

Calendar of the Close Rolls (45 vols, 1892–1954).

The Canonization of Saint Osmund from the manuscript records in the muniment room of Salisbury Cathedral ed. A.R. Malden (Wiltshire Record Society, 1901).

Caxton's Eneydos 1490 Englisht from the French Liure des Eneydes, 1483 ed. M.T. Culley and F.J. Furnivall (EETS, ES 57; 1890).

Chaucer, Geoffrey. *The Riverside Chaucer* ed. L.D. Benson (3rd edition, Oxford, 1988).

The Chronicle of Jocelin of Brakelond, concerning the acts of Samson, Abbot of the Monastery of St Edmund ed. H.E. Butler (1949).

The Chronicle of William de Rishanger of the Barons' Wars. The Miracles of Simon de Montfort ed. J.O. Halliwell (Camden Society, OS 15; 1840).

Chronicles of London ed. C.L. Kingsford (Oxford, 1905).

Chronicles of the reigns of Stephen, Henry II and Richard I ed. R. Howlett (4 vols, RS, 1884–9).

The cloud of unknowing, and the book of privy counselling ed. P. Hodgson (EETS, OS 218; 1944).

A collection of ordinances and regulations for the government of the royal household, made in divers reigns from King Edward III to King William and Queen Mary . . . ed. Anon. for the Society of Antiquaries (London, 1790).

Les collections du trésor royal sous le règne de Charles VI (1380–1422): l'inventaire de 1400 ed. P. Henwood (Paris, 2004).

Collet, A. '*Traité d'hygiène* de Thomas le Bourguignon (1286)' *Romania* 112 (1991) pp. 450–87.

Constable, G. 'The vision of Gunthelm and other visions attributed to Peter the Venerable' *Revue Bénédictine* 66 (1956) pp. 92–114.

A contemporary narrative of the proceedings against Dame Alice Kyteler; prosecuted for sorcery in 1324, by Richard de Ledrede, Bishop of Ossory ed. T. Wright (Camden Society, OS 24; 1843).

The Coronation of Richard III: the extant documents ed. A.F. Sutton and P.W. Hammond (Gloucester, 1983).

Councils and synods with other documents relating to the English Church. I. AD 871–1204 ed. D. Whitelock, M. Brett and C.N.L. Brooke (2 vols, Oxford, 1981).

Councils and synods with other documents relating to the English Church. II. AD 1205–1313 ed. F.M. Powicke and C.R. Cheney (2 vols, Oxford, 1964).

The Court and household of Eleanor of Castile in 1290: an edition of British Library, Additional Manuscript 35294 ed. J.C. Parsons (Toronto, 1977).

The Courts of the archdeaconry of Buckingham 1483–1523 ed. E.M. Elvey (Buckinghamshire Record Society, 19; 1975).

The Customary of the Benedictine abbey of Eynsham in Oxfordshire ed. A. Gransden (Siegburg, 1963).

The Cyrurgie of Guy de Chauliac ed. M.S. Ogden (EETS, OS 265; 1971).

Daniel, Walter. *The life of Ailred of Rievaulx* ed. F.M. Powicke (1950).

Dictionary of medieval Latin from British sources ed. R.E. Latham, D.R. Howlett *et al.* (1975–).

'*The doctrine of the hert,* edited from the manuscripts with introduction and notes' ed. Sister Mary Patrick Candon (Unpublished Ph.D. dissertation, Fordham University, 1963).

Doyle, A.I. '"Lectulus noster floridus": an allegory of the penitent soul' in *Literature and religion in the later Middle Ages: philological studies in honor of Siegfried Wenzel* ed. R.G. Newhauser and J.A. Alford (Binghamton, 1995) pp. 179–90.

Dugdale, W. *Monasticon anglicanum* ed. J. Caley, H. Ellis and B. Bandinel (new edition, 6 vols in 8, 1846).

The Early English versions of the Gesta Romanorum ed. S.J.H. Herrtage (EETS, ES 33; 1879).

The Eclogues of Alexander Barclay, from the original edition by John Cawood ed. B. White (EETS, OS 175; 1928).

Edwards, A.J.M. 'An early twelfth century account of the translation of St Milburga of Much Wenlock' *Transactions of the Shropshire Archaeological Society* 57 (1962) pp. 134–51.

English episcopal acta: X. Bath and Wells 1061–1205 ed. F.M.R. Ramsey (Oxford, 1995).

English mediaeval lapidaries ed. J. Evans and M.S. Serjeantson (EETS, OS 190; 1933).

English Wycliffite sermons ed. A. Hudson and P. Gradon (5 vols, Oxford, 1983–96).

La Estoire de Seint Aedward le rei, attributed to Matthew Paris ed. K. Young Wallace (ANTS, 41; 1983).

Eulogium historiarum sive temporis ed. F.S. Haydon (3 vols, RS, 1858–63).

Eynsham cartulary ed. H.E. Salter (2 vols, Oxford Historical Society, 49, 51; 1907–8).

Farmer, H. 'The canonization of St Hugh of Lincoln' *Lincolnshire Architectural and Archaeological Society Reports and Papers* 6 (1956) pp. 86–117.

Farming and gardening in late medieval Norfolk: Norwich Cathedral Priory gardeners' accounts, 1329–1530 ed. C. Noble and *Skayman's Book, 1516–1518* ed. C. Moreton and P. Rutledge (Norfolk Record Society, 61; 1997).

A fifteenth-century courtesy book ed. R.W. Chambers (EETS, OS 148; 1914).

Fleta ed. G.O. Sayles (3 vols, Selden Society, 72, 89, 99; 1953, 1972, 1984).

The floure and the leafe; the assembly of ladies; the isle of ladies ed. D. Pearsall (Kalamazoo, 1990).

Fredborg, K.M., L. Nielsen and J. Penborg. 'An unedited part of Roger Bacon's "Opus maius": "De signis"' *Traditio* 34 (1978) pp. 75–136.

Gage, J. 'Extracts from the household book of Edward Stafford, Duke of Buckingham' *Archaeologia* 25 (1834) pp. 311–41.

Geiriadur Prifysgol Cymru: a dictionary of the Welsh language (Cardiff, 1950–).

Gerald of Wales. *Opera* ed. J.S. Brewer *et al.* (8 vols, RS, 1861–91).

Gervase of Tilbury. *Otia imperialia: recreation for an emperor* ed. S.E. Banks and J.W. Binns (Oxford, 2002).

Gesta Henrici quinti: the deeds of Henry the Fifth ed. F. Taylor and J.S. Roskell (Oxford, 1975).

Gilleland, J.R. 'Eight Anglo-Norman cosmetic recipes: MS. Cambridge, Trinity College 1044' *Romania* 109 (1988) pp. 50–69.

Giuseppi, M.S. 'The wardrobe and household accounts of Bogo de Clare, AD 1284–6' *Archaeologia* 70 (1918–20) pp. 1–56.

Goering, J., and F.A.C. Mantello. 'The "Perambulauit Iudas ..." (Speculum Confessionis) attributed to Robert Grosseteste' *Revue Bénédictine* 96 (1986) pp. 125–68.

Goering, J., and F.A.C. Mantello. '"Notus in Iudea Deus": Robert Grosseteste's confessional formula in Lambeth Palace MS 499' *Viator* 18 (1987) pp. 253–73.

Goering, J., and P.J. Payer. 'The "Summa penitentie fratrum predicatorum": a thirteenth-century confessional formula' *Mediaeval Studies* 55 (1993) pp. 1–50.

Gregory the Great. *Moralia in Iob* ed. M. Adriaen (3 vols, CCSL 143, 143A, 143B; 1979–85).

Griesser, P.B. 'Die "Ecclesiastica officia Cisterciensis ordinis" des Cod. 1711 von Trient' *Analecta Sacri Ordinis Cisterciensis* 12 (1956) pp. 153–288.

Grosseteste, Robert. *Hexaëmeron* ed. R.C. Dales and S. Gieben (ABMA, 6; 1982).

Grosseteste, Robert. *De decem mandatis* ed. R.C. Dales and E.B. King (ABMA, 10; 1987).

The Harley Lyrics: the Middle English lyrics of Ms. Harley 2253 ed. G.L. Brook (4th edition, Manchester, 1968).

Henrici VI Angliae regis miracula postuma ex codice Musei Britannici Regio 13. C. VIII ed. P. Grosjean (Subsidia Hagiographica, 22; Brussels, Société des Bollandistes).

Henry of Huntingdon. *Historia Anglorum* ed. T. Arnold (RS, 1879).

Heresy trials in the diocese of Norwich, 1428–31 ed. N.P. Tanner (Camden, 4th series, 20; 1977).

Hesketh, G. 'An unpublished Anglo-Norman life of Saint Katherine of Alexandria from MS. London, BL, Add. 40143' *Romania* 118 (2000) pp. 33–82.

The Historians of the church of York and its archbishops ed. J. Raine (3 vols, RS, 1879–94).

Holdsworth, C.J. 'Eleven visions connected with the Cistercian monastery of Stratford Langthorne' *Cîteaux* 13 (1962) pp. 185–204.

Household accounts from medieval England ed. C.M. Woolgar (2 vols, British Academy, Records of Social and Economic History, NS 17–18; 1992–3).

The Household book of Queen Isabella of England for the fifth regnal year of Edward II 8th July 1311 to 7th July 1312 ed. F.D. Blackley and G. Hermansen (University of Alberta, Classical and Historical Studies, 1; 1971).

The Household of Edward IV: the Black Book and the ordinance of 1478 ed. A.R. Myers (Manchester, 1959).

Hugh of Fouilloy. *De claustro animae. PL*, 176, cols 1017–182.

Hunt, Tony. *Popular medicine in thirteenth-century England: introduction and texts* (Cambridge, 1990).

Hunt, Tony. 'Early Anglo-Norman receipts for colours' *JWCI* 58 (1995) pp. 203–9.

Innocent III. *De sacro altaris mysterio. PL*, 217, cols 773–916.

Inquests and indictments from late fourteenth century Buckinghamshire: the superior eyre of Michaelmas 1389 at High Wycombe ed. L. Boatwright (Buckinghamshire Record Society, 29; 1994).

Instructions for parish priests by John Myrc ed. E. Peacock (EETS, OS 31; 1868; second revised edition, 1902).

The Inventory of King Henry VIII: Society of Antiquaries MS 129 and British Library MS Harley 1419 ed. D. Starkey (3 vols, in progress, 1998–).

Isidore of Seville. *Isidori Hispalensis episcopi etymologiarum sive originum libri XX* ed. W.M. Lindsay (2 vols, Oxford, 1911).

Jacob's Well, an Englisht treatise on the cleansing of man's conscience ed. A. Brandeis (EETS, OS 115; 1900).

James, M.R. 'Twelve medieval ghost-stories' *EHR* 37 (1922) pp. 413–22.

Jerome. *Liber interpretationis Hebraicorum nominum* ed. P. de Lagarde in *S. Hieronymi presbyteri opera: pars I. Opera exegetica* ed. P. Antin (CCSL 72; 1959) pp. 57–161.

Le Jeu d'Adam: ordo representacionis Ade ed. W. Noomen (Paris, 1971).

John of Salisbury. *Policraticus I–IV* ed. K.S.B. Keats-Rohan (CCCM 118; 1993).

Johnstone, H. 'The wardrobe and household of Henry, son of Edward I' *Bulletin of the John Rylands Library* 7 (1922–3) pp. 384–420.

Keele, R. 'Richard Lavenham's *De causis naturalibus*: a critical edition' *Traditio* 56 (2001) pp. 113–47.

Kemp, B. 'The miracles of the hand of St James' *Berkshire Archaeological Journal* 65 (1970) pp. 1–19.

Kent heresy proceedings 1511–12 ed. N. Tanner (Kent Records, 26; 1997).

Kilwardby, Robert. *On time and imagination: De tempore, De spiritu fantastico* ed. P. Osmund Lewry (ABMA, 9; 1987).

Kirby, T.F. *Annals of Winchester College from its foundation in the year 1382 to the present time* (1892).

The knowing of woman's kind in childing: a Middle English version of material derived from the Trotula and other sources ed. A. Barratt (Turnhout, 2001).

Lancashire, I. 'Orders for Twelfth Day and Night *circa* 1515 in the Second Northumberland Household Book' *English Literary Renaissance* 10 (1980) pp. 7–45.

Lanfrank's 'Science of cirurgie' ed. R. von Fleischhacker (EETS, OS 102; 1894).

Langland, William. *The vision of Piers Plowman: a critical edition of the B-Text . . .* ed. A.V.C. Schmidt (new edition, London, 1987).

Langland, William. *Piers Plowman: a facsimile of Bodleian Library, Oxford, MS Douce 104* ed. D.Pearsall and K.Scott (Cambridge, 1992).

Lapidge, M. *The cult of St Swithun* (Winchester Studies 4.ii; Oxford, 2003).

The late medieval religious plays of Bodleian MSS Digby 133 and E Museo 160 ed. D.C. Baker, J.L. Murphy and L.B. Hall Jr. (EETS, OS 283; 1982).

Lawrence, C.H. *St Edmund of Abingdon: a study in hagiography and history* (Oxford, 1960).

Legge, M.D. 'John Pecham's *Jerarchie*' *Medium Aevum* 11 (1942) pp. 77–84.

Lewis, C.T., and C. Short. *A Latin dictionary* (Oxford, 1879).

Lewry, P.O. 'Thirteenth-century teaching on speech and accentuation: Robert Kilwardby's commentary on *De accentibus* of Pseudo-Priscian' *Mediaeval Studies* 50 (1988) pp. 96–185.

Libellus de vita et miraculis S. Godrici, heremitae de Finchale auctore Reginaldo monacho Dunelmensi. Adjicitur appendix miraculorum ed. J. Stevenson (Surtees Society, 20; 1845).

Liber exemplorum ad usum praedicantium saeculo XIII compositus a quodam fratre minore anglico de provincia Hiberniae ed. A.G. Little (British Society of Franciscan Studies, 1; 1908).

'Liber miraculorum Beati Edmundi archiepiscopi' in *Thesaurus novus anecdotorum* ed. E. Martène and U. Durand (5 vols, Paris, 1717) iii, cols. 1881–98.

Liber quotidianus contrarotulatoris garderobe anno regni regis Edwardi primi vicesimo octavo AD *MCCXCIX & MCCC* ed. Anon. for Society of Antiquaries (1787).

Liber regie capelle: a manuscript in the Biblioteca Publica, Evora ed. W. Ulmann (Henry Bradshaw Society, 92; 1961).

Libri de nativitate Mariae: Pseudo-Matthaei evangelium textus et commentarius ed. J. Gijsel (CCSA 9; 1997).

The life and miracles of St William of Norwich, by Thomas of Monmouth . . . ed. A. Jessopp and M.R. James (Cambridge, 1896).

The Life of St Catherine, by Clemence of Barking ed. W. MacBain (ANTS, 18; 1964).

Le livre de seyntz medicines: the unpublished devotional treatise of Henry of Lancaster ed. E.J. Arnould (ANTS, 2; 1940).

The Local port book of Southampton for 1439–40 ed. H.S. Cobb (Southampton Records Series, 5; 1961).

Lollard sermons ed. G. Cigman (EETS, OS 294; 1989).

Lollards of Coventry 1486–1522 ed. S. McSheffrey and N. Tanner (Camden, 5th series, 23; 2003).

Lower ecclesiastical jurisdiction in late-medieval England: the courts of the Dean and Chapter of Lincoln, 1336–1349, and the Deanery of Wisbech, 1458–1484 ed. L.R. Poos (British Academy, Records of Social and Economic History, NS 32; 2001).

La lumere as lais, by Pierre d'Abernon of Fetcham ed. G. Hesketh (3 vols, ANTS, 54–5, 56–7, 58; 1996–2000).

The macro plays: The Castle of Perseverance; Wisdom; Mankind ed. M. Eccles (EETS, OS 262; 1969).

Manières de langage ed. A.M. Kristol (ANTS, 53; 1995).

Manners and meals in olden time ed. F.J. Furnivall (EETS, OS 32; 1868).

Manners, G. 'On an apothecary's account for James I, 1622' *Proceedings of the Society of Antiquaries of London*, 2nd series, 4 (1867–70) pp. 435–7.

Manorial records of Cuxham, Oxfordshire, circa 1200–1359 ed. P.D.A. Harvey (Historical Manuscripts Commission, Joint Publications, 23; 1976).

Le Manuscrit Leyde Bibliothèque de l'Université BPL 2579, témoin principal des phases de rédaction du traité de doctrina cordis à attribuer au dominicain français Hugues de Saint-Cher (pseudo-Gérard de Liège) ed. G. Hendrix (Ghent, 1980).

Map, Walter. *De nugis curialium. Courtiers' trifles* ed. M.R. James, revised by C.N.L. Brooke and R.A.B. Mynors (Oxford, 1983).

Marbode of Rennes' (1035–1123) De lapidibus considered as a medical treatise with text, commentary and C.W. King's translation, together with text and translation of Marbode's minor works on stones ed. J.M. Riddle (Wiesbaden, Sudhoffs Archiv Zeitschrift für Wissenschaftsgeschichte, Beihefte, 20, 1977).

Materials for the history of Thomas Becket, Archbishop of Canterbury ed. J.C. Robertson (7 vols, RS, 1875–85).

Medieval English prose for women: selections from the Katherine Group and Ancrene Wisse ed. B. Millett and J. Wogan-Browne (Oxford, 1990).

Mémoires d'Olivier de la Marche ed. H. Beaune and J. d'Arbaumont (4 vols, Paris, 1883–8).

Memorials of St Anselm ed. R.W. Southern and F.S. Schmitt (ABMA, 1; 1969).

Middle English dictionary ed. H. Kurath and S.M. Kuhn *et al.* (Ann Arbor, Michigan, 1956–).

Middle English sermons edited from British Museum MS. Royal 18 B. XXIII ed. W.O. Ross (EETS, OS 209; 1940).

Middle English translation of Macer Floridus De Viribus Herbarum, A. ed. Gösta Frisk (The English Institute in the University of Upsala, Essays and studies on English language and literature, 3; 1949).

Mikkers, E. 'Un "Speculum novitii" inédit d'Étienne de Salley' *Collectanea Cisterciensis Ordinis Reformatorum* 8 (1946) pp. 17–68.

Mikkers, E. 'Un Traité inédit d'Étienne de Salley sur la psalmodie' *Cîteaux* 23 (1972) pp. 245–88.

Mirk's Festial ed. T. Erbe (EETS, ES 96; 1905).

Mirour de Seinte Eglyse (St Edmund of Abingdon's Speculum Ecclesiae) ed. A.D. Wilshere (ANTS, 40; 1982).

The monastic constitutions of Lanfranc ed. D. Knowles, revised by C.N.L. Brooke (Oxford, 2002).

Myers, A.R. 'The captivity of a royal witch: the household accounts of Queen Joan of Navarre, 1419–21' in A.R. Myers *Crown, household and Parliament in fifteenth century England* (1985) pp. 93–133.

Myers, A.R. 'The household of Queen Elizabeth Woodville, 1466–7' in A.R. Myers *Crown, household and Parliament in fifteenth century England* (1985) pp. 251–318.

Myers, A.R. 'The household of Queen Margaret of Anjou, 1452–3' in A.R. Myers *Crown, household and Parliament in fifteenth century England* (1985) pp. 135–209.

Myers, A.R. 'The jewels of Queen Margaret of Anjou' in A.R. Myers *Crown, household and Parliament in fifteenth century England* (1985) pp. 211–29.

Nicolas, N.H. 'Observations on the institution of the most noble Order of the Garter' *Archaeologia* 31 (1846) pp. 1–163.

Non-cycle plays and fragments ed. N. Davis (EETS, SS 1; 1970).

The N-town play: Cotton MS Vespasian D. 8 ed. S. Spector (2 vols, EETS, SS 11–12; 1991).

The observances in use at the Augustinian priory of S. Giles and S. Andrew at Barnwell, Cambridgeshire ed. J.W. Clark (Cambridge, 1897).

Oxford English Dictionary (2nd edition, Oxford, 1989).

Paley Baildon, W. 'The trousseaux of Princess Philippa, wife of Eric, King of Denmark, Norway, and Sweden' *Archaeologia* 67 (1916) pp. 163–88.

Paris, Matthew. *Chronica majora* ed. H.R. Luard (7 vols, RS, 1872–83).

La Passiun de Seint Edmund ed. J. Grant (ANTS, 36; 1978).

Peter Idley's instructions to his son ed. C. D'Evelyn (Modern Language Association of America, monograph series, 6; Boston and London, 1935).

Peter the Venerable. *Contra Petrobrusianos hereticos* ed. J. Fearns (CCCM 10; 1968).

The political songs of England, from the reign of John to that of Edward II ed. T. Wright (Camden Society, OS 6; 1839).

Post, J.B. 'A fifteenth-century customary of the Southwark stews' *Journal of the Society of Archivists* 5 (1974–7) pp. 418–28.

Privy purse expenses of Elizabeth of York: wardrobe accounts of Edward the Fourth, with a memoir of Elizabeth of York, and notes ed. N.H. Nicolas (1830).

The Prose Salernitan questions edited from a Bodleian manuscript (Auct. F.3.10): an anonymous collection dealing with science and medicine written by an Englishman c. 1200 with an appendix of ten related collections ed. B. Lawn (ABMA, 5; 1979).

RB 1980: the Rule of St Benedict in Latin and English ed. T. Fry *et al.* (Collegeville, 1981).

The Receyt of the Lady Kateryne ed. G. Kipling (EETS, OS 296; 1990).

Records of the wardrobe and household 1285–1286 ed. B.F. Byerly and C.R. Byerly (1977).

Records of the wardrobe and household 1286–1289 ed. B.F. Byerly and C.R. Byerly (1986).

The Registers of John de Sandale and Rigaud de Asserio, Bishops of Winchester (AD 1316–1323) ed. F.J. Baigent (Hampshire Record Society, 8; 1897).

Registrum Ricardi de Swinfield, episcopi Herefordensis, AD MCCLXXXIII–MCCCXVII ed. W.W. Capes (Canterbury and York Society, 6; 1909).

Registrum Thome de Cantilupo episcopi Herefordensis, AD MCCLXXV–MCCLXXXII ed. R.G. Griffiths and W.W. Capes (Canterbury and York Society, 2; 1907).

The regulations and establishment of the household of Henry Algernon Percy, the fifth Earl of Northumberland, at his castles of Wressle and Leconfield, in Yorkshire. Begun Anno Domini MDXII ed. T. Percy (new edition, 1905).

Reichl, K. 'An Anglo-Norman legend of Saint Margaret (MS BM. Add. 38664)' *Romania* 96 (1975) pp. 53–66.

A relation, or rather a true account, of the island of England: with sundry particulars of the customs of these people, and of the royal revenues under King Henry the Seventh, about the year 1500 ed. C.A. Sneyd (Camden Society, OS 37; 1847).

Robert de Avesbury. *De gestis mirabilibus regis Edwardi tertii* ed. E. Maunde Thompson (RS, 1889).

Robert of Brunne's 'Handlyng Synne' ed. F.J. Furnivall (EETS, OS 119, 123; 1901–3).

Robinson, J.A. 'Household roll of Bishop Ralph of Shrewsbury 1337–8' in *Collectanea I* ed. T.F. Palmer (Somerset Record Society, 39; 1924) pp. 72–174.

The roll of arms of the princes, barons and knights who attended King Edward I to the siege of Caerlaverock, in 1300 . . . ed. T. Wright (London, 1864).

Roll of the household expenses of Richard de Swinfield, Bishop of Hereford, during part of the years 1289 and 1290, A ed. J. Webb (Camden Society, OS 59, 62; 1854–5).

Rotuli parliamentorum; ut et petitiones et placita in parliamento ed. Anon. (6 vols, n.d. *c.* 1783).

Ruelle, P. *L'Ornement des dames (ornatus mulierum): texte anglo-normand du xiiie siècle* (Brussels: Université Libre de Bruxelles, Travaux de la Faculté de Philosophie et Lettres, 36; 1967).

Safford, E.W. 'An account of the expenses of Eleanor, sister of Edward III, on the occasion of her marriage to Reynald, Count of Guelders' *Archaeologia* 77 (1927) pp. 111–40.

St Modwenna ed. A.T. Baker and A. Bell (ANTS, 7; 1947).

The Saint of London: the life and miracles of St Erkenwald ed. E.G. Whatley (Medieval and Renaissance Texts and Studies, 58; Binghamton, 1989).

Select cases from the coroners' rolls AD 1265–1413 . . . ed. C. Gross (Selden Society, 9; 1896).

Select cases on defamation to 1600 ed. R.H. Helmholz (Selden Society, 101; 1985).

Six ecclesiastical satires ed. J.M. Dean (Kalamazoo, Michigan, 1991).

Södergård, Ö. 'Un Art d'aimer anglo-normand' *Romania* 77 (1956) pp. 289–330.

Southampton probate inventories 1447–1575 ed. E. Roberts and K. Parker (2 vols, Southampton Records Series, 34–5; 1992).

'The squire of low degree' in *Middle English metrical romances* ed. W.H. French and C.B. Hale (2 vols, New York, 1964) ii, pp. 721–55.

Staccini, R. 'L'inventario di una speziera del Quattrocento' *Studi Medievali* 22 (1981) pp. 377–420.

Statutes of the realm ed. A. Luders, T.E. Tomlins, J. Raithby *et al.* (11 vols, 1810–28).

Stone, L.W. 'Jean de Howden, poète anglo-normand du xiiie siècle' *Romania* 69 (1946–7) pp. 496–519.

Stubbs' Select charters from the beginning to 1307 ed. H.W.C. Davis (9th edition, Oxford, 1921).

Taylor, A.J. 'Count Amadeus of Savoy's visit to England in 1292' *Archaeologia* 106 (1979) pp. 123–32.

Theophilus. *De diversis artibus* ed. C.R. Dodwell (Oxford, 1986).

Thomas Cantimpratensis. *Liber de natura rerum* ed. H. Boese (Berlin, 1973).

Thompson, D.V., Jr. '*Liber de coloribus illuminatorum siue pictorum* from Sloane MS. No. 1754' *Speculum* 1 (1926) pp. 280–307, 448–50.

Three eleventh-century Anglo-Latin saints' lives: Vita S. Birini, Vita et miracula S. Kenelmi and Vita S. Rumwoldi ed. and trans. by R.C. Love (Oxford, 1996).

Three prose versions of the Secreta Secretorum ed. R. Steele (EETS, ES 74; 1898).

Three receptaria from medieval England: the languages of medicine in the fourteenth century ed. T. Hunt, with M. Benskin (Oxford, 2001).

The Towneley plays ed. M. Stevens and A.C. Cawley (2 vols, EETS, SS 13–14; 1994).

Treatises of fistula in ano, haemorrhoids, and clysters by John Arderne from an early fifteenth-century manuscript translation ed. D'A. Power (EETS, OS 139; 1910).

The Trotula: a medieval compendium of women's medicine ed. M.H. Green (Philadelphia, 2001).

Turner, T.H. 'Liberationes factae per executores dominae Alianorae consortis Edwardi regis Angliae primi' in *Manners and household expenses of England in the thirteenth and fifteenth centuries* ed. T.H. Turner (Roxburghe Club, 1841) pp. 93–145.

Turner, T.H. 'The will of Humphrey de Bohun, Earl of Hereford and Essex, with extracts from the inventory of his effects, 1319–1322' *Archaeological Journal* 2 (1846) pp. 339–49.

Twenty-six political and other poems (including 'Petty Job') from the Oxford MSS Digby 102 and Douce 322 ed. J. Kail (EETS, OS 124; 1904).

Two fifteenth-century cookery-books ed. T. Austin (EETS, OS 91; 1888).

Two Wycliffite texts: the sermon of William Taylor 1406; the testimony of William Thorpe 1407 ed. A. Hudson (EETS, OS 301; 1993).

La Vie de Seint Auban: an Anglo-Norman poem of the thirteenth century ed. A.R. Harden (ANTS, 19; 1968).

La Vie de Saint Laurent: an Anglo-Norman poem of the twelfth century ed. D.W. Russell (ANTS, 34; 1976).

La Vie Seint Richard evesque de Cycestre, by Pierre d'Abernon of Fetcham ed. D.W. Russell (ANTS, 51; 1995).

Visio Thurkilli relatore, ut videtur, Radulpho de Coggeshall ed. P.G. Schmidt (Akademie der Wissenschaften der DDR Zentralinstitut für alte Geschichte und Archäologie, Bibliotheca Scriptorum Graecorum et Romanorum Teubneriana, Leipzig; 1978).

Visitations of religious houses in the diocese of Lincoln ed. A.H. Thompson (3 vols, Lincoln Record Society, 7, 14, 21; 1914–29).

The Vita Wulfstani of William of Malmesbury to which are added the extant abridgements of this work and the miracles and translation of St Wulfstan ed. R.R. Darlington (Camden, 3rd series, 40; 1928).

Vitae patrum. PL, 73–4.

Walter de Bibbesworth. *Le Tretiz* ed. W. Rothwell (ANTS, Plain Texts Series, 6; 1990).

Walter of Henley and other treatises on estate management and accounting ed. D. Oschinsky (Oxford, 1971).

The Wardrobe book of William de Norwell, 12 July 1338 to 27 May 1340 ed. M. Lyon, B. Lyon and H.S. Lucas (Brussels, 1983).

Wenzel, S. 'Robert Grosseteste's treatise on confession, "Deus est"' *Franciscan Studies* 30 (1970) pp. 218–93.

The Westminster Chronicle 1381–1394 ed. L.C. Hector and B.F. Harvey (Oxford, 1982).

Wickham Legg, J. 'On an inventory of the vestry in Westminster Abbey, taken in 1388' *Archaeologia* 52 part 1 (1890) pp. 195–286.

Women's writing in Middle English ed. A. Barratt (London, 1992).

Woodruff, C.E. 'The will of Peter de Aqua Blanca Bishop of Hereford (1268)' in *Camden Miscellany XIV* (Camden, 3rd series, 37; 1926) pp. i-xi, 1–13.

The Works of John Metham, including the Romance of Amoryus and Cleopes . . . ed. H. Craig (EETS, OS 132; 1916).

York civic records, vol. III ed. A. Raine (Yorkshire Archaeological Society, Record Series, 106; 1942).

The York plays ed. R. Beadle (1982).

SECONDARY SOURCES

Ackerman, R.W. 'The debate of the body and the soul and parochial Christianity' Speculum 37 (1962) pp. 541–65.

Albala, K. Eating right in the Renaissance (Berkeley, 2002).

Appleby, D.F. 'The priority of sight according to Peter the Venerable' Mediaeval Studies 60 (1998) pp. 123–57.

Averkorn, R. 'Les Nobles, sont-ils toujours beaux? Quelques remarques sur les descriptions de personnages dans les chroniques médiévales de la Péninsule Ibérique' in Le Beau et le laid au Moyen Âge (Aix-en-Provence, 2000) pp. 27–44.

Bakka, E. 'The Alfred Jewel and sight' Antiquaries' Journal 46 (1966) pp. 277–82.

Bardsley, S. 'Sin, speech, and scolding in late medieval England' in Fama: the politics of talk and reputation in medieval Europe ed. T. Fenster and D.L. Smail (Ithaca, 2003) pp. 145–64.

Barley, N.F. 'Old English colour classification: where do matters stand ?' Anglo-Saxon England 3 (1974) pp. 15–28.

Barratt, A. 'The five wits and their structural significance in Part II of Ancrene Wisse' Medium Aevum 56 (1987) pp. 12–24.

Barratt, D.M. 'A second Northumberland Household Book' Bodleian Library Record 8 (1967–72) pp. 93–8.

Barrow, J. 'Athelstan to Aigueblanche, 1056–1268' in Hereford Cathedral: a history ed. G. Aylmer and J. Tiller (London, 2000) pp. 21–47.

Bartlett, R. England under the Norman and Angevin Kings 1075–1225 (Oxford, 2000).

Bartlett, R. The hanged man: a story of miracle, memory, and colonialism in the Middle Ages (Princeton, 2004).

Bartosz, A. 'Fonction du geste dans un texte romanesque médiéval: remarques sur la gestualité dans le première partie d'Erec' Romania 111 (1990) pp. 346–60.

Baugh, A.C., and T. Cable. A history of the English language (3rd edition, 1978).

Bell, J. 'Aristotle as a source for Leonardo's theory of colour perspective after 1500' JWCI 56 (1993) pp. 100–18.

Bennett, J.M. 'Writing fornication: medieval leyrwite and its historians' Transactions of the Royal Historical Society 6th series, 13 (2003) pp. 131–62.

Berlin, B., and P. Kay. Basic color terms: their universality and evolution (2nd edition, Berkeley, 1991).

Berryman, R.D. Use of the woodlands in the late Anglo-Saxon period (British Archaeological Reports, British Series, 271; 1998).

Biddle, M. 'Acoustic pots: a proposed study' Medieval Archaeology 6–7 (1962–3) p. 304.

Bidon, D.A., and F. Piponnier. 'Gestes et objets de la toilette aux xivème et xvème siècles' in Les Soins de beauté (Nice, 1987) pp. 211–44.

Binski, P. The Painted Chamber at Westminster (Society of Antiquaries, Occasional Paper 9, 1986).

Binski, P. 'Abbot Berkyng's tapestries and Matthew Paris's life of St Edward the Confessor' Archaeologia 109 (1991) pp. 85–100.

Blair, J. 'The 12th-century Bishop's Palace at Hereford' Medieval Archaeology 31 (1987) pp. 59–72.

Blangez, G. 'Destre et senestre, miséricorde et justice: un système de symboles' in Mélanges de langue et littérature françaises du Moyen-Âge offerts à Pierre Jonin (Aix-en-Provence, 1979) pp. 115–24.

Bloch, M. Les Rois thaumaturges: le caractère surnaturel attribué à la puissance royale, particulièrement en France et en Angleterre (Strasbourg, 1924).

Bloch, M. Feudal society (2 vols, 1965).

Bloomfield, M.W. The seven deadly sins: an introduction to the history of a religious concept, with special reference to medieval English literature (Michigan, 1952, reprinted 1967).

Bolens, L. 'Les Parfums et la beauté en Andalousie médiévale (xième–xiiième siècles)' in Les Soins de beauté (Nice, 1987) pp. 145–69.

Bolman, E.S. 'De coloribus: the meanings of color in Beatus manuscripts' Gesta 38 (1999) pp. 22–34.

Bond, E.A. 'Notices of the last days of Isabella, Queen of Edward the Second, drawn from an account of her household' Archaeologia 35 (1853–4) pp. 453–69.

Borenius, T. 'The cycle of images in the palaces and castles of Henry III' JWCI 6 (1943) pp. 40–50.

Boüard, M. de. 'Encyclopédies médiévales sur la "connaissance de la nature et du monde" au Moyen Âge' Revue des questions historiques 112 (1930) pp. 258–304.

Brakspear, H. 'The Bishop's Palace, Sonning' Berks., Bucks., and Oxon. Archaeological Journal 22 (1916) pp. 7–21.

Brault, G.J. Early blazon: heraldic terminology in the twelfth and thirteenth centuries with special reference to Arthurian literature (Oxford, 1972).

Bruce-Mitford, R.L.S. 'The Fuller Brooch' British Museum Quarterly 17 (1952) pp. 75–6.

Bruce-Mitford, R.L.S. 'Late Saxon disc-brooches' in Dark-Age Britain: studies presented to E.T. Leeds ed. D.B. Harden (1956) pp. 173–90.

Bühler, C.F. 'Prayers and charms in certain Middle English scrolls' Speculum 39 (1964) pp. 270–8.

Bullock-Davies, C. Register of royal and baronial domestic minstrels 1272–1327 (Woodbridge, 1986).

Burke, J.F. Vision, the gaze, and the function of the senses in Celestina (University Park, Pennsylvania, 2000).

Burnett, C. 'Sound and its perception in the Middle Ages' in The second sense: studies in hearing and musical judgement from Antiquity to the seventeenth century ed. C. Burnett, M. Fend and P. Gouk (Warburg Institute Surveys and Texts, 22; 1991) pp. 43–69.

Burnett, C. 'Sapores sunt octo: the medieval Latin terminology for the eight flavours' Micrologus 10 (2002) pp. 99–112.

Burrow, J.A. Gestures and looks in medieval narrative (Cambridge, 2002).

Callus, D.A. 'Introduction of Aristotelian learning to Oxford' Proceedings of the British Academy 29 (1943) pp. 229–81.

Camille, M. Mirror in parchment: the Luttrell Psalter and the making of medieval England (1998).

Camille, M. 'Before the gaze: the internal senses and late medieval practices of seeing' in Visuality before and beyond the Renaissance ed. R.S. Nelson (Cambridge, 2000) pp. 197–223.

Campbell, N. 'Aquinas' reasons for the aesthetic irrelevance of tastes and smells' British Journal of Aesthetics 36 (1996) pp. 166–76.

Carlin, M. Medieval Southwark (1996).

Carpenter, D. 'St Thomas Cantilupe: his political career' in St Thomas Cantilupe Bishop of Hereford: essays in his honour ed. M. Jancey (Hereford, 1982) pp. 57–72.

Carruthers, M.J. The book of memory: a study of memory in medieval culture (Cambridge, 1990).

Carruthers, M.J. The craft of thought: meditation, rhetoric, and the making of images, 400–1200 (Cambridge, 1998).

Carusi, P. 'Les Cinq Sens entre philosophie et médecine (Islam xᵉ–xiiᵉ siècles)' Micrologus 10 (2002) pp. 87–98.

Casagrande, C. 'Sistema dei sensi e classificazione dei peccati (secoli XII–XIII)' Micrologus 10 (2002) pp. 33–53.

Casagrande, G., and C. Kleinhenz. 'Literary and philosophical perspectives on the wheel of the five senses in Longthorpe Tower' Traditio 41 (1985) pp. 311–27.

Cassidy-Welch, M. Monastic spaces and their meanings: thirteenth-century English Cistercian monasteries (Turnhout, 2001).

Classen, C. Worlds of sense: exploring the senses in history and across cultures (1993).

Classen, C., D. Howes and A. Synnott. Aroma: the cultural history of smell (1994).

Cochet, L'Abbé. 'Acoustic pottery' The Gentleman's Magazine, NS 15 (1863) pp. 540–3.

Coldstream, N. The decorated style: architecture and ornament 1240–1360 (1994).

Colvin, H.M., et al. The history of the King's works (6 vols, 1963–82).

Corbin, A. The foul and the fragrant (1986).

Cosman, M.P. Fabulous feasts: medieval cookery and ceremony (New York, 1976).

Cox, J.C. *Churchwardens' accounts from the fourteenth century to the close of the seventeenth century* (1913).

Crane, S. 'Social aspects of bilingualism in the thirteenth century' in *Thirteenth Century England VI* ed. M. Prestwich, R.H. Britnell and R. Frame (Woodbridge, 1997) pp. 103–15.

Craun, E.D. '"Inordinata locutio": blasphemy in pastoral literature, 1200–1500' *Traditio* 39 (1983) pp. 135–62.

Craun, E.D. *Lies, slander, and obscenity in medieval English literature: pastoral rhetoric and the deviant speaker* (Cambridge, 1997).

Curta, F. 'Colour perception, dyestuffs, and colour terms in twelfth-century French literature' *Medium Aevum* 73 (2004) pp. 43–65.

Dean, R.J., and M.B.M. Boulton. *Anglo-Norman literature: a guide to texts and manuscripts* (ANTS, Occasional Publications Series, 3; 1999).

Denery, D.G., II. *Seeing and being seen in the later medieval world: optics, theology and religious life* (Cambridge, 2005).

Denholm Young, N. 'The *cursus* in England' in idem, *Collected papers on medieval subjects* (Oxford, 1946) pp. 26–55.

Dronke, P. 'Tradition and innovation in medieval Western colour imagery' in *The realms of colour* ed. A. Portmann and R. Ritsema (Eranos 1972 Yearbook, 41: Leiden, 1974) pp. 51–107.

Dronke, P. 'Les Cinq Sens chez Bernard Silvestre et Alain de Lille' *Micrologus* 10 (2002) pp. 1–14.

Eastwood, B.S. 'Galen on the elements of olfactory sensation' *Rheinisches Museum für Philologie* 124 (1981) pp. 268–90.

Eco, U., R. Lambertini, C. Marmo and A. Tabarroni. 'On animal language in the medieval classification of signs' in *On the medieval theory of signs* ed. U. Eco and C. Marmo (Amsterdam, 1989) pp. 3–41.

Edgerton, S.Y., Jr. 'Alberti's colour theory: a medieval bottle without Renaissance wine' *JWCI* 32 (1969) pp. 109–34.

Ellis, P., *et al. Ludgershall Castle Wiltshire: a report on the excavations by Peter Addyman, 1964–1972* (Wiltshire Archaeological and Natural History Society Monograph Series, 2; 2000).

Emery, A. *Greater medieval houses of England and Wales 1300–1500* (3 vols, Cambridge, 1996–2006).

Evans, J. *Magical jewels of the Middle Ages and the Renaissance particularly in England* (Oxford, 1922).

Evans, J.W., and R. Turner. *St David's Bishop's Palace; St Non's Chapel* (revised edition, Cardiff, 1999).

Febvre, L. *The problem of unbelief in the sixteenth century: the religion of Rabelais* trans. B. Gottlieb (Cambridge, Mass., 1982).

Finucane, R.C. *Miracles and pilgrims: popular beliefs in medieval England* (1977).

Finucane, R.C. 'The Cantilupe–Pecham controversy' in *St Thomas Cantilupe Bishop of Hereford: essays in his honour* ed. M. Jancey (Hereford, 1982) pp. 103–23.

Flandrin, J.L. 'Internationalisme, nationalisme et régionalisme dans la cuisine des xive et xve siècles: le témoignage des livres de cuisine' in *Manger et boire au Moyen Âge* ed. D. Menjot (2 vols, Nice, 1984) ii, pp. 75–91.

Flandrin, J.L. 'Soins de beauté et recueils de secrets' in *Les Soins de beauté* (Nice, 1987) pp. 13–29.

Flandrin, J.L. 'Seasoning, cooking and dietetics in the late Middle Ages' in *Food: a culinary history* ed. J.L. Flandrin and M. Montanari (New York, 1999) pp. 313–27.

Fleming, P. *Family and household in medieval England* (Basingstoke, 2001).

Flint, V.I.J. 'Space and discipline in early medieval Europe' in *Medieval practices of space* ed. B.A. Hanawalt and M. Kobialka (Minneapolis, 2000) pp. 149–66.

Forest-Hill, L. *Transgressive language in medieval English drama: signs of challenge and change* (Aldershot, 2000).

Frank, G. *The memory of the eyes: pilgrims to living saints in Christian late Antiquity* (Berkeley, 2000).

Gage, J. *Colour and culture: practice and meaning from Antiquity to abstraction* (1993).

Gage, J. *Colour and meaning: art, science and symbolism* (1999).

Gaimster, D., and B. Nenk. 'English households in transition *c.* 1450–1550: the ceramic evidence' in *The age of transition: the archaeology of English culture 1400–1600* ed. D. Gaimster and P. Stamper (Society for Medieval Archaeology Monograph Series 15; 1997) pp. 171–95.

Gannon, A. *The iconography of Anglo-Saxon coinage: sixth to eighth centuries* (Oxford, 2003).

Gerschel, L. 'Couleur et teinture chez divers peuples indo-européens' *Annales ESC* 21 (1966) pp. 608–31.

Glaenzer, A. 'La Tenture de La Dame à la licorne, du *Bestiaires d'amours* à l'ordre des tapisseries' *Micrologus* 10 (2002) pp. 401–28.

Golding, L. *The Jewish problem* (1938).

Gougaud, L. *Anciennes coutumes claustrales* (Abbaye Saint-Martin de Ligugé (Vienne); Moines et monastères, 8; 1930).

Greatrex, J. 'Horoscopes and healing at Norwich Cathedral Priory in the later Middle Ages' in *The Church and learning in later medieval society: essays in honour of R.B. Dobson. Proceedings of the 1999 Harlaxton Symposium* ed. C.M. Barron and J. Stratford (Donington, 2002) pp. 170–7.

Gregory, R. (ed.). *The Oxford companion to the mind* (2nd edition, Oxford, 2004).

Gurevich, A. *Medieval popular culture: problems of belief and perception* trans. J.M. Bak and P.A. Hollingsworth (Cambridge, 1988).

Hahn, C. 'Visio Dei: changes in medieval visuality' in *Visuality before and beyond the Renaissance* ed. R.S. Nelson (Cambridge, 2000) pp. 169–96.

Hamburger, J.M. 'The visual and the visionary: the image in late medieval monastic devotions' *Viator* 20 (1989) pp. 161–83.

Hamilakis, Y. 'The past as oral history: towards an archaeology of the senses' in *Thinking through the body: archaeologies of corporeality* ed. Y. Hamilakis, M. Pluciennik and S. Tarlow (New York, 2002) pp. 121–36.

Hanawalt, B.A. *The ties that bound: peasant families in medieval England* (Oxford, 1986).

Hanawalt, B.A. *Growing up in medieval London* (Oxford, 1993).

Harriss, G.L. 'The court of the Lancastrian kings' in *The Lancastrian court* ed. J. Stratford (Harlaxton Medieval Studies, 13; 2003) pp. 1–18.

Harvey, B.F. 'The aristocratic consumer in England in the long thirteenth century' in *Thirteenth-Century England VI* ed. M. Prestwich, R.H. Britnell and R. Frame (Woodbridge, 1997) pp. 17–37.

Harvey, B.F. 'Monastic pittances in the Middle Ages' in *Food in medieval England: diet and nutrition* ed. C.M. Woolgar, D. Serjeantson and T. Waldron (Oxford, 2006) pp. 215–27.

Harvey, E.R. *The inward wits: psychological theory in the Middle Ages and the Renaissance* (Warburg Institute Surveys, 6; 1975).

Harvey, P.D.A., and A. McGuinness. *A guide to British medieval seals* (1996).

Henig, M. 'Archbishop Hubert Walter's gems' *Journal of the British Archaeological Association* 136 (1983) pp. 56–61.

Henry, A. 'Un Texte oenologique de Jofroi de Waterford et Servais Copale' *Romania* 107 (1986) pp. 1–37.

Hieatt, C.B. 'Medieval Britain' in *Regional cuisines of medieval Europe: a book of essays* ed. M. Weiss Adamson (New York, 2002) pp. 19–45.

Hill, T.D. 'Enide's colored horse and Salernitan color theory: *Erec et Enide*, lines 5268–81' *Romania* 108 (1987) pp. 523–7.

Hilton, R.H. *The English peasantry in the later Middle Ages* (Oxford, 1975).

Hinton, D.A. *Gold and gilt, pots and pins: possessions and people in medieval Britain* (Oxford, 2005).

Holdsworth, W. *A history of English law* (7th edition, 17 vols, 1956–72).

Hope, W. St. J. 'The Sarum consuetudinary and its relation to the cathedral church of Old Sarum' *Archaeologia* 68 (1916–17) pp. 111–26.

Horvath, R. 'Romancing the word: *fama* in the Middle English *Sir Launfal* and *Athelston*' in *Fama: the politics of talk and reputation in medieval Europe* ed. T. Fenster and D.L. Smail (Ithaca, 2003) pp. 165–86.

Howard, H. *Pigments of English medieval wall painting* (2003).

Howes, D. (ed.). *The varieties of sensory experience: a sourcebook in the anthropology of the senses* (Toronto, 1991).

Hudson, A. 'Middle English' in *Editing medieval texts: English, French and Latin written in England. Papers given at the twelfth annual Conference on Editorial Problems, University of Toronto 5–6 November 1976* ed. A.G. Rigg (New York, 1977) pp. 34–57.

Hudson, A. *The premature reformation: Wycliffite texts and Lollard history* (Oxford, 1988).

Hughes, J. *Arthurian myths and alchemy: the kingship of Edward IV* (Far Thrupp, 2002).

Hunt, R.W. 'English learning in the late twelfth century' *Transactions of the Royal Historical Society* 4th series, 19 (1936) pp. 19–42.

Iradiel, P. 'Cuidar el cuerpo, cuidar la imagen: los paradigmas de la belleza femenina en la Valencia bajomedieval' in *Les Soins de beauté* (Nice, 1987) pp. 61–86.

Izbicki, T.M. 'Forbidden colors in the regulation of clerical dress from the Fourth Lateran Council (1215) to the time of Nicholas de Cusa (d. 1464)' *Medieval Clothing and Textiles* 1 (2005) pp. 105–14.

Jenner, M.S.R. 'Civilization and deodorization? Smell in early modern English culture' in *Civil histories: essays presented to Sir Keith Thomas* ed. P. Burke, B. Harrison and P. Slack (Oxford, 2000) pp. 127–44.

Johnstone, H. 'The Queen's household' in T.F. Tout *Chapters in the administrative history of mediaeval England: the wardrobe, the chamber and the small seals* (6 vols, Manchester, 1920–33) v, pp. 231–89.

Johnstone, P. *High fashion in the Church: the place of church vestments in the history of art from the ninth to the nineteenth century* (Leeds, 2002).

Jones, C. 'Interpreting the perceptions of past people' in *The archaeology of perception and the senses* ed. C. Jones and C. Hayden (special number of *Archaeological Review from Cambridge* 15: 1 (1998)) pp. 7–22.

Jones, P.M. 'Music therapy in the later Middle Ages: the case of Hugo van der Goes' in *Music as medicine: the history of music therapy since Antiquity* ed. P. Horden (Aldershot, 2000) pp. 120–44.

Jones, P.M., and L.T. Olsan. 'Middleham Jewel: ritual, power, and devotion' *Viator* 31 (2000) pp. 249–90.

Joubert, F. *La Tapisserie médiévale au musée de Cluny* (Paris, 1987).

Jütte, R. *A history of the senses: from Antiquity to cyberspace* trans. J. Lynn (Oxford, 2005).

Kemp, S. 'A medieval controversy about odor' *Journal of the History of the Behavioural Sciences* 33 (1997) pp. 211–19.

Kieckhefer, R. *Magic in the Middle Ages* (Cambridge, Canto edition, 2000).

Kipling, G. *Enter the King: theatre, liturgy, and ritual in the medieval civic triumph* (Oxford, 1998).

Kowaleski, M. *Local markets and regional trade in medieval Exeter* (Cambridge, 1995).

Lachaud, F. 'Embroidery for the court of Edward I (1272–1307)' *Nottingham Medieval Studies* 37 (1993) pp. 33–52.

Lachaud, F. 'Vêtement et pouvoir à la cour d'Angleterre sous Philippa de Hainaut' in *Au Cloître et dans le monde: femmes, hommes et sociétés (IXᵉ–XVᵉ siècles). Mélanges en l'honneur de Paulette l'Hermite-Leclercq* ed. P. Henriet and A.-M. Legras (Paris, 2000) pp. 217–33.

Lachaud, F. 'Dress and social status in England before the sumptuary laws' in *Heraldry, pageantry and social display in medieval England* ed. P. Coss and M. Keen (Woodbridge, 2002) pp. 105–23.

Langlois, C.V. *La Connaissance de la nature et du monde au Moyen Âge d'après quelques écrits français à l'usage des laïcs* (Paris, 1911).

Largey, G.P., and D.R. Watson. 'The sociology of odors' *American Journal of Sociology* 77 (1972) pp. 1021–34.

Larmat, J. 'Les Bains dans la littérature française du Moyen Âge' in *Les Soins de beauté* (Nice, 1987) pp. 195–210.

Lauer, H.H. 'Böser Blick' *Lexikon des Mittelalters* (10 vols, Munich, 1981–3) ii, cols 470–2.

Laurioux, B. *Le Règne de Taillevent: livres et pratiques culinaires à la fin du Moyen Âge* (Paris, 1997).

Lawn, B. *The Salernitan questions: an introduction to the history of medieval and Renaissance problem literature* (Oxford, 1963).

Lawson, G., C. Scarre, I. Cross and C. Hills, 'Mounds, megaliths, music and mind: some thoughts on the acoustical properties *and purposes* of archaeological spaces' in *The archaeology of perception and the senses* ed. C. Jones and C. Hayden (special number of *Archaeological Review from Cambridge* 15:1 (1998)) pp. 111–34.

Leclercq, J. *The love of learning and the desire for God: a study of monastic culture* trans. Catharine Misrahi (New York, 1961).

Lester, G. 'Idle words: stereotyping by language in the English mystery plays' *Medieval English Theatre* 11 (1989) pp. 129–33.

Leveque-Agret, I. 'Les Parfums à la fin du Moyen Âge: les différentes formes de fabrication et d'utilisation' in *Les Soins de beauté* (Nice, 1987) pp. 135–44.

Lewis, C.S. *The discarded image: an introduction to medieval and Renaissance literature* (Cambridge, 1964).

Lohr, C.F. 'Medieval Latin Aristotle commentaries: authors A–F' *Traditio* 23 (1967) pp. 313–413.

Lorcin, M.-T. 'Rides et cheveux gris dans les ouvrages de Roger Bacon' in *Les Soins de beauté* (Nice, 1987) pp. 253–9.

Maddicott, J.R. 'Follower, leader, pilgrim, saint: Robert de Vere, Earl of Oxford, at the shrine of Simon de Montfort, 1273' *EHR* 109 (1994) pp. 641–53.

Mandrou, R. *Introduction to modern France 1500–1640: an essay in historical psychology* trans. R.E. Hallmark (1975).

Marks, R. *Stained glass in England during the Middle Ages* (1993).

Marner, D. *St Cuthbert: his life and cult in medieval Durham* (2000).

Martin, N.D.S. 'The life of St Thomas of Hereford' in *St Thomas Cantilupe Bishop of Hereford: essays in his honour* ed. M. Jancey (Hereford, 1982) pp. 15–19.

Martineau-Genieys, C. 'Modèles, maquillage et misogynie, à travers les textes littéraires français du Moyen Âge' in *Les Soins de beauté* (Nice, 1987) pp. 31–50.

Matthew, H.C.G., and B. Harrison (eds). *Oxford Dictionary of National Biography* (61 vols, Oxford, 2004).

McGlashan, N.D., and R.E. Sandell. 'The Bishop of Salisbury's house at his manor of Potterne' *Wiltshire Archaeological and Natural History Magazine* 69 (1974) pp. 85–96.

McGuire, J.R. 'L'Onguent et l'initiative féminine dans *Yvain*' *Romania* 112 (1991) pp. 65–82.

McVaugh, M.R. 'Smells and the medieval surgeon' *Micrologus* 10 (2002) pp. 113–32.

Merceron, J.E. 'Cooks, social status, and stereotypes of violence in medieval French literature and society' *Romania* 116 (1998) pp. 170–87.

Michael, M.A. 'A manuscript wedding gift from Philippa of Hainault to Edward III' *Burlington Magazine* 127 (1985) pp. 582–9.

Milroy, J. 'Middle English dialectology' in *The Cambridge history of the English language* ed. R.M. Hogg *et al.* (6 vols, Cambridge, 1992–2001) ii, pp. 156–206.

Morley, B., and D. Gurney. *Castle Rising Castle, Norfolk* (East Anglian Archaeology Report, 81; Dereham, 1997).

Morris, C. *The discovery of the individual 1050–1200* (1972).

Mütherich, F. 'An illustration of the five senses in mediaeval art' *JWCI* 18 (1955) pp. 140–1.

Newton, S.M. *Fashion in the age of the Black Prince. A study of the years 1340–1365* (Woodbridge, 1980).

Nilson, B. *Cathedral shrines of medieval England* (Woodbridge, 1998).

Nordenfalk, C. 'Les Cinq Sens dans l'art du Moyen Âge' *Revue de l'Art* 34 (1976) pp. 17–28.

Nordenfalk, C. 'The five senses in late medieval and Renaissance art' *JWCI* 48 (1985) pp. 1–22.

North, J.D. 'Some Norman horoscopes' in *Adelard of Bath: an English scientist and Arabist of the early twelfth century* ed. C. Burnett (Warburg Institute Surveys and Texts, 14; 1987) pp. 147–61.

O'Connor, E. 'Hell's pit and Heaven's rose: the typology of female sights and smells in Panormita's *Hermaphroditus' Medievalia et Humanistica* NS, 23 (1996) pp. 25–51.

Orme, N. 'The cathedral school before the Reformation' in *Hereford Cathedral: a history* ed. G. Aylmer and J. Tiller (London, 2000) pp. 565–78.

Orme, N. *Medieval children* (New Haven, 2001).

Ormrod, W.M. 'The personal religion of Edward III' *Speculum* 64 (1989) pp. 849–77.

Owst, G.R. *Preaching in medieval England: an introduction to sermon manuscripts of the period c. 1350–1450* (Cambridge, 1926).

Owst, G.R. *Literature and pulpit in medieval England: a neglected chapter in the history of English letters and of the English people* (Cambridge, 1933).

Owst, G.R. 'Sortilegium in English homiletic literature of the fourteenth century' in *Studies presented to Sir Hilary Jenkinson* ed. J.C. Davies (1957) pp. 272–303.

Page, C. *The owl and the nightingale: musical life and ideas in France, 1100–1300* (1989).

Page, C. 'Music and medicine in the thirteenth century' in *Music as medicine: the history of music therapy since Antiquity* ed. P. Horden (Aldershot, 2000) pp. 109–19.

Palmer, R. 'In bad odour: smell and its significance in medicine from Antiquity to the seventeenth century' in *Medicine and the five senses* ed. W.F. Bynum and R. Porter (Cambridge, 1993) pp. 61–8, 285–7.

Paravicini Bagliani, A. 'En Guise d'introduction: le pouvoir pontifical a-t-il besoin des cinq sens?' *Micrologus* 10 (2002) pp. ix–xiv.

Parkes, M.B. *Pause and effect: an introduction to the history of punctuation in the West* (Aldershot, 1992).

Parsons, J.C. 'Ritual and symbol in the English medieval queenship to 1500' in *Women and sovereignty* ed. L.O. Fradenburg (Edinburgh, 1992) pp. 60–77.

Pasnau, R. *Theories of cognition in the later Middle Ages* (Cambridge, 1997).

Pastoureau, M. 'Du Bleu au noir. Éthiques et pratiques de la couleur à la fin du Moyen Âge' *Médiévales* 14 (1988) pp. 9–21.

Pastoureau, M. 'L'Église et la couleur des origines à la Réforme' *Bibliothèque de l'École des Chartes* 147 (1989) pp. 203–30.

Pastoureau, M. *Blue: the history of a color* (Princeton, 2001).

Pastoureau, M. *The Devil's cloth: a history of stripes and striped fabric* (New York, 2001).

Pastoureau, M. 'Le Bestiaire des cinq sens (xiie–xvie siècles)' *Micrologus* 10 (2002) pp. 133–45.

Pastoureau, M. 'Les Teinturiers médiévaux: histoire sociale d'un métier réprouvé' in M. Pastoureau *Une Histoire symbolique du Moyen Âge occidental* (Paris, 2004) pp. 173–95.

Pastoureau, M. 'L'Homme roux: iconographie médiévale de Judas' in M. Pastoureau *Une Histoire symbolique du Moyen Âge occidental* (Paris, 2004) pp. 197–209.

Pearsall, D., and E. Salter *Landscapes and seasons of the medieval world* (1973).

Pegg, M.G. *The corruption of angels: the great inquisition of 1245–1246* (Princeton, 2001).

Phillips, K.M. 'The invisible man: body and ritual in a fifteenth-century noble household' *Journal of Medieval History* 31 (2005) pp. 143–62.

Piponnier, F. *Costume et vie sociale: la cour d'Anjou xive–xve siècles* (Paris, 1970).

Piponnier, F., and P. Mane. *Dress in the Middle Ages* trans. C. Beamish (New Haven, 1997).

Platt, C. *Medieval Southampton: the port and trading community, AD 1000–1600* (1973).

Polo de Beaulieu, M.-A. 'La Condamnation des soins de beauté par les prédicateurs du Moyen Âge (xiiième–xvème siècles)' in *Les Soins de beauté* (Nice, 1987) pp. 297–309.

Prestwich, M. *York civic ordinances, 1301* (Borthwick papers, 49; 1976).

Prestwich, M. 'The piety of Edward I' in *England in the thirteenth century: proceedings of the 1984 Harlaxton Symposium* ed. W.M. Ormrod (Grantham, 1985) pp. 120–8.

Prestwich, M. *Armies and warfare in the Middle Ages: the English experience* (New Haven, 1996).

Prestwich, M. *Edward I* (new edition, New Haven, 1997).

Prestwich, M. *Plantagenet England 1225–1360* (Oxford, 2005).

Price, P. *Bells and man* (Oxford, 1983).

Rastall, R. 'The sounds of Hell' in *The iconography of Hell* ed. C. Davidson and T.H. Seiler (Medieval Institute Publications, Kalamazoo, Early Drama, Art and Music Monograph Series, 17; 1992) pp. 102–31.

Rawcliffe, C. *The Staffords, Earls of Stafford and Dukes of Buckingham 1394–1521* (Cambridge, 1978).

Rawcliffe, C. *Medicine & society in later medieval England* (Far Thrupp, 1995).

Rawcliffe, C. *Medicine for the soul. The life, death and resurrection of an English medieval hospital: St Giles's, Norwich, c.1249–1550* (Far Thrupp, 1999).

Reeve, M.M. 'The former painted cycle of the life of Edward I at the Bishop's Palace, Lichfield' *Nottingham Medieval Studies* 46 (2002) pp. 70–83.

Renevey, D. 'Household chores in *The doctrine of the hert*: affective spirituality and subjectivity' in *The medieval household in Christian Europe, c.850–c.1550 managing power, wealth, and the body* ed. C. Beattie, A. Maslakovic and S. Rees Jones (Turnhout, 2003) pp. 167–85.

Renfrew, C., and E.B.W. Zubrow (eds). *The ancient mind: elements of cognitive archaeology* (Cambridge, 1994).

Richardson, A. 'Gender and space in English royal palaces *c.*1160–*c.*1547: a study in access analysis and imagery' *Medieval Archaeology* 47 (2003) pp. 131–65.

Rivlin, R., and K. Gravelle. *Deciphering the senses: the expanding world of human perception* (New York, 1985).

Robinson, D. (ed.) *The Cistercian abbeys of Britain: far from the concourse of men* (1998).

Rosen, E. 'The invention of eyeglasses' *Journal of the History of Medicine and Allied Sciences* 9 (1956) pp. 13–46, 183–218.

Roth, C. *A history of the Jews in England* (3rd edition, Oxford, 1964).

Rouse, E.C. *Longthorpe Tower, Cambridgeshire* (London, 1987).

Rouse, E.C., and A. Baker. 'The wall-paintings at Longthorpe Tower near Peterborough, Northants' *Archaeologia* 96 (1955) pp. 1–57.

Roussel, C. 'Le Jeu des formes et des couleurs: observations sur "la beste glatissant"' *Romania* 104 (1983) pp. 49–82.

Rowe, C. 'Conceptions of colour and colour symbolism in the Ancient World' in *The realms of colour* ed. A. Portmann and R. Ritsema (Eranos 1972 Yearbook, 41: Leiden, 1974) pp. 327–64.

Rowland, B. 'Bishop Bradwardine on the artificial memory' *JWCI* 41 (1978) pp. 307–12.

Ruddock, A.A. *Italian merchants and shipping in Southampton 1270–1600* (Southampton Records Series, 1; 1951).

Russell, H.G. 'Lollard opposition to oaths by creatures' *American Historical Review* 51 (1946) pp. 668–84.

Sabine, E.L. 'Latrines and cesspools of mediaeval London' *Speculum* 9 (1934) pp. 303–21.

Sandler, L.F. *Omne bonum: a fourteenth-century encyclopedia of universal knowledge, British Library MSS Royal 6 E VI–6 E VII* (2 vols, 1996).

Santi, F. 'Il senso del ragno. Sistemi a confronto' *Micrologus* 10 (2002) pp. 147–61.

Saul, N. *Richard II* (New Haven, 1997).

Saussure, F. de. *Course in general linguistics* trans. R. Harris (1983).

Saxl, F. 'A spiritual encyclopaedia of the later Middle Ages' *JWCI* 5 (1942) pp. 82–142.

Scattergood, J. 'Fashion and morality in the late Middle Ages' in *England in the fifteenth century: proceedings of the 1986 Harlaxton symposium* ed. D. Williams (Woodbridge, 1987) pp. 255–72.

Schmitt, J.-C. *La Raison des gestes dans l'Occident médiéval* (Paris, 1990).

Schmitt, J.-C. 'The rationale of gestures in the West: third to thirteenth centuries' in *A cultural history of gesture: from Antiquity to the present day* ed. J. Bremmer and H. Roodenburg (Cambridge, 1991) pp. 59–70.

Schmitt, J.-C. *Les Revenants: les vivants et les morts dans la société médiévale* (Paris, 1994).

Schofield, J. *Medieval London houses* (New Haven, 1994).

Schofield, P.R. 'Peasants and the manor court: gossip and litigation in a Suffolk village at the close of the thirteenth century' *Past and Present* 159 (May 1998) pp. 3–42.

Sears, E. 'The iconography of auditory perception in the early Middle Ages: on psalm illustration and psalm exegesis' in *The second sense: studies in hearing and musical judgement from Antiquity to the seventeenth century* ed. C. Burnett, M. Fend and P. Gouk (Warburg Institute Surveys and Texts, 22; 1991) pp. 19–42.

Sears, E. 'Sensory perception and its metaphors in the time of Richard of Fournival' in *Medicine and the five senses* ed. W.F. Bynum and R. Porter (Cambridge, 1993) pp. 17–39, 276–83.

Seiler, T.H. 'Filth and stench as aspects of the iconography of Hell' in *The iconography of Hell* ed. C. Davidson and T.H. Seiler (Medieval Institute Publications, Kalamazoo, Early Drama, Art and Music Monograph Series, 17; 1992) pp. 132–40.

Serjeantson, D., and C.M. Woolgar. 'Fish consumption in medieval England' in *Food in medieval England: diet and nutrition* ed. C.M. Woolgar, D. Serjeantson and T. Waldron (Oxford, 2006), pp. 102–30.

Short-title catalogue of books printed in England, Scotland, and Ireland, and of English books printed abroad 1475–1640 (2nd edition, 3 vols, 1976–91).

Short-title catalogue of books printed in the German-speaking countries and German books printed in other countries from 1455 to 1600 now in the British Museum (1962).

Skard, S. 'The use of color in literature: a survey of research' *Proceedings of the American Philosophical Society* 90 (1940) pp. 163–249.

Smedick, L.K. 'Parallelism and pointing in Rolle's rhythmical style' *Mediaeval Studies* 41 (1979) pp. 404–67.

Smith, B.R. *The acoustic world of early modern England* (Chicago, 1999).

Spencer, B.W. 'Medieval pilgrim badges: some general observations illustrated mainly from English sources' in *Rotterdam papers: a contribution to medieval archaeology* ed. J.G.N. Renaud (Rotterdam, 1968) pp. 137–53.

Staniland, K. 'Clothing and textiles at the court of Edward III 1342–1352' in *Collectanea Londiniensia. Studies in London archaeology and history presented to Ralph Merrifield* ed. J. Bird, H. Chapman and J. Clark (London and Middlesex Archaeological Society, special paper, 2; 1978) pp. 223–34.

Steane, J. *The archaeology of the medieval English monarchy* (1993).

Stevens, D. 'A Somerset coroner's roll, 1315–21' *Somerset and Dorset Notes and Queries* 31 (1985) pp. 451–72.

Stock, A. 'A sounding vase at Fountains Abbey?' *Cistercian Studies* 23 (1988) pp. 190–1.

Stoller, P. 'Sound in Songhay cultural experience' *American Ethnologist* 11 (1984) pp. 559–70.

Stoller, P. *The taste of ethnographic things: the senses in anthropology* (Philadelphia, 1989).

Strohm, P. *Hochon's arrow: the social imagination of fourteenth-century texts* (Princeton, 1992).

Summerson, H. 'Attitudes to capital punishment in England, 1200–1350' *Thirteenth Century England VIII* ed. M. Prestwich, R. Britnell and R. Frame (Woodbridge, 2001) pp. 123–33.

Sutton, A.F., L. Visser-Fuchs and P.W. Hammond *The reburial of Richard, Duke of York, 21–30 July 1476* (1996).

Tachau, K.H. *Vision and certitude in the age of Ockham: optics, epistemology and the foundations of semantics* (Leiden, 1988).

Taylor, P.B. 'The Canon Yeoman's breath: emanations of a metaphor' *English Studies* 60 (1979) pp. 380–8.

Thomas, K. *Religion and the decline of magic: studies in popular beliefs in sixteenth- and seventeenth-century England* (Harmondsworth, 1991).

Thomas, K. 'Introduction' in *A cultural history of gesture: from Antiquity to the present day* ed. J. Bremmer and H. Roodenburg (Cambridge, 1991) pp. 1–14.

Thompson, D.V. *The materials and techniques of medieval painting* (New York, 1956).

Thompson, M. *Medieval bishops' houses in England and Wales* (Aldershot, 1998).

Thomson, J.A.F. *The later Lollards 1414–1520* (Oxford, 1965).

Thorndike, L. 'Sanitation, baths, and street-cleaning in the Middle Ages and Renaissance' *Speculum* 3 (1928) pp. 192–203.

Thorndike, L. 'Chiromancy in mediaeval Latin manuscripts' *Speculum* 40 (1965) pp. 674–706.

Thurley, S. *The royal palaces of Tudor England: architecture and court life 1460–1547* (New Haven, 1993).

Thurley, S. *Whitehall Palace: an architectural history of the royal apartments, 1240–1698* (New Haven, 1999).

Tilliette, J.Y. 'Le Symbolisme des cinq sens dans la littérature morale et spirituelle des xie et xiie siècles' *Micrologus* 10 (2002) pp. 15–32.

Toynbee, P. 'The colour *perse* in Dante and other medieval writers' in P. Toynbee *Dante studies and researches* (1902) pp. 307–14.

Trease, G.E. 'The spicers and apothecaries of the royal household in the reigns of Henry III, Edward I and Edward II' *Nottingham Medieval Studies* 3 (1959) pp. 19–52.

Turner, R. *Lamphey Bishop's Palace; Llawhaden Castle; Carswell medieval house; Carew Cross* (2nd edition, Cardiff, 2000).

Twycross, M., and S. Carpenter. 'Masks in the medieval English theatre' *Medieval English Theatre* 3 (1981) pp. 7–44, 69–113.

Vandenberghe, S. 'Une Aquamanile ornitomorphe trouvée à Gistel (Flandre occidentale) [an ornithomorphic aquamanile from Gistel]' in *Everyday and exotic pottery in Europe c. 600–1900: studies in honour of J.G. Hurst* ed. D.R.M. Gaimster and M. Redknap (Oxford, 1992) pp. 336–9.

Veale, E.M. *The English fur trade in the later Middle Ages* (2nd edition, London Record Society, 38; 2003).

Verkerk, D.H. 'Black servant, black demon: color ideology in the Asburnham Pentateuch' *Journal of Medieval and Early Modern Studies* 31 (2001) pp. 57–77.

Vince, A.G. 'The Saxon and medieval pottery of London: a review' *Medieval Archaeology* 29 (1985) pp. 25–93.

Vinge, L. *The five senses: studies in a literary tradition* (Lund, 1975).

Vinyoles i Vidal, T.M. 'Els sons, els colors i les olors de Barcelona pels volts del 1400' in *Història Urbana del Pla de Barcelona: Actes del II Congrès d'Història del Pla de Barcelona celebrat a l'Institut Municipal d'Història els dies 6 i 7 de desembre de 1985* ed. A.M. Adroer i Tasis (2 vols, Barcelona, 1989) i, pp. 133–44.

Waddell, C. 'The early Cistercian experience of the liturgy' in *Rule and life: an interdisciplinary symposium* ed. M.B. Pennington (Shannon: Cistercian studies series, 12; 1971) pp. 77–116.

Wagner, A.R. *Heralds and heraldry in the Middle Ages: an inquiry into the growth of the armorial function of heralds* (1939).

Walker Bynum, C. 'Wonder' *American Historical Review* 102 (1997) pp. 1–26.

Ward, J.C. *English noblewomen in the later Middle Ages* (Harlow, 1992).

Watson, L. *Jacobson's organ and the remarkable nature of smell* (1999).

Weiss Adamson, M. 'The role of medieval physicians in the diffusion of culinary recipes and cooking practices' in *Du Manuscrit à la table: essais sur la cuisine au Moyen Âge et répertoire des manuscrits médiévaux contenant des recettes culinaires* ed. C. Lambert (Montreal, 1992) pp. 69–80.

Welander, D. *The history, art and architecture of Gloucester Cathedral* (Far Thrupp, 1991).

Whitehead, C. *Castles of the mind: a study of medieval architectural allegory* (Cardiff, 2003).

Wickham Legg, J. *Notes on the history of liturgical colours* (1882).

Williams, G. *Henry de Gower, bishop and builder* (St David's, n.d. *c.* 1980).

Williams, J.M. 'Synaesthetic adjectives: a possible law of semantic change' *Language* 52 (1976) pp. 461–78.

Winstone, J.H. 'The Bishop's Palace at Wookey' *Somerset Archaeology and Natural History* 141 (1998) pp. 91–101.

Wolfson, H.A. 'The internal senses in Latin, Arabic, and Hebrew philosophic texts' *Harvard Theological Review* 28 (1935) pp. 69–133.

Woolgar, C.M. 'Diet and consumption in gentry and noble households: a case study from around the Wash' in *Rulers and ruled in late medieval England: essays presented to Gerald Harriss* ed. R.E. Archer and S. Walker (1995) pp. 17–32.

Woolgar, C.M. *The great household in late medieval England* (New Haven and London, 1999).

Woolgar, C.M. 'Group diets in late medieval England' in *Food in medieval England: diet and nutrition* ed. C.M. Woolgar, D. Serjeantson and T. Waldron (Oxford, 2006) pp. 191–200.

Woolgar, C.M. 'Fasting and feasting: food and taste in the Middle Ages' in *Food: the history of taste* ed. P. Freedman (forthcoming).

Woolgar, C.M., D. Serjeantson and T. Waldron (eds) *Food in medieval England: diet and nutrition* (Oxford, 2006).

INDEX

Lightning Source UK Ltd.
Milton Keynes UK
UKHW030645291021
393035UK00006B/457